13.90

AF593803

BERNARD MICHEL

# ICE MECHANICS

LES PRESSES DE L'UNIVERSITÉ LAVAL
Québec, 1978

© 1978, Bernard Michel, Québec.
All rights reserved. Printed in Canada.
Legal deposit (Quebec), fourth quarter 1978.
DISTRIBUTOR: LES PRESSES DE L'UNIVERSITÉ LAVAL
I.S.B.N. 0-7746-6876-8

*To Mariette*

# *PREFACE*

*Ice mechanics is an engineering science whose object is to understand the basic mechanical behavior of different types of ice and solve the problems of ice interaction with natural or artificial boundaries, including engineering structures.*

*A very large amount of research work has been done in the last hundred years on the physical properties of ice. At the beginning of the century most of the engineering research was aimed at solving the problems caused by ice obstruction during the winter months, for the operation of hydro power stations. This led to a first book on "Ice Engineering" by H.T. Barnes in 1928. This book was not followed by others in the engineering fields and ice properties continued to the extensively studied for their own merits by reputed glaciologists.*

*A new impetus was given to the field of ice mechanics at the beginning of the seventies because of its impact on exploration for oil and gas in the Canadian Arctic and along the Coast of Alaska. One major problem was the determination of forces exerted by a moving ice field on offshore structures. The question of the bearing capacity of a floating ice cover also became of paramount interest for its use as a supporting surface for transportation and for drilling purposes. At the same time, the design of icebreaking ships took a new turn after the maiden trip of the "Manhattan" through the North-West passage in the Canadian Arctic Archipelago and the conquest of the North Pole by the U.S.S.R. Arktika.*

*There has been in the last few years a very large increase in the number of published articles and research reports on the subject of ice engineering. Amongst others, two international forums have been extensively used to dissiminate the research results: the "Ice Symposiums" of the International Association for Hydraulic Research and the "Conferences of Port and Ocean Engineering under Arctic Conditions" (POAC). Unfortunately the literature is so abundant and so diverse that it is extremely difficult for a newcomer to get acquainted quickly with ice engineering and even for a seasoned researcher to obtain a good in-depth understanding of the numerous subjects that are discussed.*

*This book is thus the first one to deal with the basics of ice mechanics and some of its applications to engineering problems. The major part of the fundamental knowledge presented in the book origins from the research work done at the Ice Mechanics Laboratory of the Civil Engineering Department of Laval University. Without the experimental studies and analysis done at this Laboratory, pertaining to the formation of ice, its crystallographic structure and above all giving general explanations to the basic mechanical behavior of ice under brittle and ductile conditions; such a book could never have been written. The applications to ice engineering then comes naturally from the understanding of the fundamental behavior of ice as a material.*

*We have not tried to produce a handbook on ice engineering covering all the papers which are published at this time. This field is in a state of rapid evolution and only some major applications have been discussed. There are certainly many important studies which I have overlooked and to the authors of these works I extend my apologies.*

*Because of its fundamental aspect we hope that the book will serve first as an introduction for students in ice mechanics but will also help engineers and scientists already engaged in the field.*

*The writer sincerely appreciates the support of Professor A. Côté of the Civil Engineering Department of Laval University and Professor L. Lliboutry of the Laboratoire de Glaciologie at the University of Grenoble, for their help during a sabbatical leave when the main parts of this book were written.*

*The assistance of Professor C.E. Deslauriers in proof-reading of the manuscript is gratefully acknowledged. The help of J. Parent and J.C. Pugno for the drawings and the patience of Diane Dussault in typing the text is greatly appreciated.*

*Bernard* MICHEL

*July 1978.*

## PARTIAL LIST OF SYMBOLS

e – void ratio
E – Young's modulus of ice
g – acceleration of gravity
G – rigidity modulus of ice
h – ice thickness
$\ell$ – characteristic length of ice plate
L – latent heat of fusion of ice
p – pressure
Q – water discharge
R – universal gaz constant
t – time
y – water depth
$\gamma$ – shear strain
$\epsilon$ – axial strain
$\theta$ – temperature
$\mu$ – coefficient of friction
$\upsilon$ – Poisson's ratio
$\rho$ – density of water
$\rho'$ – density of ice
$\sigma$ – normal stress
$\tau$ – shear stress

## SI UNITS AND CONVERSION FACTORS

The units used in this book conform to the Système International d'Unités (i.e. the SI system) which is now the internationally accepted form of the metric system. The SI units for mass, length, time, and electric current are the metre (m), kilogram (kg), second (s), and ampere (A). Temperatures are measured in Kelvins (K).

Listed below are some derived SI units and other units acceptable in the SI system which are used in this book.

| *Quantity* | *Name of unit* | *Symbol for unit* | *Definition of unit* |
|---|---|---|---|
| Energy | joule | J | $kg\ m^2\ s^{-2}$ |
| Force | newton | N | $kg\ m\ s^{-2} = J\ m^{-1}$ |
| Power | watt | W | $kg\ m^2\ s^{-3} = J\ s^{-1}$ |
| Electric charge | coulomb | C | A s |
| Electric potential difference | volt | V | $kg\ m^2\ s^{-3}\ A^{-1} = J\ A^{-1}\ s^{-1}$ |
| Electric resistance | ohm | Ω | $kg\ m^2\ s^{-3}\ A^{-2} = V\ A^{-1}$ |
| Electric capacitance | farad | F | $A^2\ s^4\ kg^{-1}\ m^{-2} = A\ s\ V^{-1}$ |
| Frequency | hertz | Hz | $s^{-1}$ |
| Celcius temperature ($\theta_c$) | degree Celcius | °C | $\theta_c$ (°C) = $\theta$ (K)-273.15 |
| Temperature interval | degrees | deg | No specification of °C or K needed |

*Acceptable:*

| | | | |
|---|---|---|---|
| Volume | litre | l | $10^{-3}$ $m^3$ |
| Pressure | bar | bar | $10^5$ N $m^{-2}$ |
| Dynamic viscosity | poise | P | $10^{-1}$ kg $m^{-1}$ $s^{-1}$ |
| Energy | electronvolt | eV | 1.6021 x $10^{-19}$ J |

The following prefixes are used to construct decimal multiples of units.

| *Multiple* | *Prefix* | *Symbol* | *Multiple* | *Prefix* | *Symbol* |
|---|---|---|---|---|---|
| $10^{-1}$ | deci | d | 10 | deca | da |
| $10^{-2}$ | centi | c | $10^2$ | hecto | h |
| $10^{-3}$ | mili | m | $10^3$ | kilo | k |
| $10^{-6}$ | micro | $\mu$ | $10^6$ | mega | M |
| $10^{-9}$ | nano | n | $10^9$ | giga | G |
| $10^{-12}$ | pico | p | $10^{12}$ | tera | T |
| $10^{-15}$ | femto | f | | | |
| $10^{-18}$ | atto | a | | | |

The values of some common physical constants in SI units include:

| | |
|---|---|
| Avogadro number, $N_A$ | 6.023 x $10^{-23}$ $mol^{-1}$ |
| Boltzmann constant, k | 1.3805 x $10^{-23}$ J/K |
| Planck constant, h | 6.626 x $10^{-34}$ J s |
| Stefan-Boltzmann constant, $\sigma$ | 5.6697 x $10^{-8}$ W/$m^2$ $K^4$ |
| Standard temperature and pressure (s.t.p.) | 273.15 K and 1.013 x $10^5$ N/$m^2$ |
| Volume of 1 kmol of ideal gas at s.t.p. | 22.41 $m^3$ |
| Gravitational acceleration | 9.807 m/s |
| Universal gas constant, R | 8.314 J/mol K |

Conversion factors for some common units are listed below. An asterisk (*) denotes an exact relationship.

| | | |
|---|---|---|
| Length | *1 in | 25.4 mm |
| | *1 ft | 0.304 8 m |
| | *1 yd | 0.914 4 m |
| | 1 mile | 1.609 3 km |
| | *1 Å (ångstrom) | $10^{-10}$ m |
| Time | *1 min | 60 s |
| | *1 h | 3.6 ks |
| | *1 day | 86.4 ks |
| | 1 year | 31.5 Ms |
| Area | *1 $in^2$ | 645.16 $mm^2$ |
| | 1 $ft^2$ | 0.092 903 $m^2$ |
| | 1 $yd^2$ | 0.836 13 $m^2$ |
| | 1 acre | 4046.9 $m^2$ |
| | 1 hectare | 10 000 $m^2$ |
| | 1 $mile^2$ | 2.590 $km^2$ |
| Volume | 1 $in^3$ | 16.387 $cm^3$ |
| | 1 $ft^3$ | 0.028 32 $m^3$ |
| | 1 $yd^3$ | 0.764 53 $m^3$ |
| | 1 UK gal | 4546.1 $cm^3$ |
| | 1 US gal | 3785.4 $cm^3$ |
| Mass | 1 oz | 28.352 g |
| | 1 grain | 0.064 80 g |
| | *1 lb | 0.453 592 37 kg |
| | 1 cwt | 50.802 3 kg |
| | 1 ton | 1016.06 kg |

| | | |
|---|---|---|
| Force | 1 pdl | 0.138 26 N |
| | 1 lbf | 4.448 2 N |
| | 1 kgf | 9.806 7 N |
| | 1 tonf | 9.964 0 kN |
| | *1 dyn | $10^{-5}$ N |
| Temperature difference | *1 degF (degR) | 5/9 degC (K) |
| Energy (work, heat) | 1 ft lbf | 1.355 8 J |
| | 1 ft pdl | 0.042 14 J |
| | *1 cal (internat. table) | 4.186 8 J |
| | 1 erg | $10^{-7}$ J |
| | 1 Btu | 1.055 06 kJ |
| | 1 chu | 1.899 1 kJ |
| | 1 hp h | 2.684 5 MJ |
| | *1 k W h | 3.6 MJ |
| | 1 therm | 105.51 MJ |
| | 1 thermie | 4.185 5 MJ |
| Calorific value (volumetric) | 1 Btu/ft$^3$ | 37.259 kJ/m$^3$ |
| | 1 chu/ft$^3$ | 67.067 kJ/m$^3$ |
| | 1 kcal/ft$^3$ | 147.86 kJ/m$^3$ |
| | 1 kcal/m$^3$ | 4.186 8 kJ/m$^3$ |
| | 1 therm/ft$^3$ | 3.726 0 GJ/m$^3$ |
| Velocity | 1 ft/s | 0.304 8 m/s |
| | 1 ft/min | 5.080 0 mm/s |
| | 1 ft/h | 84.667 $\mu$m/s |
| | 1 mile/h | 0.44704 m/s |
| Volumetric flow | 1 ft$^3$/s | 0.028 316 m$^3$/s |
| | 1 ft$^3$/h | 7.865 8 cm$^3$/s |
| | 1 UK gal/h | 1.262 8 cm$^3$/s |
| | 1 US gal/h | 1.051 5 cm$^3$/s |

| | | |
|---|---|---|
| Mass flow | 1 lb/h | 0.126 00 g/s |
| | 1 ton/h | 0.282 24 kg/s |
| Mass per unit area | 1 lb/in$^2$ | 703.07 kg/m$^2$ |
| | 1 lb/ft$^2$ | 4.882 4 kg/m$^2$ |
| | 1 ton/sq mile | 392.30 kg/km$^2$ |
| Density | 1 lb/in$^3$ | 27.680 g/cm$^3$ |
| | 1 lb/ft$^3$ | 16.019 kg/m$^3$ |
| | 1 lb/UK gal | 99.776 kg/m$^3$ |
| | 1 lb/US gal | 119.83 kg/m$^3$ |
| Pressure | 1 lbf/in$^2$ | 6.894 8 kN/m$^2$ |
| | 1 tonf/in$^2$ | 15.444 MN/m$^2$ |
| | 1 lbf/ft$^2$ | 47.880 N/m$^2$ |
| | 1 kgf/m$^2$ | 9.806 7 N/m$^2$ |
| | *1 standard atm | 101.325 kN/m$^2$ |
| | *1 at (1 kgf/cm$^2$) | 98.066 5 kN/m$^2$ |
| | *1 bar | $10^5$ N/m$^2$ |
| | 1 ft water | 2.989 1 kN/m$^2$ |
| | 1 in water | 249.09 N/m$^2$ |
| | 1 inHg | 3.386 4 kN/m$^2$ |
| | 1 mmHg (1 torr) | 133.32 N/m$^2$ |
| Power (heat flow) | 1 hp (British) | 745.70 W |
| | 1 hp (metric) | 735.50 W |
| | 1 erg/s | $10^{-7}$ W |
| | 1 ft lbf/s | 1.355 8 W |
| | 1 Btu/h | 0.293 08 W |
| | 1 Btu/s | 1.055 1 kW |

| | | |
|---|---|---|
| | 1 chu/h | 0.527 54 W |
| | 1 chu/s | 1.899 1 kW |
| | 1 kcal/h | 1.163 0 kW |
| | 1.163 0 kW | |
| | 1 ton of refrigeration | 3516.9 W |
| Moment of inertia | 1 lb $ft^2$ | 0.042 140 kg $m^2$ |
| Momentum | 1 lb ft/s | 0.138 26 kg m/s |
| Angular momentum | 1 lb $ft^2$/s | 0.042 140 kg $m^2$/s |
| Viscosity, dynamic | *1 poise (1g/cm s) | 0.1 N s/$m^2$ (0.1 kg/m s) |
| | 1 lb/ft h | 0.413 38 mN s/$m^2$ |
| | 1 lb/ft s | 1.488 2 N s/$m^2$ |
| Viscosity, kinematic | *1 stokes (1 $cm^2$/s) | $10^{-4}$ $m^2$/s |
| | 1 $ft^2$/h | 0.258 06 $cm^2$/s |
| Surface energy (surface tension) | 1 erg/$cm^2$ (1 dyn/cm) | $10^{-3}$ J/$m^2$ ($10^{-3}$ N/m) |
| Surface per unit volume | 1 $ft^2$/$ft^3$ | 3.280 8 $m^2$/$m^3$ |
| Surface per unit mass | 1 $ft^2$/lb | 0.204 82 $m^2$/kg |
| Mass flux density | 1 lb/h $ft^2$ | 1.356 2 g/s $m^2$ |
| Heat flux density | 1 Btu/h $ft^2$ | 3.154 6 W/$m^2$ |
| | *1 kcal/h $m^2$ | 1.163 W/$m^2$ |
| Heat transfer coefficient | 1 Btu/h $ft^2$ °F | 5.678 W/$m^2$ K |
| Specific enthalpy (latent heat, etc.) | *1 Btu/lb | 2.326 kJ/kg |
| Heat capacity (specific heat) | *1 Btu/lb °F | 4.186 8 kJ/kg K |

| | | |
|---|---|---|
| Thermal conductivity | 1 Btu/h ft °F | 1.730 7 W/m K |
| | 1 kcal/h m °C | 1.163 W/m K |

The reader may find the following two other conversion factors useful.

1 bar = 0.9868 atmosphere = 750.06 mm Hg
1 eV per molecule = 23.06 kcal $mol^{-1}$

From: Mullin J.W. (1971) — "Crystallisation" 2nd Edition, CRC Press.

# CONTENTS

PARTIAL LIST OF SYMBOLS .............................. viii

SI UNITS AND CONVERSION FACTORS .................. ix

1. FORMATION AND TYPES OF ICE .................... 1

1.1. The ice crystal .............................. 1

1.1.1. Ice structures .............................. 1
1.1.2. Molecular arrangement in ordinary ice ........ 2
1.1.3. Crystallographic structure representation ....... 4
1.1.4. Basic lattice dimensions ...................... 5
1.1.5. The hydrogen bonds ......................... 6
1.1.6. Theoretical density of pure ice ............... 7

1.2. Mecanisms of ice nucleation ........................ 8

1.2.1. Modes of nucleation .......................... 8
1.2.2. Theory of homogeneous nucleation............ 10
1.2.3. Heterogeneous nucleation..................... 13
1.2.4. Surface nucleation ........................... 19
1.2.5. Secondary nucleation......................... 22

1.3. River and lake ice ................................ 22

1.3.0. Introduction................................. 22
1.3.1. Ice appearance in natural water bodies........ 23

1.3.1.1. Ice nucleation at the surface ......... 23
1.3.1.2. Frazil nucleation .................... 25
1.3.1.3. Snow fall ........................... 27

1.3.2. Static ice cover formation .................... 28

1.3.2.1. Border ice .......................... 28
1.3.2.2. Plate ice ........................... 29
1.3.2.3. Growth of ice ...................... 30

1.3.3. Dynamic ice formation...................... 32

1.4. Sea ice and pressure ridges .......................... 38

1.4.1. Phase equilibrium in sea ice .......................... 38
1.4.2. Formation of sea ice .......................... 40
1.4.3. Salinity profiles in young sea ice .......................... 42
1.4.4. Polar ice .......................... 44
1.4.5. Pressure ridges .......................... 47

1.5. Glaciers and icebergs .......................... 48

1.5.0. Introduction .......................... 48
1.5.1. Morphology of glaciers .......................... 49
1.5.2. Equilibrium limit of a glacier .......................... 51
1.5.3. Glacier flow .......................... 53
1.5.4. Crystalline structure of a temperate glacier .... 58
1.5.5. Formation of icebergs .......................... 60

1.6. Classification of ice structures .......................... 62

1.6.0. Introduction .......................... 62
1.6.1. Basic ice layers .......................... 62
1.6.2. Structure and texture of ice covers .......................... 63
1.6.3. Classification .......................... 65
1.6.4. Inclusions and impurities .......................... 75

2. MECHANICAL PROPERTIES OF ICE .......................... 81

2.0. Introduction .......................... 81
2.1. Brittle behavior .......................... 81

2.1.0. Introduction .......................... 81
2.1.1. Brittle behavior of monocrystals .......................... 82

2.1.1.1. Elastic modulii .......................... 82
2.1.1.2. Fracture and crack propagation under tension .......................... 86
2.1.1.3. Fracture under compression .......................... 93

2.1.2. Brittle behavior of polycrystalline ice .......... 95

2.1.2.0. Introduction .......................... 95
2.1.2.1. Elastic modulii ....................... 96
2.1.2.2. Brittle crack formation and propagation 100
2.1.2.3. Tensile strength ..................... 102
2.1.2.4. Crushing strength .................... 109
2.1.2.5. Ice strength under a combined state of stress .............................. 114

2.2. Ductile behavior ................................ 116

2.2.0. Introduction .................................. 116
2.2.1. Movement of dislocations in ice .............. 117
2.2.2. Basic law of creep of an ice crystal .......... 122
2.2.3. Classification of creep processes ............. 125

2.2.3.1. Typical creep curves under a constant shear stress ........................ 129
2.2.3.2. Typical creep curves under a constant shear strain ........................ 131

2.2.4. Mechanical model of creep in polycrystalline ice 133

2.2.4.1. Basic model .......................... 133

2.2.5. Creep of ice under a constant load ........... 139

2.2.5.1. Equations of $\epsilon\alpha$ and $\nu$ creep under a constant stress ......................... 139
2.2.5.2. Equations of $\alpha\nu$ creep under a constant stress .............................. 144
2.2.5.3. Creep curves for monocrystals ........ 146
2.2.5.4. Creep curves for polycrystalline ice ... 148

2.2.6. Creep of ice at constant speed .............. 155

2.2.6.1. Equations of $\epsilon\alpha$ and $\nu$ creep under a constant strain rate ..... 155
2.2.6.2. Creep curves for monocrystals ..... 157
2.2.6.3. Creep curves for polycrystalline ice ... 160

2.2.7. Recrystallization and glacier flow ..... 161
2.2.8. Yield strength of ice ..... 164

2.2.8.0. Introduction ..... 164
2.2.8.1. Yield strength representation ..... 165
2.2.8.2. Transition from ductile to brittle failure 169
2.2.8.3. Failure criterion in the ductile range.. 170

3. THE BEARING CAPACITY OF ICE ..... 177

3.0. Introduction ..... 177
3.1. Mechanical behavior of floating ice ..... 178

3.1.1. Flexural strength of ice beams ..... 178
3.1.2. Behavior of ice covers under load ..... 191
3.1.3. Criteria for the bearing capacity of ice covers . 193

3.2. Dynamic behavior of ice covers ..... 197

3.2.1. Equations of plates on elastic foundations ..... 197

3.2.1.1. Solution for a concentrated load ..... 203
3.2.1.2. Solution for a uniform load over a circular area ..... 204
3.2.1.3. Solution for a thick plate ..... 206
3.2.1.4. Effect of transverse shear strains ..... 208

3.2.1.5. Arbitrary load distribution on an infinite plate ..... 209
3.2.1.6. Solution for a semi-infinite plate ..... 209
3.2.1.7. Solution for a narrow wedge ..... 212

3.2.2. Speed of vehicles ..... 215
3.2.3. Experimental and field investigations ..... 218

3.3. Static behavior of ice covers ..... 223
3.4. Design considerations ..... 228

3.4.0. Introduction ..... 228
3.4.1. Use of natural ice covers ..... 230
3.4.2. Built-up ice surfaces ..... 238

4. FORCES EXERTED BY ICE ON STRUCTURES ..... 246

4.0. Introduction ..... 246
4.1. Force of an expanding ice sheet ..... 248

4.1.0. Introduction ..... 248
4.1.1. Coefficient of thermal expansion ..... 250
4.1.2. Thermal exchange ..... 252
4.1.3. Ductile behavior of ice submitted to increasing temperatures ..... 258
4.1.4. Computation of ice thrust ..... 264
4.1.5. Protection against static ice pressure ..... 271
4.1.6. Other effects ..... 274

4.2. Impact of drifting ice ..... 276

4.2.0. Introduction ..... 276
4.2.1. Movement of ice floes ..... 277
4.2.2. Indentation by a vertical face structure ..... 281

4.2.2.0 Introduction .......................... 281
4.2.2.1. General relationships ................. 281
4.2.2.2. Crushing of ice against a vertical face 283
4.2.2.3. Buckling of ice fields ................ 299
4.2.2.4. Splitting of ice floes ................ 302

4.2.3. Failure of ice floes on an inclined structure ... 305

4.2.3.0 Introduction .......................... 305
4.2.3.1. Relations between the vertical and horizontal components .................. 307
4.2.3.2. Simplified analysis (Korzhavin's formulas) .............................. 310
4.2.3.3. Crushing failure before ride-up ....... 316
4.2.3.4. Analysis of bending on a cone with an elasto-plastic material and foundation . 316
4.2.3.5. Impact of a multiyear ice ridge ...... 319
4.2.3.6. Dimensional analysis ................ 321

4.2.4. Field verifications .......................... 323
4.2.5. Abrasive action of moving ice ............... 339

4.3. Ice pressure by unconsolidated accumulations ......... 341

4.3.0. Introduction ............................... 341
4.3.1. Equilibrium of an ice accumulation ........... 342
4.3.2. Stability of an idealized ice jam .............. 349
4.3.3. Maximum thrust of unconsolidated ice accumulations ...................................... 352
4.3.4. Ice pile-ups on structures .................... 354

4.4. Vertical forces exerted by ice ........................ 358

4.4.0. Introduction ............................... 358
4.4.1. Strength of adhesion of ice to structures ...... 359
4.4.2. Vertical forces on a straight wall ............. 361
4.4.3. Lifting force on isolated structures ........... 364

5. ICEBREAKERS ........ 377

5.0. Introduction ........ 377
5.1. Ice conditions and modes of icebreaking ........ 380
5.2. Ship resistance in unconsolidated ice ........ 381
5.3. The mechanics of continuous icebreaking in uniform solid ice ........ 388
5.4. Icebreaking by ramming ........ 399
5.5. Maneuverability in ice ........ 410
5.6. Characteristics of icebreaking ships ........ 413

5.6.1. Icebreakers ........ 413
5.6.2. Commercial icebreaking ships ........ 421

5.7. Special methods to break ice ........ 432

6. ICE MODELING ........ 437

6.0. Introduction ........ 437
6.1. Laws of similitude ........ 438

6.1.1. Basic relations ........ 438
6.1.2. Complete similitude ........ 445
6.1.3. Distorted similitude ........ 447

6.1.3.1. General equations ........ 447
6.1.3.2. Distorted hydraulic models ........ 449
6.1.3.3. Unconsolidated ice accumulations ........ 449
6.1.3.4. Distortion with a solid ice cover ........ 451

6.2. Model techniques ........ 453

6.2.0. Introduction ........ 453
6.2.1. Model materials ........ 455

6.2.1.1. Solid particles and blocks ........ 455
6.2.1.2. Real ice ........ 457
6.2.1.3. Weak breakable materials ........ 458

6.2.2. Models of ice jams .......................... 461
6.2.3. Models of ice covers in rivers ................ 464
6.2.4. Models of structures in ice ................... 467
6.2.5. Models of ships in ice ........................ 470

NAME INDEX ........................................... 479

INDEX OF SUBJECTS .................................... 485

# CHAPTER 1

# FORMATION AND TYPES OF ICE

## 1.1 - THE ICE CRYSTAL

### 1.1.1 - ICE STRUCTURES

Three types of crystallographic structures are presently known for ice:

- the hexagonal structure
- the cubic structure
- the amorphous form

The hexagonal structure is the only one which is found in nature. It is stable down to very low temperatures at normal pressures.

Other structures of ice have been observed in the laboratory. At least seven allotropic varieties have been observed at pressures exceeding 200 atmospheres. At very low temperatures and atmospheric pressures ice of cubic structure and of amorphous form has been identified. Because we are dealing only with natural ice in this text we will discuss only the behavior of hexagonal ice of type 1. For those interested by other ice types a phase diagram with discussion is given in Whalley (1969).

## 1.1.2 - MOLECULAR ARRANGEMENT IN ORDINARY ICE

It has been known since Bragg (1922) that the oxygen atoms in natural ice are arranged in a tetrahedral pattern, each oxygen atom being surrounded by four equally spaced oxygen atoms at the vertices of the tetrahedron as shown in Fig. 1.1. If the tetrahedron were perfect, the angles between the direction of all oxygen atoms would be $109^{\circ}$ 28' which is extremely close to what has been observed for ice.

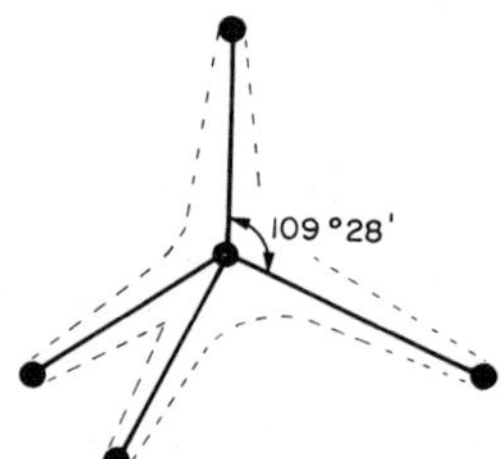


*Fig. 1.1. Tetrahedral arrangement of oxygen atoms in ice.*

The continuous assembly of oxygen atoms in an ice crystal has been described by Barnes (1929) and confirmed by numerous measurements. It is shown in Fig. 1.2.

A number of important features of the molecular structure of ice can be seen from these figures. Firstly, it is apparent that the tetrahedral coordination of the oxygen atoms gives rise to a crystal structure possessing hexagonal symmetry. The hexagonal shape exhibited by many ice crystals is clearly related to the hexagonal symmetry of the molecular arrangement. Another important characteristic of the molecular structure is that the molecules are all concentrated close to a series of parallel planes known as the basal planes. The normal to the basal planes is referred to as the c-axis of the crystal. Finally, it is clear that ice has a very open structure and this is reflected in its low density compared to liquid water.

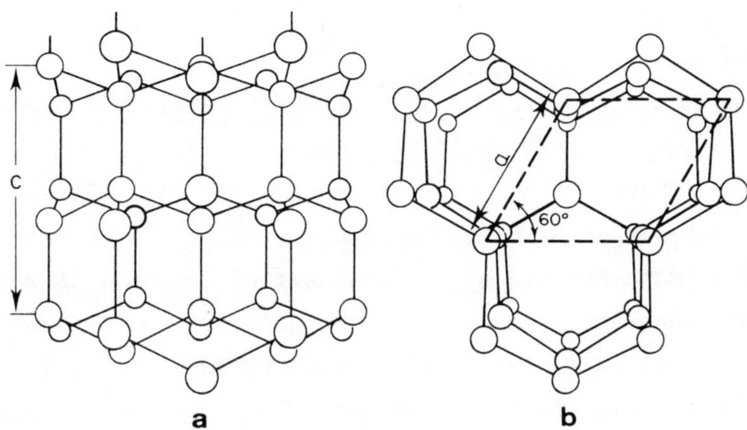


*Fig. 1.2. Continuous molecular structure of ice. The lines joining the oxygen atoms represent bounds. The hydrogen atoms are not shown.*
*a) View perpendicular to the c-axis whose length is shown.*
*b) View along the c-axis where the distance a in the basal plane is shown.*

It can be seen from Fig. 1.2 that the fundamental building block of the crystallographic structure of ice is the assemblage of oxygen atoms indicated by the parallepiped drawn in dashed lines in Fig. 1.2. The region of space defined by this group of atoms is termed a unit cell. The complete crystallographic structure may be constructed by stacking identical unit cells face to face in perfect alignment in three dimensions. In this unit cell each of the atoms at the eight vertices is shared between eight adjacent cells, each of the four atoms on the edges is shared between four cells, and the remaining two atoms are completely within the unit cell. Hence, the total number of oxygen atoms and therefore the total number of water molecules in the unit cell is: $(8/8) + (4/4) + 2 = 4$.

### 1.1.3 - CRYSTALLOGRAPHIC STRUCTURE REPRESENTATION

The various symmetry elements exhibited by the complete lattice of oxygen atoms in ice are represented in crystallography by the Hermann-Mauguin space group symbol $P6_3$ mmc: P refers to the fact that the space lattice is primitive; $6_3$ indicates that the principal axis of symmetry, which is the c-axis, is a hexagonal screw axis. This implies that when a group of oxygen atoms in the crystal is translated along the c-axis and is at the same time rotated about this axis at such a rate that the group makes one complete revolution after travelling a distance c in the direction of the c-axis, the atoms in the group have travelled a distance along the c-axis equal to c/2 (Fig. 1.3a).

The first m in the space group symbol indicates that there are mirror planes M normal to the principal axis such that each molecule in a mirror image position above the plane (Fig. 1.3a). The remaining two symbols m and c indicate that there are two sets of symmetry planes parallel to the c-axis and that one of these is a set of mirror planes M and the other a set of glide planes G (Fig. 1.3b). The term glide plane in this context means that if the mirror image of a particular molecule is formed in the glide plane and is then translated a distance c/2 along the c-axis, it will coincide with another molecule (Fig. 1.3c).

Finally, it should be noted that the molecular structure of ice is centrosymmetric, for a point can always be located within the lattice such that every molecule which occurs at a given distance from this point in one direction has a corresponding molecule at the same distance from the point in the opposite direction.

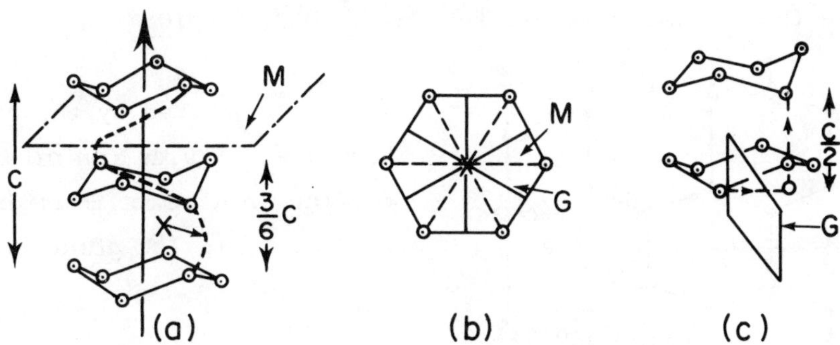


*Fig. 1.3. The symmetry elements of ice: (a) the screw axis $6_3$ and a mirror plane M normal to the c-axis. (b) the mirror planes M and the glide planes G parallel to the c-axis. (c) the operation of a glide plane. From Hobbs (1974).*

## 1.1.4 - BASIC LATTICE DIMENSIONS

The projection of the ice lattice on a plane through the c-axis is shown in Fig. 1.4. The dark circles indicate oxygen atoms in the plane of the cut and the white circles are in an intermediate offset plane. The lattice points of the unit cell are shown as A B C D in the plane. The fundamental dimensions are r, the distance between two atoms, a the dimension in the basal plane and c the cell length along the c-axis.

If there is a perfect tetrahedral coordination with equal distance r between bonded molecules, the values of c can be related to a and r by the geometric relations:

$$a = 2\ r \sin 54^{\circ}\ 44'$$
$$c = 2\ r\ (1 + \sin 19^{\circ}\ 28') = 1.634\ a \qquad (1\text{-}1)$$

Measurements by Megaw (1934) with X-ray diffraction, which are still valid, show that the basic lattice parameters have the following values for ice at $0^{\circ}C$.

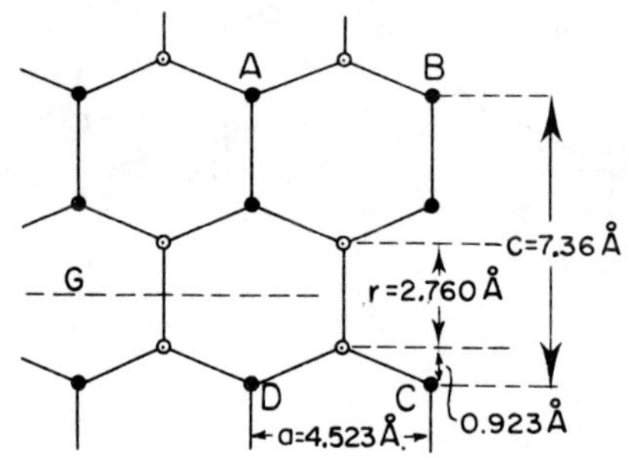


*Fig. 1.4. Projection of the ice lattice on a plane through the c-axis. G is the glide plane.*

$r = 2.760$ Å $\qquad c = 7.367$ Å

$a = 4.523$ Å $\qquad c/a = 1.629$

Kumai (1968) deduced from electron diffraction measurements that the ratio of c/a was equal to the theoretical value, for ice at a temperature of $-110^{\circ}C$.

## 1.1.5 - THE HYDROGEN BONDS

The position of the hydrogen atoms in the ice molecule is not as well ascertained as that of the oxygen atoms. At this time it is still thought that the best representation is that given by Pauling (1935) statistical model. This model is based on the following assumptions:

(*i*) Each oxygen atom has two hydrogen atoms attached to it at distances of about 0.95 Å, thereby forming a water molecule.

(*ii*) Each water molecule is oriented so that its two hydrogen atoms are directed approximately towards two of the four oxygen atoms which sur-round it tetrahedrally.

(iii) The orientations of adjacent water molecules are such that only one hydrogen atom lies between each pair of oxygen atoms.

(iv) Under ordinary conditions ice can exist in any one of a large number of configurations, each corresponding to a certain distribution of the hydrogen atoms with respect to the oxygen atoms.

Furthermore it has been found that the hydrogen atom will jump from one position close to one oxygen atom to the other position next to the other one. This is particularly helped by point defects in the structure of the ice. Riehl (1965) has found that the hydrogen protons would oscillate along the bonds with a frequency of $2 \times 10^{11}\ s^{-1}$.

### 1.1.6 - THEORETICAL DENSITY OF PURE ICE

One of the first observation concerning the structure of ice is that it has a very open lattice; the spacing between the oxygen atoms are quite large compared with the size of the atoms themselves.

From the lattice constants we can compute the density of ice by selecting a representative volume of the ice and counting the number of molecules in this volume.

Let us consider the volume of a unit cell shown in Fig. 1.2, which is:

$$v = ca^2 \sin 60^\circ \qquad (1\text{-}2)$$

With $a = 4.523$ Å and $c=7.367$ Å this gives a volume of 130.52 Å$^3$. The unit cell will contain 4 water molecules. The number of molecules in one mole of ice is given by Avogadro's number: $6.025 \times 10^{23}$. Thus the volume of a mole of ice is:

$$\frac{1}{4} \times 6.025 \times 10^{23} \times 130.52 \times 10^{-24} = 19.66 \text{ cm}^3$$

The mass of a mole of ice is 18.02 g so the density of ice at $0^{\circ}$C is predicted to be:

$$\rho = \frac{18.02}{19.66} = 0.9166 \text{ g cm}^{-3}$$

An experimental value of $\rho' = 0.9167$ g cm$^{-3}$ has been found by Ginnings and Corrucini (1947). This is a very good agreement.

## 1.2 - MECANISMS OF ICE NUCLEATION

### 1.2.1 - MODES OF NUCLEATION

If we cool a liquid and measure the temperature at which crystallization starts spontaneously, we find the nucleation temperature to be always lower than the melting point.

Fig. 1.5 shows the cooling curve of pure water when heat is being removed at a constant rate. It may be noticed that a certain amount of supercooling is required to nucleate the first ice crystal. The second stage, liquid and solid, corresponds to crystal growth and the last to the cooling of ice itself.

The condition of supercooling alone is not a sufficient cause for a system to begin to crystallize. Before crystals can grow there must exist in the solution a number of minute centers of crystallization known as seeds, embryos or nuclei. Nucleation may occur spontaneously or it may be induced artificially. It is not always possible, however, to decide whether a system has nucleated of its own or whether it has done so under the influence of some external stimulus.

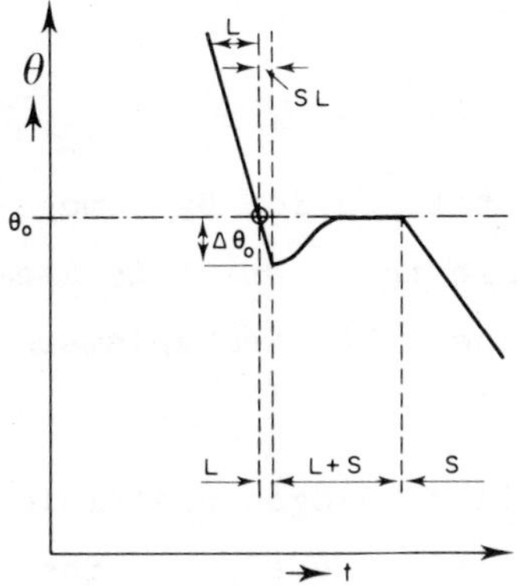


*Fig. 1.5. Schematic cooling curve of a liquid with heat being removed at a constant rate.* $\theta$ *is the temperature;* $\theta_o$ *the temperature of phase equilibrium,* $\Delta\theta_o$ *the degree of supercooling;* t *is the time;* L, S *and* SL *are respectively the liquid, solid and supercooled liquid states.*

We may distinguish the following modes of nucleation:

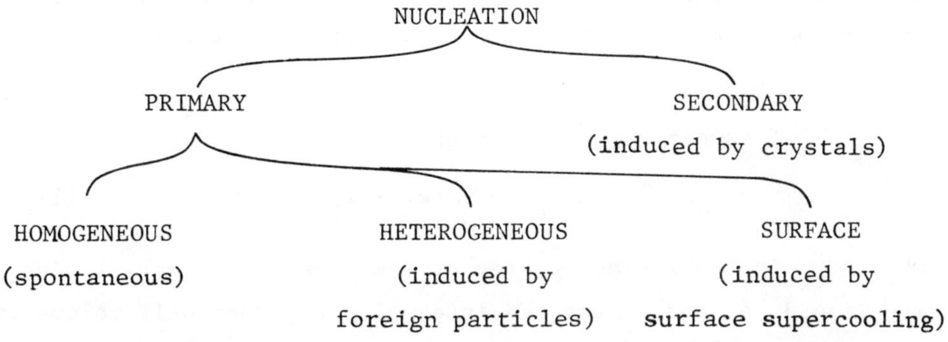


How a crystal nucleus is found within a homogeneous fluid is not known with any degree of certainty. Not only do the fluid molecules have to agglomerate to form a nucleus but they also have to become oriented into a fixed lattice. It takes about 80 molecules to form such a water (ice) nucleus and it is most improbable that such an event would happen at normal freezing temperature. This is why it takes such a large amount of supercooling (-40 ±2)$^{\circ}$C before the appearance of ice crystals in pure water.

Ice in nature, can then be nucleated only by other processes or by secondary nucleation.

## 1.2.2 - THEORY OF HOMOGENEOUS NUCLEATION

Although the mechanism of crystallization by homogeneous nucleation is more or less of academic interest it is useful to analyse as a reference from which the other mechanisms could be discussed.

The formation of a crystal within a homogeneous fluid demands the expenditure of a certain quantity of energy in the creation of the solid surface. Therefore the total quantity of work W, required to form a stable crystal nucleus is equal to the work required to form the surface Ws (a positive quantity) and the bulk of the particle Wv (a negative quantity).

$$W = Ws + Wv \tag{1-3}$$

This can also be written:

$$W = a\,\sigma_{c\ell} - v\,(e_{\ell} - e_{s}) \tag{1-4}$$

where $\sigma_{c\ell}$ is the surface energy of the crystal in contact with the liquid, $e_{\ell}$ and $e_{s}$ are the internal energy per unit volume in the liquid and crystal respectively, a and v are the surface area and volume of the crystal.

If this is applied to a nucleus forming a spherical crystal of radius r, then Eq. (1-4) becomes:

$$W = 4\,\pi\,r^2\,\sigma_{c\ell} - \frac{4}{3}\,\pi\,r^3\,(e_{\ell} - e_{s}) \tag{1-5}$$

The work required to form a nucleus of radius r is usually called the excess free energy, or the Gibb's free energy $\Delta G$, as it was first derived for the formation of water droplets from the vapor phase and is refered to a unit volume of the fluid. On Fig. 1.6, the variation of the free energy in function of the radius of the embryos is represented and is the sum of the surface energy and of the volume free energy. It can be seen that this function has a maximum. This maximum value $\Delta G_c$ corresponds to the critical nucleus $r_c$ where a spherical cluster of molecules forming an embryo will be transformed into a viable crystal. It can be obtained by maximizing Eq. (1-5):

$$\frac{d\Delta G}{dr} = 8\,\pi\,r\,\sigma_{c\ell} - 4\,\pi\,r^2\,(e_\ell - e_s) = 0 \qquad (1\text{-}6)$$

therefore

$$r_c = \frac{2\,\sigma_{c\ell}}{(e_\ell - e_s)} \qquad (1\text{-}7)$$

$$\Delta G_c = \frac{16\,\pi\,\sigma_{c\ell}^{\,3}}{3\,(e_\ell - e_s)^2} = \frac{4\,\pi\,\sigma_{c\ell}\,r_c^{\,2}}{3} \qquad (1\text{-}8)$$

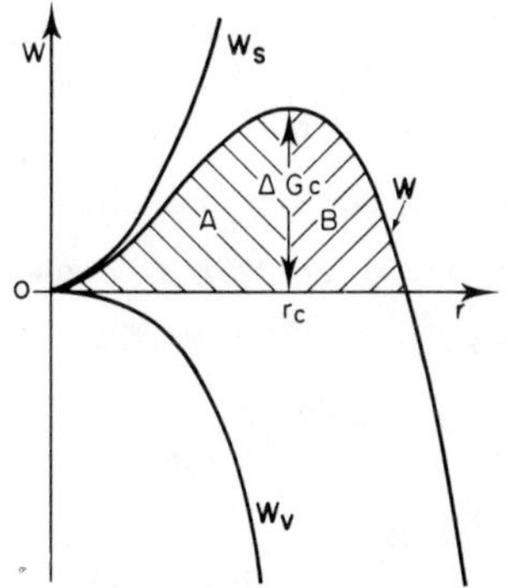


*Fig. 1.6. Free energy change W in a drop of liquid as a function of radius. In zone A only unstable embryos can exist. Over the maximum free energy change $\Delta G_c$, stable nuclei are formed in zone B.*

The behaviour of a newly created crystalline lattice structure in a supercooled solution depends on its size: it can either grow or redissolve but the process which it undergoes should result in a decrease in the free energy of the particle.

The critical size $r_c$, therefore, represents the minimum size of a stable nucleus. Particles smaller than $r_c$ will dissolve or dissappear in the supercooled liquid, because only in this way can the particle achieve a reduction in its free energy. Similarly, particles larger than $r_c$ will continue to grow as crystals.

Although it can be seen why a particle of size greater than $r_c$ is stable, it does not explain how the amount of energy $\Delta G_c$ necessary to form a stable nucleus is produced. This may be explained as follows. The energy of a fluid system at constant temperature and pressure is constant, but this does not mean that the energy level is the same in all parts of the fluid. There will be fluctuations in the energy about the constant mean value, i.e. there will be a statistical distribution of energy, or molecular velocity, in the molecules constituing the system and in these supercooled regions where the energy level rises temporarily to the critical value, nucleation will be favoured.

The rate of nucleation J, e.g. the number of nuclei formed per unit time per unit volume can be expressed in the form of the Arrhenius'equation which is the rate equation commonly used for a thermally activated process:

$$J = A \exp(-\Delta G_c/k\theta^*) \tag{1-9}$$

where k is the Boltzmann gas constant, $1.3805 \times 10^{-23}\ JK^{-1}$ and $\theta^*$ the absolute temperature.

The volume free energy $(e_\ell - e_s)$ may be expressed in function of the temperature of supercooling and the latent heat of fusion of the ice, L:

$$(e_\ell - e_s) = \frac{L\ \Delta\theta_o}{\theta^*_o} \tag{1-10}$$

where $\theta^*_o$ is the solid-liquid equilibrium temperature, 273° Kelvin and $\Delta\theta = \theta^*_o - \theta^*$ is the supercooling of the liquid.

With Eqs (1-7), (1-8) and (1-9) this gives:

$$r_c = \frac{2\ \sigma_{c\ell}\ \theta^*_o}{L\ \Delta\theta_o} \tag{1-11}$$

$$\Delta G_c = \frac{16\ \pi\ \sigma_{c\ell}^{\ 3}\ \theta^{*2}_o}{3\ L^2\ \Delta\theta_o^{\ 2}} \tag{1-12}$$

$$J = A\ \exp\left[\frac{-16\ \pi\ \sigma_{c\ell}^{\ 3}\ \theta^*_o}{3\ k\ L^2\ \Delta\theta^2_o}\right] \tag{1-13}$$

For a value of $\sigma_{c\ell}$ of 33 mJ/m², L = 333.84 k J/kg and an amount of supercooling of -40°C, the critical radius of an embryo will be $1.35 \times 10^{-9}$ m and it will contain 80 ice molecules.

Eq. (1-13) shows the dominant effect of supercooling on the nucleation rate.

### 1.2.3 - HETEROGENEOUS NUCLEATION

We will now consider the mechanism of heterogeneous nucleation which is basically the more important one in nature, either for the formation of ice crystals in clouds or the nucleation of supercooled water. In both cases the presence of foreign particles significantly reduces the required supercooling below that needed for homogeneous nucleation. There are a large number of heteronuclei in nature. Aqueous solutions as normally prepared in the laboratory may contain $10^6$ to $10^8$ solid particles per cm³. It is virtually impossible to achieve a solution completely free of foreign bodies although careful filtration may reduce the numbers to $10^3$ $cm^{-3}$ and render water more difficult to nucleate.

Because the surface free energy of an ice embryo will generally be a minimum when the shape of the embryo is some polyhedron we will consider the model proposed by Fletcher (1970), as shown on Fig. 1.7.

The embryo C is a prismatic column of height H, inscribed radius r, and contains n prism faces. It rests on a substratum S which is required for heterogeneous nucleation.

The surface energies are respectively $\sigma_{s\ell}$ between the substratum and the liquid, $\sigma_{sc}$ between the substratum and the crystal and $\sigma_{c\ell}$ between the crystal and liquid.

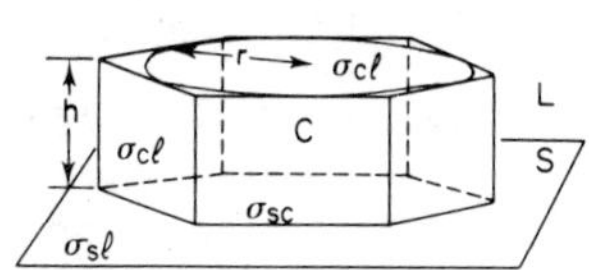


*Fig. 1.7. A prismatic embryo growing upon a perfect plane substrate. Subscript c, ℓ and s refer to the crystal, liquid and substrate.*

With this configuration of the crystal and the notation used previously, the increase in free energy of the system is:

$$\Delta G = \pi r^2 \zeta h (e_s - e_\ell) + \pi r^2 \zeta(\sigma_{sc} - \sigma_{s\ell}) + (\pi r^2 \zeta + 2\pi r \zeta h)\sigma_{c\ell} \quad (1\text{-}14)$$

with the value:

$$\zeta = \frac{n}{\pi} \tan \frac{\pi}{n} \quad (1\text{-}15)$$

The surface energies can be represented by a quantity:

$$\chi = (\sigma_{c\ell} - \sigma_{s\ell} + \sigma_{sc})/\sigma_{c\ell} \quad (1\text{-}16)$$

or more generally:

$$\chi = 1 - \cos\gamma \tag{1-17}$$

with

$$\cos\gamma = (\sigma_{s\ell} - \sigma_{sc})/\sigma_{c\ell} \tag{1-18}$$

$\gamma$ being usually taken as the angle of contact between ice and the nucleating material. For complete affinity (cf. complete wetting) of the substrate with ice, $\gamma = 0$ and $\chi = 0$ and for complete non-affinity $\gamma = 180^o$ and $\chi = 2$.

With these notations, Eq. (1-14) can be simplified to:

$$\Delta G = \pi\, r^2\, \zeta h\, (e_s - e_\ell) + \pi\, r^2\, \chi\, \zeta\, \sigma_{c\ell} + 2\,\pi\, r\, \zeta h\, \sigma_{c\ell} \tag{1-19}$$

As for the case of homogeneous nucleation, the value of $\Delta G$ initially increases with increasing values of r and h, but after a certain critical size is reached, the ice embryo grows freely, producing a decrease in the free energy of the system. The critical size of the embryo is given by:

$$\frac{\partial \Delta G}{\partial h} = 0 \text{ and } \frac{\partial \Delta G}{\partial r} = 0 \tag{1-20}$$

which lead to:

$$r_c = \frac{2\,\sigma_{c\ell}}{(e_\ell - e_s)} \tag{1-21}$$

$$h_c = \chi r_c \tag{1-22}$$

$$\Delta G_c = 4\,\pi\,\chi\,\zeta\,\frac{\sigma_{c\ell}^{\,3}}{(e_\ell - e_s)^2} \tag{1-23}$$

With the expression of the volume free energy in (1-10) we also have:

$$\Delta G_c = \frac{4\ \pi\ \chi\ \zeta\ \sigma_{c\ell}^{\ 3}\ \theta_o^{*2}}{L^2\ \Delta\theta_o^{\ 2}} \tag{1-24}$$

or

$$\Delta\theta_o = \sqrt{\frac{4\ \pi\ \chi\ \zeta}{\Delta G_c}} \times \frac{\sigma_{c\ell}^{\ 3/2}\ \theta_o^*}{L} \tag{1-25}$$

It is most interesting to discuss, in this last expression, the value of the temperature of supercooling required to nucleate a water body if it is of sufficient size to contain heteronuclei.

Let us start with the case of reference of homogeneous nucleation. Then $\gamma = 180^{\circ}$, and $\chi = 2$. This gives:

$$\Delta\theta_o = 2\sqrt{\frac{2\ \pi\ \zeta}{\Delta G_c}} \times \frac{\sigma_{c\ell}^{\ 3/2}\ \theta_o^*}{L} \tag{1-26}$$

For the same critical free energy of nucleation and similar crystal form, the amount of supercooling will then be reduced with an heteronucleus (n) to:

$$\frac{\Delta\theta_n}{\Delta\theta_o} = \sqrt{\frac{\chi}{2}} = \sqrt{\frac{1 - \cos\gamma}{2}} \tag{1-27}$$

At the limit, when nucleation is initiated with an ice crystal $\gamma = 0$ and $\Delta\theta_n = 0$. This could be expected as it is well known that no supercooling is required to grow ice from any existing ice nucleus.

It is usually considered that usual inorganic nucleators in static conditions have a temperature threshold of about $-3.5^{\circ}$C. If we compare this to $-40^{\circ}$C for homogeneous nucleation we can compute with (1-27) that the value of $\gamma$ is $10^{\circ}$ for these best nucleators. All others, of course, must have an higher angle of contact with water and ice.

In nature, there are however, nucleators that are much more efficient than inorganic nucleators. Devik (1949), Höhne (reported by Geiger 1965), Schnell (1972) and Katsaros *et al.*, (1974) reported nucleating thresholds $\Delta\theta_n$ between $1^\circ C$ and $1.5^\circ C$ for natural water surfaces. Michel (1977) made some experiences in a pan cooled by emitted radiations at night. He also found surface supercooling between $-0.5^\circ C$ to $-1.0^\circ C$ before the apparition of long needles of ice at the surface. Robert (1978) made systematic tests in a cold room with a small tank of agitated water, isolated from the outside atmosphere. For ordinary tap water he found an average nucleating temperature of $-2.5^\circ C$. With a large amount of silt particles in the water the threshold would increase between $-1.0^\circ C$ to $-1.5^\circ C$. Leaves of aulne and floating pine needles would also increase the temperature of nucleation around $-1.0^\circ C$.

Most of the research on heterogeneous nucleation has been done on water droplets in clouds. In that case, the presence of an active ice nucleus is not a certainty and must be determined on a probability basis. A model assumes that every drop nucleates at a temperature determined by the most effective ice nucleus it contains. Therefore no freezing event would occur at a fixed temperature. If we assume that the concentration n $(\theta)$ of ice nuclei which become active between $\Delta\theta_n$ and $\Delta\theta$ in $^\circ C$, is given by:

$$n = n_c \exp \alpha \, [\Delta\theta - \Delta\theta_n] \qquad (1\text{-}28)$$

where $n_c$ is the concentration of active particles at the critical temperature of supercooling $\Delta\theta_n$ above which no foreign particle is active in the bulk and $\alpha$ is the affinity of nucleation of the foreign particles. Then, if a drop contains a random distribution of ice nuclei, the probability p that a droplet of volume v will contain one active nucleus when cooled to $\Delta\theta$ is given by Poisson statistics as:

$$p = 1 - \exp[-nv] \qquad (1\text{-}29)$$

when $p = \frac{1}{2}$, that is when 50% of the droplets are frozen at a temperature $\theta$ for a volume v, we obtain:

$$|\Delta\theta - \Delta\theta_n| = (1/\alpha)(.366 + \ln n_c v) \qquad (1\text{-}30)$$

This relation applies well to experimental data. In Fig. 1.8 each curve corresponds to a different concentration of foreign particles, the highest on the graph corresponding to the highest concentration n. The slope of these curves depends on the affinity of nucleation of the particles $\alpha$. Fig. 1.9 shows without any doubt that above a critical supercooling $\Delta\theta_n$, no particle in the droplets is active as a nucleus in larger volume of water as confirmed by Dorsey (1948).

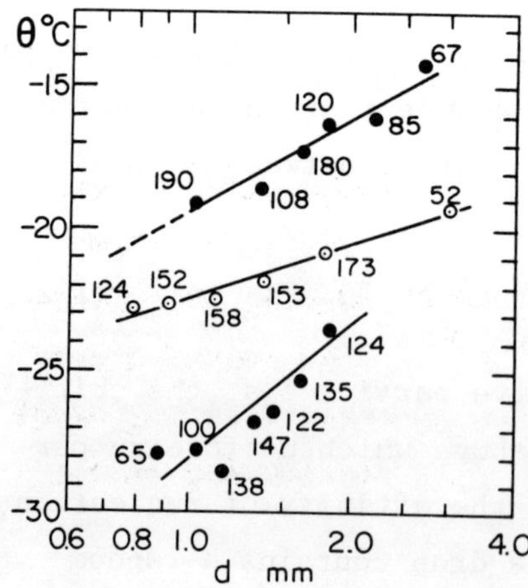


*Fig. 1.8. Freezing of groups of water droplets of various water qualities. Figures indicate total number of tests for a given diameter d. From Bigg (1953).*

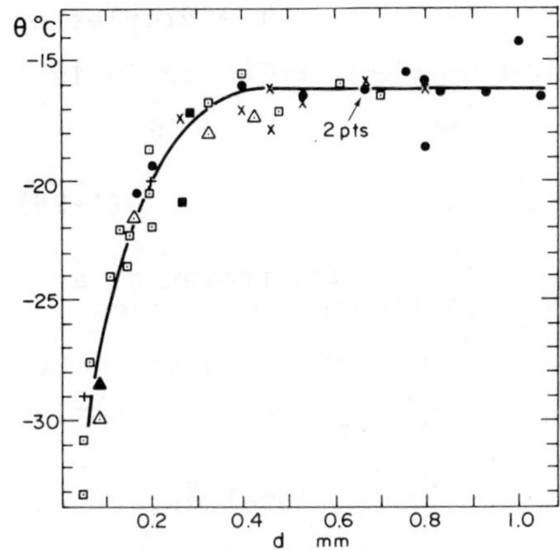


*Fig. 1.9. Spontaneous freezing point of water droplets as a function of droplet size d. Source of water in droplet: • sublimation Δ1% sucrose solution, Δ1% NaCl solution, + distilled water, boiled water, cooled boiled water, × condensated water. From Heverly (1949).*

## 1.2.4 - SURFACE NUCLEATION

It is quite surprising to observe that in large bodies of turbulent water in nature, the amount of supercooling which is measured before the appearance of ice crystals is only of the order of a few hundredths of a $^{\circ}$C, at most a few tenths. We think that surface nucleation may well explain this phenomenon (Michel 1967, 1977).

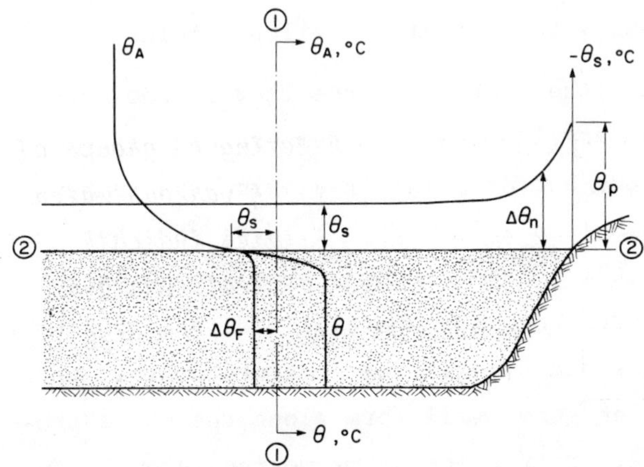


*Fig. 1.10. Cross section of flow showing the three-dimensional tempature boundary layer in flowing water.*

Let us consider on Fig. 1.10 the temperature distribution in flowing water. There are two boundary limits, one at the water-air interface and the other at the water-shore interface.

If we consider first, the temperature distribution along a vertical section 1.1, we note that there is a strong temperature gradient between the water temperature in the bulk, $\theta$, and the air temperature $\theta_a$. Most of the temperature difference occurs in the air but there is also a boundary layer with a smaller temperature difference in the top layer of water. This temperature difference in the top water boundary layer depend on the turbulence of the flow. It is large for calm water and laminar flow, it gets smaller as the turbulence increases. If the top water boundary layer is large enough the surface temperature will attain first the nucleating threshold for natural water bodies (around $-1^{o}C$) and germs of ice will be nucleated at the surface.

In calm water of lakes or laminar flow, ice needles will be nucleated at the surface and plate ice will grow from them. In slightly turbulent flow the ice germs will be entrained within the whole water mass to nucleate frazil particles.

Let us consider the case where the flow is too much turbulent to form a boundary layer with a temperature differential of the order of $1^{o}C$ at the surface. If we examine the temperature along the flow surface on line 2-2 of Fig. 1.10 we see that there will also be another boundary layer as we approach the shore. The flow will be less turbulent, the shore will also be colder so that the nucleating threshold $\Delta\theta_n$ will be attained at this three-dimensional boundary. Ice germs will form along the shore producing either border ice, or free ice germs moving into the turbulent flow.

The presence of ice germs nucleated at the top is sufficient to nucleate all the particles of frazil. Some definition of terms is however prerequisite before we describe the full nucleating process. One nucleus is a foreign particle of favorable structure which serves as a substratum for the eventual formation of a frazil particle. One embryo is a nucleus which has absorbed a water layer in a close packed form characteristic of the structure of ice. A germ is an ice particle coming from any source including a nucleated embryo.

Forced nucleation occurs when an embryo is nucleated by contact with a germ. The amount of supercooling that is needed for forced nucleation of a particle of radius $r_\ell$ can be computed from Eq. (1-11). It is:

$$\Delta\theta_F = \frac{2\,\sigma_{c\ell}\,\theta^*_o}{r_\ell\,L} \qquad (1\text{-}31)$$

where $\Delta\theta_F$ is the threshold for forced nucleation. The best frazil nucleators in flowing water will be the ones having the smallest diameters. With clay particles of the order of 0.05 μm in diameter and a value of $\sigma_{c\ell} = 33$ mJ/m$^2$, this gives an approximate threshold of $\Delta\theta_F \simeq 0.03^{o}$C, which is the value that is normally observed at time of frazil nucleation in the bulk of the water.

Thus the processes of frazil nucleation is completely elucidated. Germs originating from surface nucleation or border ice come in contact with embryos in slightly supercooled water. Once the threshold of forced nucleation is attained this produces a myriad of germs of ice which propagate quickly to the whole supercooled water body. This process of multiplication of ice crystals from an ice germ was extensively studied by Altberg (1936) in the laboratory, from which he explained frazil formation.

### 1.2.5 - SECONDARY NUCLEATION

Probably the best method for inducing crystallization is to inoculate or seed a supersaturated solution with small particles of the material to be crystallized. Deliberate seeding is frequently employed in industrial crystallization to effect a control over the product size and size distribution.

As shown by Eq. (1-27) no supercooling is needed to form ice in this manner as it just grows from the existing crystal.

In nature, this mechanism may be important when ice clouds are formed over large water bodies and when snowfall is falling into the water. It must be remembered however that the threshold for ice nucleation in water drops is much lower than for surface nucleation; so if the ice particles are not already present in the atmosphere when water attains the freezing point, then surface nucleation may occur first.

## 1.3 - RIVER AND LAKE ICE

### 1.3.0 - INTRODUCTION

In order to determine the various properties of ice that are significant for the solution of the various problems of interaction of ice with engineering structures it is first a necessity to have some fundamental knowledge on the formation of the solid part of ice covers.

It then becomes apparent that ice is not the simple material that it is thought to be at first sight and, because of the large number of processes that may intervene in its formation and growth from the liquid phase, it may take many different crystalline structures that will account for a wide diversity of physical and mechanical properties. It is not surprising to see, for instance, some natural solid ice as strong as concrete that will be incorporated in an ice cover with another type of solid ice as weak as kitchen cookies.

A discussion on the properties of ice that would not specify the ice types could then only be very qualitative in character as a general discussion on soils or rocks. To get to useful engineering data you have to specify which type of ice you are dealing with as for the case of these other natural materials.

## 1.3.1 - ICE APPEARANCE IN NATURAL WATER BODIES

There are essentially three ways in which the first ice may appear in a water body; either in a river, a lake or in the sea. The first one is the heterogeneous nucleation at the surface of a calm or slowly moving water body, the second the nucleation of frazil particles which appear through a fast moving water mass and the last one is simply the freeze up of snow or atmospheric ice nuclei falling into the water.

### 1.3.1.1 - ICE NUCLEATION AT THE SURFACE

The first ice crystals might appear right at the surface of a water body in calm water or in laminar flow, as described by Michel (1971). Because, there will be an important temperature gradient with depth, a thin supercooled water layer will appear at the water-air interface and along the banks of the water body. When the top surface of this supercooled layer will attain the natural temperature of surface nucleation the first ice crystal will appear.

Usually this will happen first along the banks where the water cooling will be faster, but ice may also nucleate on the surface away from the banks.

*Fig. 1.11. Formation of ice needles and dendrites on a calm free water surface with a small temperature gradient.*

The concentration of crystals that are formed at the beginning of this type of nucleation is an important parameter to determine the texture of the solid ice cover that will follow. Most generally, the initial crystals that are formed are needles because the supercooling is greater than 0.3°C. An example is shown on Fig. 1.11. But the number of crystals depends very much on the number of active nucleus of foreign substances present in the water body and on the rate of cooling. If there is a small temperature gradient this will cause a slow crystallization. The initial needles (which have an optical c-axis perpendicular to the geometric axis and at random orientation to the water surface, (as shown by Shumskii, 1964) will tend to grow more along the basal plane into the water. When the growth in the basal plane is long enough these crystals may turn over and float with the c-axis taking a vertical orientation. On the other hand if there is a large temperature gradient on a calm water surface, the solidification will proceed much more quickly and there will be a large increase in nucleation centers. The initial needles will interlock with each other before they have time to develop plate like crystals and the resulting structure will be more complex, the c-axis of each individual crystal having a random orientation.

#### 1.3.1.2 - FRAZIL NUCLEATION

Frazil formation is a peculiar phenomena of ice appearance in rivers and occasionally in lakes or at sea when surface currents induces a turbulent mass and heat exchange.

Let us examine, Fig. 1.12, the evolution of water temperature along the axis of a section of a river subject to temperature below freezing and a constant rate of heat transfer per unit of surface over the section. A given volume of water will be submitted to a rate of cooling until it is supercooled to a point where frazil crystals start to appear. Since in turbulent flow, water molecules intermix, the water temperature and, consequently, the crystallization process takes place in depth over the whole cross-section of the flow. Because frazil particles are formed by forced nucleation the supercooling is very small, of the order of only a few hundredths of a degree Celcius. An example is shown on Fig. 1.13.

When the large number of ice crystals begin to appear, the rate of temperature change shows a decrease and frazil is produced until the water temperature comes back to $0^{o}C$ at a decreasing rate. During that time, individual crystals grow to take the form of small circular disks that agglomerate to form porous flocks whose size depends on the state of turbulence of the flow. During the frazil formation period a considerable reserve of "cold" exists in the supercooled water. The potential of particle growth is high and that is why the particles form rapidly and agglomerate into flocks and slush.

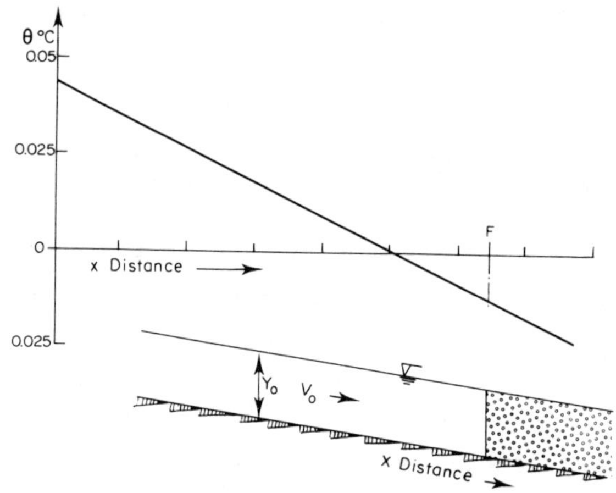


*Fig. 1.12. Water depth, temperature, distance and frazil front during frazil formation. From Michel (1965).*

In this ideal case, frazil is produced indefinitely in the same zone and the point of origin is fixed, this zone of formation extending only over a few hundred meters. In nature the principal difference comes from the continuous variation in the heat transfer process because of large atmospheric changes. A continuous lowering of air temperature provokes an upstream displacement of the frazil production front. If it is warm enough during the day to increase the water temperature and cold enough at night to decrease it again, the same section of a river will produce a certain amount of frazil every night. If energy losses due to friction at the foot of a rapid maintain the temperature above $0^{\circ}C$, the zone of frazil formation may oscillate around this rapid so that the supercooling of the water and frazil production may last many days in the same river stretch.

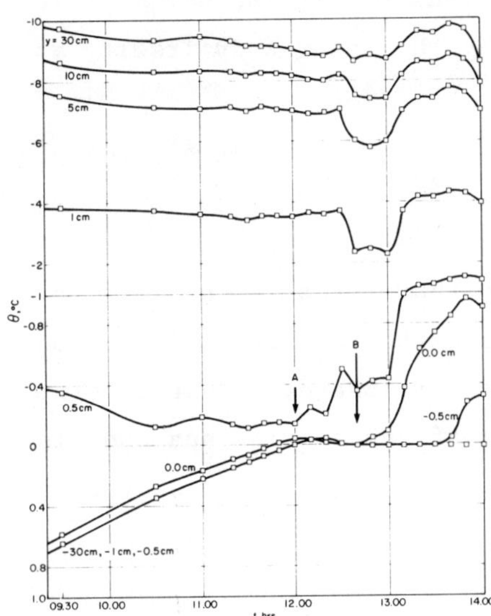


*Fig. 1.13. Evolution of the temperature above and below the surface of an agitated tank of water at time of frazil appearance. A corresponds to the time of frazil appearance and B to the time when the water temperature is back to zero. From Michel et al., (1975).*

## 1.3.1.3 - SNOW-FALL

One of the simplest way to initiate an ice cover on a water body is by secondary nucleation from ice particles coming from the atmosphere, when the water temperature is close to the freezing point. In a lake or a calm water body, a heavy snowfall will often initiate the ice cover when the floating snow particles will refreeze at the top; even if the top water temperature is not exactly at the freezing point during the snow fall.

However in a stream where the flow is turbulent the snow particles will melt very quickly if the water temperature is not at the freezing point, so that the frozen precipitation has to coincide with the water being at the freezing point. The frozen precipitation might take various forms, including those coming from surface mist produced by the water body itself, as shown by Osterkamp (1971).

### 1.3.2 - STATIC ICE COVER FORMATION

A static ice cover is one that forms essentially in place without ice advection and where the thickening is caused mainly by thermal exchange with the atmosphere. These covers are found essentially in calm water of lakes or at sea and also in areas of slow laminar flow in rivers or along sea shores and they have been described by Michel (1971).

#### 1.3.2.1 - BORDER ICE

Border ice is usually the first type of ice to appear in a river or along shorelines in areas of laminar flow along the banks of lakes and in the seas. Because in laminar flow or calm water, there is no intermixing of the top layer with the lower ones, the temperature differences are important both in the vertical direction and horizontally away from the banks. The top layer adjacent to the bank goes through sensible super-cooling while the average temperature of water in the middle of the river may still be far above the freezing point. Ice is then nucleated first along the colder material of the banks. This nucleation propagates on the surface toward the middle of the flow forming, at least in the beginning phases, a clear and solid ice sheet.

The edge of this ice sheet finally comes in contact with the turbulent water and its further progress depends only on the thermal atmospheric exchange as compared to the temperature and turbulence of the water.

The shore ice growth process is a slow one that can be somewhat accelerated when the water is carrying frazil or snow slush. The slush clings in parallel rows to the growing dendrites at the ice boundary and forms successive clearly marked layers in the solid ice sheet.

In rivers and along the sea shore ice grows not only from the shore but also around emerging boulders or from anchor ice which is emerging from the bottom. This latter type of formation is very apparent in small brooks.

### 1.3.2.2 - PLATE ICE

In calm water of lakes, in the sea and in very slow moving water, ice crystals will nucleate right at the water surface in the form of ice needles, as described previously. The individual needles will grow in the supercooled top water layer and form a continuous ice plate at the surface. These plates might be more or less extensive in plan but it has been often observed that they will cover the whole surface of a small lake during a cold night at time of ice cover formation.

The initial ice plate may also originate from a snowfall in calm water close to freezing point, followed by cold enough weather to refreeze the snow at the top.

The structure of the surface layer of the ice cover

and the succeeding ones which will grow underneath depend very much on the conditions of formation of this first ice plate.

### 1.3.2.3 - GROWTH OF ICE

The growth of solid continuous ice in a river or lake, from heat exchange with the atmosphere, is not as simple as it looks to be at first sight.

If we consider the simple case of vertical static growth of ice from an original ice surface, that we take into account the thickness of the existing solid ice sheet and that of a snow layer on top of the ice and we assume that the temperature at the snow surface is that of the air, we have the basic heat transfer equation:

$$\frac{-K_i \theta_s \, dt}{h} = \frac{K_s (\theta_s - \theta_a) \, dt}{h_s} = g\rho' L \, dh \qquad (1\text{-}33)$$

The equation leads to:

$$\frac{dh}{dt} = \frac{1}{g\rho' L}\left(\frac{1}{h/K_i + h_s/K_s}\right)\frac{dS}{dt} \qquad (1\text{-}34)$$

with

$$S = \int_{to}^{t} [-\theta_a] \, dt \qquad (1\text{-}35)$$

in consistent units, where:

o' - density of solid ice

g - acceleration of gravity

$K_i$ - coefficient of thermal conductivity of that ice

$K_s$ - average coefficient of thermal conductivity of the snow

L - latent heat of fusion of the ice

h - thickness of the solid ice sheet at time **t**

$h_s$ - depth of snow

S - number of degree-days of frost

$\theta_a$ and $\theta_s$ - air temperature and temperature of the top surface of the ice in $^oC$.

It is obvious that the above formula will give ice thickness values in excess of what will be found in nature for many reasons. Solar radiations will be absorbed in the ice sheet and they are not included in the heat budget as well as heat advected at the bottom from the underlaying water body. Furthermore, a boundary layer at the upper snow-air surface makes the surface temperature at this location higher than the air temperature.

But the major difference in nature comes from the fact this is not the way that most static ice covers grow in rivers and lakes of northern regions (Michel, 1971). The above simplified formula would apply mainly to Arctic regions where snow accumulations are very small on the ice. In most cases, the weight of snow falls on the ice is adequate to drown the solid ice cover so that water will move through cracks (thermal or others) to flood the ice.

In fact, a rough computation shows that a very small amount of deposited snow is sufficient to stop the growth of solid ice underneath: snow about half the thickness of the ice will suffice. In many areas, a first snowstorm floods the ice of a river or a lake and a new type of ice called snow-ice, (whitish in color) begins to form in the water saturated snow layer.

There are then many possible modes of growth of the combination of snow ice and black ice (transparent solid ice) depending on the amount and timing of the successive snow falls.

An example is shown on Fig. 1.14. Because of the nearly impossible task of predicting the snow-ice and black ice growth in many rivers and lakes, and engineering formula proposed by Assur (1956) is often used for the total growth of solid ice:

$$\frac{dh}{dt} = \frac{\alpha^2 K_i}{\rho' g L h} \frac{dS}{dt} \qquad (1\text{-}36)$$

where $\alpha$ is a coefficient always smaller than one that characterizes the ice growth for each particular water body.

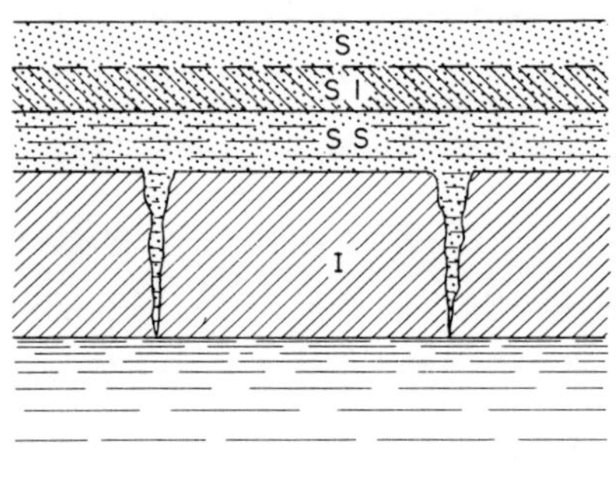


*Fig. 1.14. Flooded snow lying on a fractured ice sheet with layers of snow (S), snow ice (SI), snow slush (SS) and black ice (I).*

## 1.3.3 - DYNAMIC ICE FORMATION

The more usual way of ice cover formation in larger northern rivers is the dynamic accumulation of ice slush and floes and the progression against the flow of the upstream edge of the cover. This process also accounts for some ice cover formation in lakes and along sea shores when the ice pieces are being pushed along a solid ice edge by the wind and a water surface current is set up in the epilimnion.

Let us consider first the frazil evolution process.

We have said previously that during the period of nucleation and active production of frazil, the ice crystals agglomerate to form frazil flocks of very low ice content. Once the water comes back to $0^{\circ}C$, the flocks of inactive frazil oscillate in the turbulent flow and those reaching the surface serve as a frame-work for the regular growth of ice from heat exchange with the atmosphere.

This inactive frazil evolves in a river in a manner shown on Fig. 1.15. As soon as the flocks are formed they have a tendency to concentrate at the surface of the flow to form the usual frazil slush. Because of the continuing heat exchange with the atmosphere these slush balls become denser and in the case of low velocity flow, they stay at the surface long enough so that a continuous layer of ice is formed in their upper part to produce ice pancakes with their peculiar saucer like appearance because of collisions with neighbours. If these initial floes can travel over long distances larger ice floes may develop by sintering of individual pancake floes together. Sometimes large floes passing over rapid reaches, or peculiar rivers singularities, will break again in smaller floes.

The frazil particles which are formed not only stick together but they adhere and grow on certain material of favorable crystalline structure. They also deposit in separated no-flow areas in front of, and behind obstacles, in flowing water. So they adhere to projection like stones and weeds on the river bottom to form what is known as anchor ice. These deposits will grow in the supercooled water and form transparent ice plates (Schaefer, 1950), or in high flow velocities, the frazil is shingled in a snow-like crust.

When the ice drifts over a long distance, a sizeable area of the water surface becomes covered with floating slush, ice pans and floes. In deep sections with low surface velocity, in braided channels or other singularities, the coverage may reach 100% of the surface and the ice floes will press together and freeze to form a continuous ice bridge. The river reaches susceptible to formation of an ice bridge can be determined from the continuity equation.

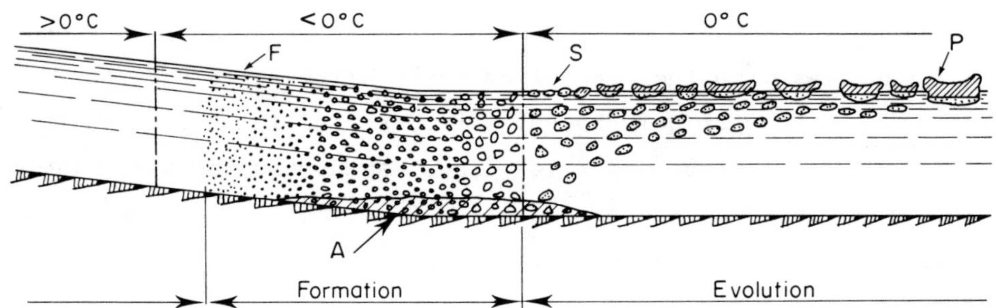


*Fig. 1.15. Development of frazil in a river, from frazil discoids (F), to frazil slush (S) and pancakes (P). Anchor ice (A) is also shown. From Michel (1965).*

Once an ice cover has started to form it may progress very quickly by simple juxtaposition of incoming frazil slush and ice pans. In larger rivers, as in low velocity stretches of the St. Lawrence river the ice pack may advance as much as 25 miles a day.

When a cover is formed in a river by accumulation of incoming ice floes which freeze on top at a short distance from the leading edge, the total thickness may be obtained from the following analysis.

Let us consider the stability of an ice cover formed by packing of incoming ice floes and slush in front of a solid ice crust as shown on Fig. 1.16.

The Bernoulli equation between section (1) and (2) gives:

$$H = Y + V^2/2g = y + U^2/2g + eh + \xi h + \frac{(1 - e)\ \rho' h}{\rho} \tag{1-37}$$

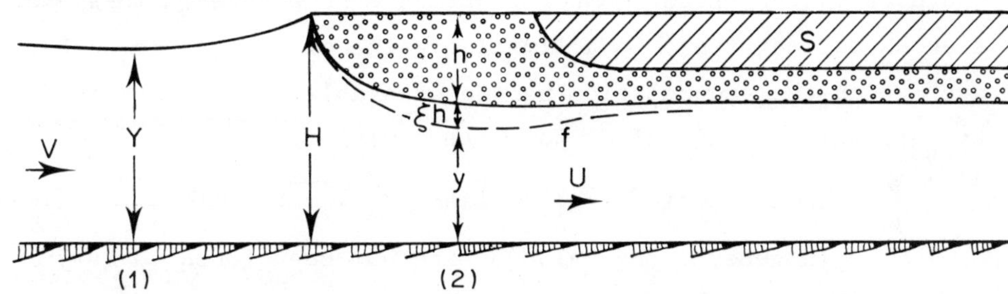


*Fig. 1.16. Stability of a progressing ice pack; S-solid ice crust, f-line of flow separation.*

The condition of non submersion of the frontal edge is:

$$H = y + h\ (1 + \xi) \tag{1-38}$$

The continuity equation is:

$$VY = Uy \tag{1-39}$$

where Y and H are the water depth and total head in front of the cover, h is the total ice cover thickness, y the flow depth at the restricted section under the cover, V and U are the flow velocities respectively in front and under the cover, $\rho$ and $\rho'$ the densities of water and solid ice and g is the acceleration of gravity.

The parameter e is the porosity of the ice accumulation, i.e., the ratio of the volume of voids filled with water to the total volume of ice and $\xi h$ is the thickness of the separated flow layer.

Eqs (1-37), (1-38) and (1-39) lead to the condition of equilibrium of the ice cover.

$$Fr = V/\sqrt{gY} = \sqrt{2\ (1 - \rho'/\rho)\ (1 - e)\ (h/y)\ [1-(1+\xi)\frac{h}{y}]} \quad (1\text{-}40)$$

where Fr is the Froude number in front of the cover. This equation shows that there is a limiting value of the Froude number over which the accumulation becomes unstable. It is obtained by derivation and it gives for $\rho'/\rho = 0.916$.

$$Frc = 0.158\ \sqrt{1 - e} \Big/ \sqrt{1 + \xi} \quad (1\text{-}41)$$

In general the value of Frc has been found to vary between the limits:

$$0.06 < Frc < 0.12 \quad (1\text{-}42)$$

It depends very much on the porosity of the accumulation and the types of ice floes. The smallest value is applicable to small individual frazil particles and the highest one for well defined ice blocks plunging under the cover with little slush. These values are documented by Kivisild (1959), Pariset *et al.*, (1961), Michel (1965) and Uzuner *et al.*, (1972).

For deep water, particularly for ice covers formed in the sea with surface currents, the Froude number cannot be used.

Eq. (1-40) then reduces to:

$$V = \sqrt{2g\ (1 - e)\ (1 - \rho'/\rho)\ h'} \qquad (1\text{-}43)$$

where h' is the equivalent thickness of the ice accumulation taking into account the added thickness of the separated water layer.

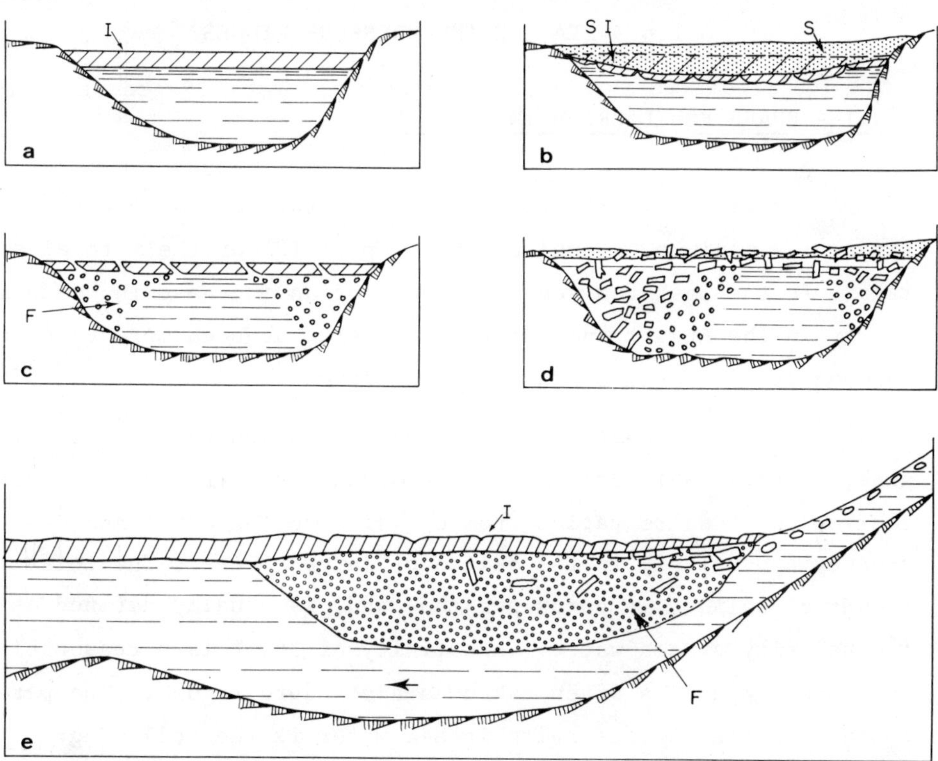


*Fig. 1.17. Various types of ice covers in a river: a) pure ice cover, b) with snow partly flooded on top, c) with frazil accumulation underneath, d) hummocked ice cover with floes, snow and frazil, e) underhanging dam with frazil, slush and floes.*

When the ice cover progression stops, the incoming floes pass underneath and form an underhanging dam which fills backward under the cover and equilibrates with the velocity of deposit of the ice pieces which are carried underneath. Fig. 1.17 shows various types of ice covers in rivers.

## 1.4 - SEA ICE AND PRESSURE RIDGES

### 1.4.1 - PHASE EQUILIBRIUM IN SEA ICE

Sea ice is a very important feature of the ocean and at times, its maximum extension may be up to 12% of their total surface. In the last few years, because of an increase in human activity in the Arctic and Antarctic, there has been more research done on its properties and characteristics.

The composition of sea water, from which the ice origins, is remarkably uniform in the oceans even if the total concentration of salts varies from one site to the other and decreases appreciably in the deltas of large rivers. The total amount of solids, in parts per thousand, is usually defined as the salinity of sea water. A salinity of 34.5% is a reasonably good average and is often taken as a standard figure. The percentage of the various salts in sea water is the following:

| NaCl | $MgCl_2$ | $Na_2SO_4$ | $CaCl_2$ | KCl | $NaHCO_3$ | Other | Total |
|---|---|---|---|---|---|---|---|
| 68.10 | 14.44 | 11.37 | 3.19 | 1.91 | 0.55 | 0.44 | 100 |

Percentage composition of salt in sea water

It must be added that this table give the composition of dried sea salt. In solution, the various salts are almost completely ionized and the composition of sea water is best described in terms of ionic concentration $Cl^-$ = 18.98%, $Na^+$ =

10.56%, $SO_4$ = 2.65% etc...

The major difference between sea ice and fresh water ice is the amount of salts, in the form of concentrated brine pockets, that are squeezed within the ice structure. Therefore, the first prerequisite for the study of the structure of sea ice is a knowledge of the phase diagram. Such a diagram is shown on Fig. 1.18. It gives the relative volumes of ice, brine and solid salts that will be in equilibrium at a given temperature.

Because of the salt content in sea water, the freezing point will be depressed below zero $^{\circ}$C as shown on Fig. 1.18. For standard sea water, this temperature of freezing is -1.8$^{\circ}$C. For any temperature, the diagram gives the amount of brine that will coexist in equilibrium with the ice. The salts in sea ice precipitate in the following order of decreasing temperatures (Assur, 1958):

| Temperature $^{\circ}$C | |
|---|---|
| - 8.2 | $Na_2SO_4 \cdot 10\ H_2O$ |
| -22.9 | $NaCl \cdot 2\ H_2O$ |
| ~-36.8 | KCl |
| ~-43.2 | $Mg\ Cl_2 \cdot 8\ H_2O$ |
| | $Mg\ Cl_2 \cdot 12\ H_2O$ |
| ~-54.0 | $CaCl_2 \cdot 6\ H_2O$ |

Experimental data for this multiphase system is limited below -30$^{\circ}$C. An important fact is that apparently sea ice always contains some liquid at any temperature found naturally on earth.

In the first part of the diagram, before $NaCl_2 \cdot 2H_2O$ precipitates, the phase diagram can be approximated to give the brine volume $\nu$, in parts per thousand, in function of the ice salinity $s$ in the same units and the temperature $\theta$ in $^{\circ}$C (Frankenstein *et al.*,1967).

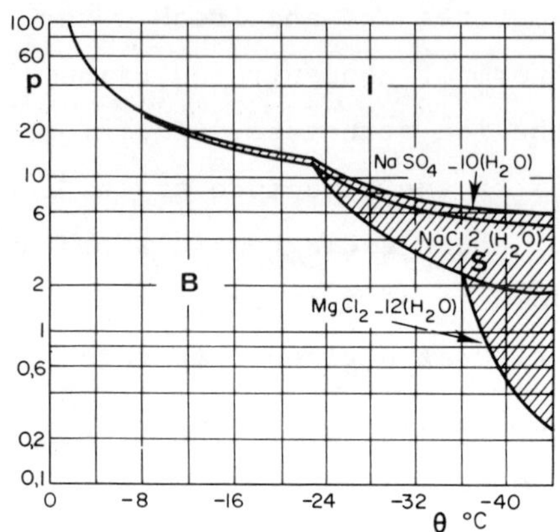


*Fig. 1.18. Phases diagram in the solidification of sea water showing the percentage p in weight of the various phases. One area represents the solid ice (I), the other the brine (B) and the third one the precipitated salts (S).*

$$\nu = s\ (0.532 - 49.12/\theta) \qquad \text{for} \quad 0.5^{\circ}C < \theta < 22.9^{\circ}C \qquad (1\text{-}44)$$

## 1.4.2 - FORMATION OF SEA ICE

The appearance of a first ice skim on sea water is very similar to what is observed in rivers and lakes, which has been described previously. If the surface is very calm and the temperature gradient small, a few crystals will appear on the supercooled layer of water at the surface to form plate ice of uniform texture. Most generally, however, wind and waves will mix the water on top of the sea and the supercooling will extend to a certain depth in water. Surface or secondary nucleation will initiate the formation of frazil particles having the form of small discoids. As they grow and agglomerate in slush they will float to the surface and refreeze to form the first layer of ice. In polar regions, small ice nuclei coming from the atmosphere are being deposited almost continuously on the water surface to initiate frazil formation. Little or no supercooling is then observed on the top water surface.

Once this initial skim of ice is formed, the salt in sea water plays a major role in the structure of the ice formed by thermal growth and it becomes quite different from fresh water ice. One major difference is the nonuniform growth of sea ice at the ice-water boundary. This is illustrated on Fig. 1.19 which shows a cross-polaroid photograph of a growing sea ice layer where the lower 1 or 2 cm of the ice consist of pure ice platelets with layers of trapped brine in between. Within the same ice crystal, the platelets coincide with the basal plane and they are accurately parallel one to the other. It can also be noticed that these planes are vertical or very close to the vertical. As freezing progresses, the platelets thicken up slightly and ice bridges develop between neighbouring ones, gradually forming an almost solid structure. The brine which is trapped behind the links is forming elongated cells between the platelets, which are called brine pockets. These pockets are aligned in rows separating the platelets. They can be seen as black spots on Fig. 1.19. The brine cells shrink in size as the ice cools forming parts of long vertical cylinders of almost microscopic cross-sections.

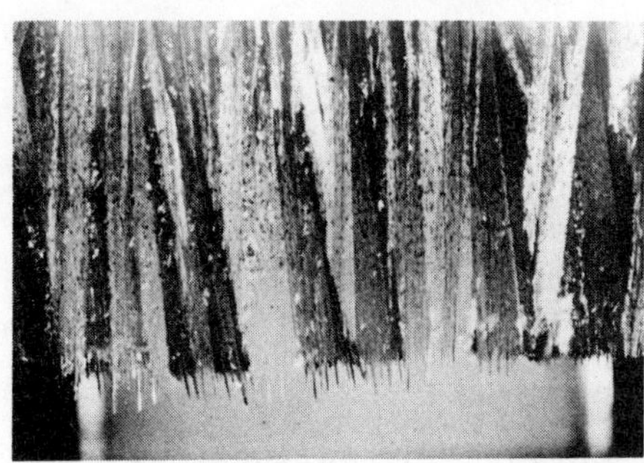

*Fig. 1.19. A crossed-polaroid photograph of a sea ice sheet growing in a test cell. The thickness of the ice from top to bottom is about 15 cm. From Eide and Martin (1975).*

Thus, each single crystal of sea ice is composed of a stack of parallel ice plates and each plate is separated by an array of brine pockets. One model of the shape of the brine inclusions is that of Assur (1958) and is shown on Fig. 1.20.

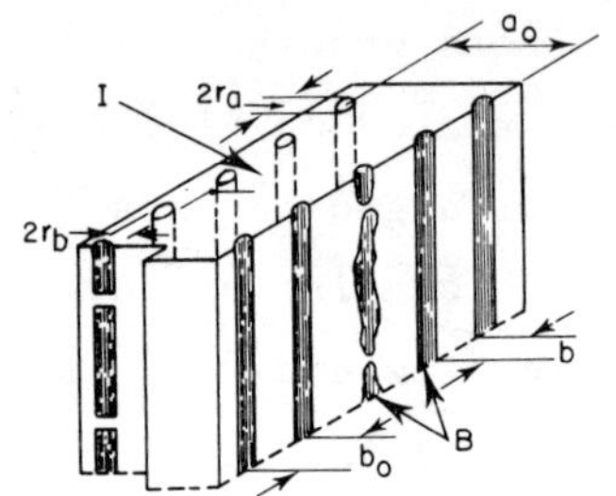


*Fig. 1.20. Idealized diagram of the shape of the brine inclusions (B) in sea ice (I). (Assur, 1958).*

In this model, the brine channels have elliptical cross-sections. The platelet thickness is $a_o$. The distance between the channels is $b_o$ and major and minor radius of the ellipses are $r_b$ and $r_a$. The relative brine volume $\nu$ for a layer of unit thickness is:

$$\nu = \frac{\pi \, r_b \, r_a}{a_o \, b_o} \tag{1-45}$$

And the relative brine area, in the brine plate; which is the plane of weakness of this type of sea ice, is:

$$\varepsilon = 2 \, r_b / b_o = \sqrt{\frac{4 \, a_o \, r_b \, \nu}{\pi \, b_o \, r_a}} \tag{1-46}$$

## 1.4.3 - SALINITY PROFILES IN YOUNG SEA ICE

The physical properties of sea ice depend very much on the brine content $\nu$, and this parameter varies with time because of both temperature and salinity changes. The salinity of sea ice changes in two ways; by brine drainage and by brine migration.

At least at temperatures above -15°C, Pounder (1965) says that brine cells have interconnections so that brine may drain slowly through the ice under the influence of gravity. Fig. 1.21 shows a sketch of the brine drainage system which was observed by Eide *et al.*,(1975) in young sea ice. The main channel had a diameter of the order of 1 cm and the orifice at the neck decreases to a diameter of about 2-3 mm. It was found that the brine in the channels would oscillate and thus be replaced with some of the underlying sea water.

*Fig. 1.21. Sketch of a drain for brine in young sea ice.*

Pounder says that brine drainage is quite rapid as ice approaches its melting point. If a block of ice is removed from contact with the sea, as being pushed on shore, it loses salt very rapidly during the warmer months of spring and summer.

A much slower process is that of migration of the brine cells from the colder end to the warmer end of the ice sheet. Because the concentration of brine within a cell is uniform, the warmer end of it is too concentrated and will dissolve to reduce its concentration. At the colder end more ice freezes to increase the brine concentration. The net effect is to move the entire cell of brine along the temperature gradient, which is downward in a natural sea ice cover.

Both brine drainage and brine migration reduce considerably the salinity of sea ice compare to that of the surrounding sea. These processes will depend somewhat on the texture of the ice. In refrozen sea water, in snow or frazil slush and in agglomerate ice, the drainage channels cannot follow the vertical so easily and the drainage process is certainly much slower. Because of these factors, the salinity profiles in sea ice will depend very much on the ice structure and will vary considerably with time, mainly from one season to the other. An example of salinity profiles for growing sea ice of columnar structure is shown on Fig. 1.22.

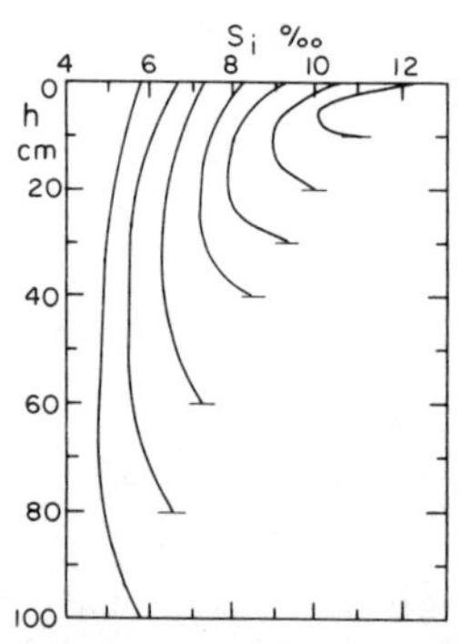


*Fig. 1.22. Series of schematic salinity profiles Si for sea ice of various thickness h. From Weeks and Assur (1967).*

## 1.4.4 - POLAR ICE

The Arctic ocean and part of the adjoining seas, as well as the regions surrounding the Antarctic continent are almost entirely covered the year round with polar ice which is several years of age. When there is no snow cover, the ice appears to be pale blue in color in contrast with the greyish white of first year ice. A polar floe is not level but is covered with gently rounded hummocks of the order of 1 m in height, spaced 30 to 40 m apart.

In the Arctic, the ice formation can be described as including the Arctic Pack, the fast ice and the drift ice (Transehe, 1928). A sketch of fast ice formation is shown on Fig. 1.23 including pressure ridges, one of which is usually grounded at the edge of the fast ice.

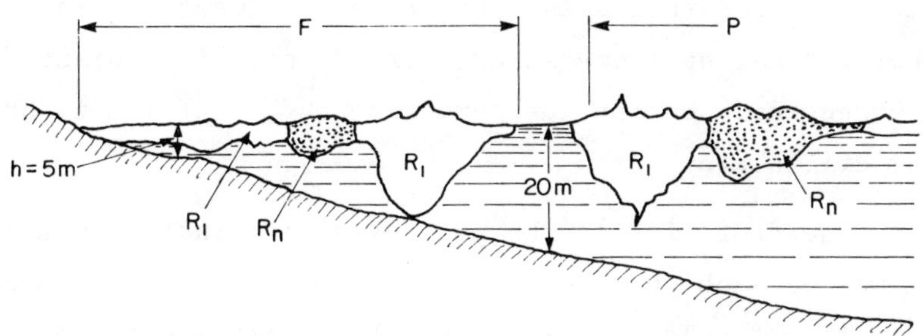


*Fig. 1.23. Sketch of sea ice in the Arctic: P is the Polar Pack ice, F the fast ice, R1 are first year partly consolidated pressure ridges and Rn are multi-year consolidated ridges. There is an open lead between the Pack and the fast ice.*

The Arctic Pack floats in the center of the Arctic ocean, covering two thirds of it. It is a permanent feature and the ice in it is progressively renewed. In July 1954, Shumskii (1964), measured the following ice thicknesses: 10 cm of snow, 40 cm of regelation ice formed by refreezing of melt water and corresponding to four years of accumulation, 110 cm of ice having more than four years but with low salinity and finally 130 cm of sea ice having less than four years of age. The total thickness was 3 m which appears to be the approximate equilibrium limit for ice in the Arctic Pack.

The Arctic Pack turns in a clockwise direction and although the velocities depend very much on location and period of the year, a typical velocity at the periphery is of the order of 2 km/year.

The fast ice in the Arctic is the ice that forms each winter between the coasts and the Pack itself. It covers approximately 5% of the Arctic regions. It has a smooth surface interrupted by pressure ridges. Under the influence of the tides and other forces, open cracks are formed, some in contact with the grounded ice. During the summer months, fast ice is broken, melts or drifts away.

The drift ice is the multi-year ice that is not completely melted each year and that accumulates with wind and currents at particular locations which are different than the Arctic Pack.

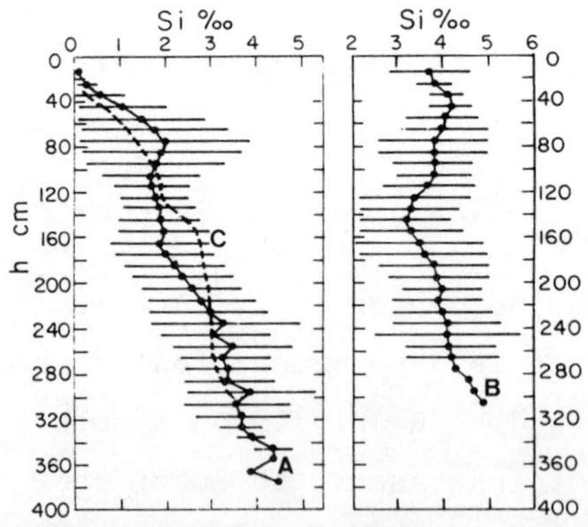


*Fig. 1.24. Average salinity profile in the Polar Pack. Curves A and B are the average salinities under hummocks or depressions, respectively from Cox and Weeks (1974) and C is the average given by Schwarzacher (1950). The bars denote the standard deviation of the average values.*

Multi-year polar ice attains an equilibrium thickness because the ice and snow which melt or evaporate at the top during the summer months is compensated by sea ice growth underneath during the colder periods. Because the melt water is essentially fresh and refreezes inside the brine drainage system, the salinity of polar ice is low and it is fairly strong. Salinity measurements by Cox and Week's (1974) and Schwarzacher (1950) are shown on Fig. 1.24.

## 1.4.5 - PRESSURE RIDGES

With the exception of ice islands and icebergs, the more massive appearances of ice in the Arctic are the pressure ridges, i.e. those linear or irregular accumulations of ice caused by the crushing and shear interaction between large ice fields. The highest free floating ice ridge reported by Kovacs *et al.*,(1973) had a sail of 12.8 m above sea level. Ridges of that size are rare and ridge sails rarely exceed 5 m.

The bulk properties of geometrically similar ridges may vary greatly. Ridges which have just been formed are made up of individual ice pieces which are very poorly bounded. In a multi-year ridge, on the other hand, the melt water produced during the summer months will drain downward into the core of the ridge. Then the water will refreeze, cementing the ice blocks together into a solid, resistant mass of low salinity ice. It is thought that at least two melt seasons are required for the voids between the ice pieces to become filled with ice.

Fig. 1.25a shows two transverse and the longitudinal cross-section of a multi-year pressure ridge that was measured in the Beaufort sea by Kovacs *et al.*, (1973). It has a maximum sail of 4 m above sea level and a keel of 13 m whose cross-section can be described as roughly semi-circular. The soundings in this ridge revealed no cavity, the mean in density being

0.90 gr $cm^{-3}$. There were neither wide cracks or leads that transited the ridge.

Fig. 1.25b gives the salinity and temperature profiles within the ridge. The temperature were measured in March when the ice temperature would be expected to be close to the annual minimum.

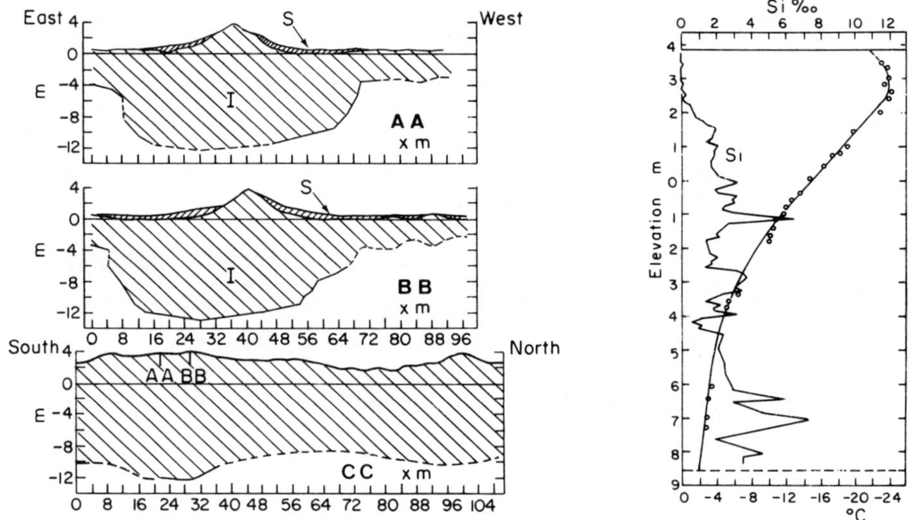


*Fig. 1.25a. Transverse cross-section AA and BB and longitudinal section CC of a multi-year pressure ridge in the Beaufort Sea. The limits of snow (S) and ice (I) are given.*

*Fig. 1.25b. Salinity and temperature profile of the ice of a pressure ridge. From Kovacs et al., (1973).*

## 1.5 - GLACIERS AND ICEBERGS

### 1.5.0 - INTRODUCTION

The definition of a glacier by Lliboutry (1965) is any large mass of natural ice, originating from snow, which is permanent at the human scale. This is quite a broad definition which includes not only the usual Alpine type glacier but also the indlandsis which are covering the huge continental land of Antarctica (13,000,000 km$^2$), that of Groenland (1,725,000 km$^2$) and also smaller ice sheets in polar and sub-polar regions. Also included would be the very small firns, without visible flow, which might or might not be transformed into ice at the lower levels.

This section is a very short summary of some of the topics treated by Lliboutry in his extensive work "Traité de Glaciologie" tome 2 (1965), to which the reader is referred to for more complete treatment.

### 1.5.1 - MORPHOLOGY OF GLACIERS

A glacier has essentially two zones; an accumulation basin where solid precipitations are accumulated and an ablation zone where the loss of ice by various means is greater than the gain by precipitations. Any glacier can then be described according to the characteristics of these two zones.

A simple glacier is one where there is one accumulation basin followed by one ablation zone, at a lower level, which has more or less the form of an ice tongue.

A composite glacier is made up of many simple glaciers where the evacuating tongues join together in a single ice stream. There are many such glaciers in the Himalayas and some in other large mountain ranges.

A third type of glacier is the opposite of the previous one. It is the multiple emissaries glacier where only one accumulation area distributes the ice in many glacier tongues at lower levels, in separate locations. These glaciers are more evident in the Antarctic, in Groenland and in many mountain ranges.

The surface of the accumulation zone of a glacier might have various forms. The more frequent ones are the dome, the highland plateau, the regular slope and the circus. Indeed, each form applies only to one accumulation of adequate size.

The most variable feature of a glacier is that of the ablation zone. Most often it is in the form of a tongue which is much longer than wider. In general it flows into a valley like a real ice stream. In the case of large glacier sheets, like the Groenland inlandsis, the ice streams, that can be detected by the crevasses they produce with a lowering of the ice level, take their origin very far inside the main ice sheet and not at its immediate periphery. They are ice streams within the whole ice mass, similar to large water streams within the ocean.

The ablation zone can also be in the form of a regular ice stream. This is the case when a glacier, leaving the mountain, spreads evenly on a flat ground. This is usually called a piedmont glacier.

The lower ends of the Antarctic ice sheet are floating on the sea, and they are called shelfs like the Ross ice shelf, although they are not necessarily all in the ablation zone of the indlandsis. The name of barrier is used for the ice cliff that limits the ice and the open sea. It might be 30 m high or more.

Lliboutry gives an interesting example of various forms of glacier that are located in the Akuliarusq peninsula in the western side of Groenland. This is shown on Fig. 1.26.

*Fig. 1.26. Sketch of various types of glaciers on the occidental edge of the Indlandsis in Groenland: A) circus glacier of Alpine type, B) Piedmont glacier, C) High valley glacier, D) Rock glacier, E) Plateau glacier of Norwegian type, F) Emissary glacier. (From Lliboutry, 1965).*

## 1.5.2 - EQUILIBRIUM LIMIT OF A GLACIER

In mountain ranges of temperate climate, the snow edge varies with the seasons. In winter, glaciers are covered with new snow but in summer, the snow cover edge moves up so there appears on the glacier surface, an area where the bare ice is uncovered and an area still covered with snow. If there has not been any snowfall for a long time, there will appear, in general, at the periphery of the old snow a more or less extended border of regelation ice. At the end of the summer, when the bare ice has its greatest extension, the part of the glacier containing the old snow is called the firn and its limit, the firn line.

If we neglect the regelation ice, the irregular movement of the glacier and the annual climatic fluctuations, the firn line is the equilibrium limit of a glacier between the accumulation zone in the firn and the ablation zone on the bare ice.

The first complete sampling of an Alpine glacier, at the Massif du Mont Blanc (Vallon *et al.*,1976) has shown that the firn in the accumulation zone was 30 m thick over solid glacier ice. On this glacier, 6 to 8 m of light, cold powder snow would accumulate in the winter. Owing to the low thermal conductivity of the snow, it was cold only in the first 5 to 7 m; beyond 12 to 15 m the firn from the previous years always stayed at $0^{\circ}$C.

The model shown on Fig. 1.27 represents the equilibrium conditions of the firn at the time of maximum melting. It applies roughly to any type of glacier, either temperate or cold.

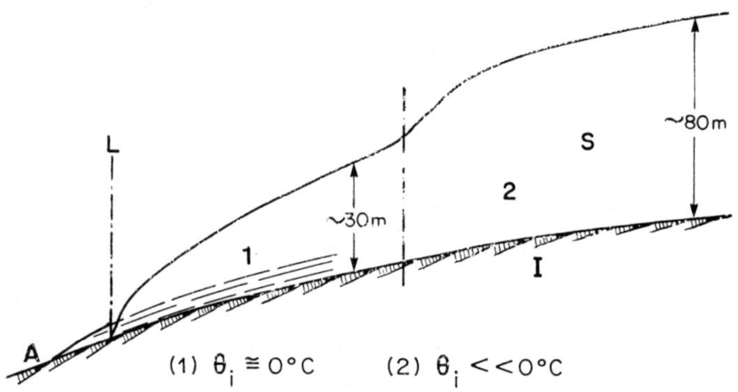


*Fig. 1.27. Equilibrium conditions of the firn over a glacier: A is the ice surface of the ablation zone, L is the firn line, S is the snow accumulation zone and I the glacier ice. In section (1) the water infiltrates to the glacier surface and the glacier is temperate, in section (2) there is no snow melting and the glacier is cold.*

Below the firn line on this figure is the ablation zone of the bare ice, when it melts or evaporates. Some water flows from the melt of the saturation zone, over the impermeable top of the glacier ice when the firn becomes soaked with melting water.

In the soaked ice layer up to the dry snow transition,

summer melting is inadequate to soak all of past year's accumulation but percolation in drainage channels lead the water to form a continuous layer on top of the impermeable ice. The maximum thickness of firn in this zone is of the order of 30 m and the ice underneath is that of a temperature glacier; very close to $0^{o}C$.

After a transition, the area above the dry snow line would correspond to glaciers where there is no summer melting in this part of the accumulation zone. Because melting is negligible, the snow remains permanently dry and the impermeable glacier ice is formed by densification at a depth of around 80 m (Bader, 1961). The temperature of the snow in the upper layers is within $1^{o}C$ to $2^{o}C$ of the mean air temperature and the ice remains cold underneath for what is called a cold polar glacier.

In general, the firn line is not exactly the equilibrium limit of a glacier. Following faster glacier flow or in a year of excess accumulation, firn having more than one year of age might appear downstream of the equilibrium limit.

### 1.5.3 - GLACIER FLOW

One of the most striking feature of a glacier is the movement under the effect of gravity of a huge ice mass down a mountain slope. Although the velocities are very low (of the order of 10 m/year under normal conditions) this movement accounts greatly for the morphology of the glacier and also for the characteristics of the ice itself.

The typical flow of a glacier can be visualized from measurements made by Lewis (1960) on the movement of a small glacier, the Skautbre in Norway, which was resting on a rock bed with approximately circular form. The velocities were measured on the surface and in an horizontal tunnel throughout the base of the glacier. The results plus reconstitution of the velocity field with flow lines are shown on Fig. 1.28.

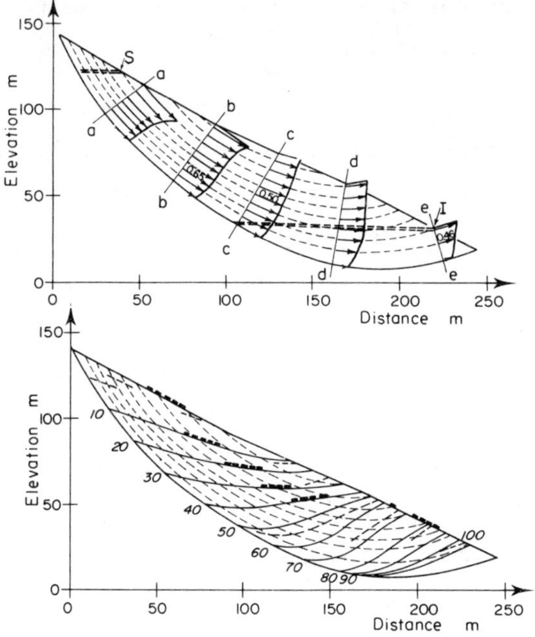


*Fig. 1.28. Flow lines, ablation surfaces and velocity profile of a circus glacier as measured and extrapolated. The dashed lines are flow lines and continuous lines the ablation surfaces. The bars indicate the position of debris every ten years. Velocities are in meters for 2500 days. From Lewis et al., (1960).*

It can first be observed that the velocities are relatively constant along a perpendicular to the bed and of the order of 2 m/year. They increase a little in the firn because of compaction but they keep a relatively high value even very close to the bed. This fact indicates that the glacier mass behaves much more like a perfectly plastic substance, that glides at the boundary than as a viscous substance with a gradually increasing velocity profile. This is why theories based on plastic flow and first proposed by Nye (1951) have been the most successful in predicting the flow of glacier ice.

Another interesting observation is the movement of

debris and atmospheric dust that had settled previously, in the accumulation zone, during the summer, on top of the layer of snow of the previous year. The movement of this strata of debris can be traced in function of time until it emerges at the surface of the zone of ablation. It appears to follow closely the planes of maximum shear stresses within the ice.

For a two dimensional glacier, Nye has given a simple analysis of glacier flow. Consider an element of a perfectly plastic material as shown in Fig. 1.29 that is under the equilibrium of hydrostatic forces and a constant limit plastic tangential stress $\tau_c$.

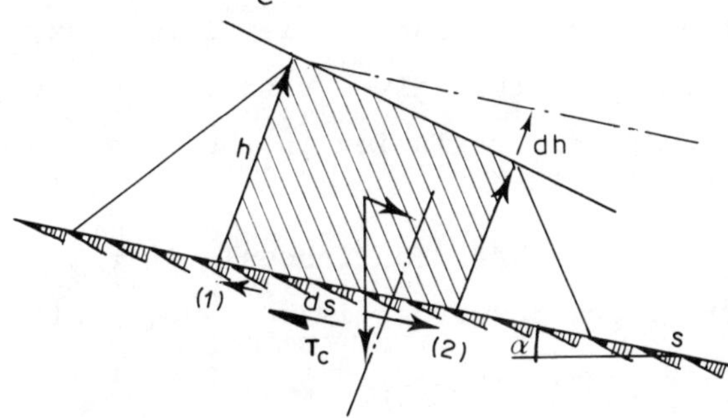


*Fig. 1.29. Equilibrium of forces in glacier flow.*

The resultant of the hydrostatic pressure of faces (1) and (2) is:

$$\frac{\rho' gh^2}{2} - \frac{\rho' g\ (h+dh)^2}{2} \simeq \rho' ghdh \qquad (1\text{-}47)$$

The component of gravity in the direction of motion s is:

$$\rho' gds\ h \sin \alpha \qquad (1\text{-}48)$$

With the tangential friction force $\tau_c$, this leads to the equation of equilibrium.

$$\frac{dh}{ds} = \sin \alpha - \frac{\tau_c}{\rho' gh} \qquad (1\text{-}49)$$

For a plane bed where sin $\alpha$ is a constant, this equation is easily integrated by setting:

$$\tau_c/\rho' g \sin\alpha = h_\infty \quad (1\text{-}50)$$

With a convenient origin for the s axis, the profile of the glacier surface is given by:

$$s \sin\alpha = h + h_\infty \log (1 - h/h_\infty) \quad (1\text{-}51)$$

We can see that there are three types of glacier flow:

1°) $h = h_\infty$, this gives a uniform flow of constant ice thickness;

2°) $h < h_\infty$, the thickness decreases continuously in the downstream direction and the surface of the glacier is convex. The upstream depth is the limit and this gives rise to a compressive flow as in the case of a piedmont glacier in the lower ablation zone;

3°) $h > h_\infty$, the thickness increases continuously in the downstream direction and the glacier surface is concave. This is what is called an extended flow, usually in the zone of ice accumulation or at the foot of a steep slope.

One major problem in glacier flow is to find the value of $\tau_c$ at the bed-boundary and its relation with the flow velocity of the glacier. This critical shear stress can be obtained from measurements of glacier slopes and depths, assuming the equation for uniform flow.

$$\tau_c = \rho' g\, h \sin\alpha \quad (1\text{-}52)$$

The usual values which are found, vary from 50 to 180 K Pa. A good average value for a temperate glacier would be 100 K Pa.

The velocity of a glacier is related to the critical shear stress at the bed boundary. Under usual conditions, a model showing full contact between glacier ice and bedrock (Lliboutry, 1965) gives values of the sliding velocities smaller than about 10 m/year. However, many higher sliding velocities are observed particularly when a glacier surges, i.e. when there is a brutal advance of the glacier front in a short period of time. Sliding velocities of the order of 100 m/year are then measured. Lliboutry explains these higher flow velocities by the mechanism of cavitation. Cavitation occurs when a cavity forms at low pressure in the lee of an obstacle on the bed. In front of the obstacle the increased pressure of the moving ice makes it melt. The melt water flows and fills the cavity at lower pressure behind the obstacle. The pressures within the cavities then adjust themselves in order to minimize the friction so that the glacier velocity can increase. In these cavities, full of water, Lliboutry (1976) distinguishes between an autonomous hydraulic regime and an interconnected hydraulic regime. In the latter case, the isolated cavities increase in size with time and become interconnected, so that the same piezometric pressure is establish within this system. A new parameter has to be taken into account $N = (p_i - p)$ where $p_i$ is the mean pressure of ice on the bed, p is the water pressure and N is the Terzhagi's effective pressure. The friction law must then be written:

$$\tau_c = \mu N \qquad (1\text{-}53)$$

As $\mu$ is a dimensionless friction parameter depending only on the bedrock shape, there is a large reduction in the critical shear stress with increasing water pressure and a corresponding increase in glacier velocity. With a sudden interconnection of large areas of cavities at a higher water pressure, the glacier may become instable and move quickly downstream.

## 1.5.4 - CRYSTALLINE STRUCTURE OF A TEMPERATE GLACIER

According to Lliboutry (1965) there are many types of ice in a glacier that can be differentiated by their air bubble content, mud or debris content and by the form, size and crystallographic orientation of the grains. In general, glacier ice is made up of a pile-up of layers of different types of ice which he calls **ice foliations.** In the Massif du Mont Blanc, Vallon (1976) found three kinds of ice: full of bubbles but without foliation (white ice); foliated with alternate bands of blue ice (without bubble) and white ice and finally only blue ice. Rare layers of debris were found in the ice but their orientation was concordant with those of the foliations. The distinction between foliated ice and non-foliated ice was clear enough to identify the annual layers of ice with certainty.

The stratified layers of foliated ice move with the glacier, faster in the center than on the sides and they follow with depth, a trajectory similar to that shown on Fig. 1.28. Then, after a certain displacement, the foliation take more or less the form of a spoon with the concavity on the usptream side. In the ablation zone of the glacier, they appear on the surface forming an arc from side to side. The layer has an important vertical inclination within the ice.

The measurements of Vallon (1976) have shown that the firn of one year was essentially made of round crystals with an average 1 mm cross-section. The growth of the ice is very slow, and 6 years later, at a depth of 30 m, the average crystal cross-section is about 2 mm, the largest ones reaching 10 mm.

A very strong anisotropy was found in the firn and all

stereograms showed a single sub-vertical preferential orientation of the optical axis of the crystals. Below a depth of 20 to 30 m, the vertical orientation was no longer rigorously vertical but inclined $10^{o}$ to $20^{o}$ which is probably the direction normal to the stratified layers in the snow.

The densities of the different seasonal horizons are shown on Fig. 1.30. The metamorphosis is faster in the winter horizons, because they retain more water, settle more quickly and reach the density of ice at a smaller depth.

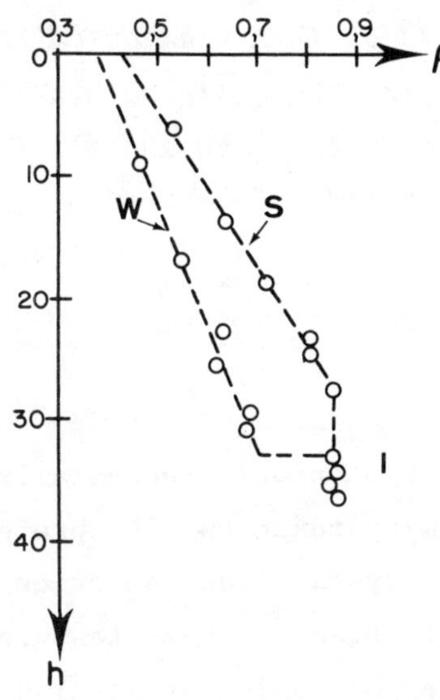


*Fig. 1.30. Density of firn versus depth in spring-autumn (S) and winter (W). At the lowest depth is found the density of glacier ice (I). From Vallon et al., (1976).*

For glacier ice itself, the average size of the crystals is shown on Fig. 1.31 in function of depth. At a depth beyond 100 m the size of the crystals does not increase, except very close to bedrock, where at least 7 m of blue ice is made of very large crystals with an average cross-section of 10 cm. This blue ice layer appears to be a common feature of temperate glaciers.

Initially round in the firn, the crystals acquire a very complex form with numerous re-entrants as they move downwards. The very strong subvertical preferential c-axis orientation was not found, as expected, in the glacier ice itself. In general, in the larger crystals, there are several preferential c-axis orientation which are clustered around the normal to the foliation planes.

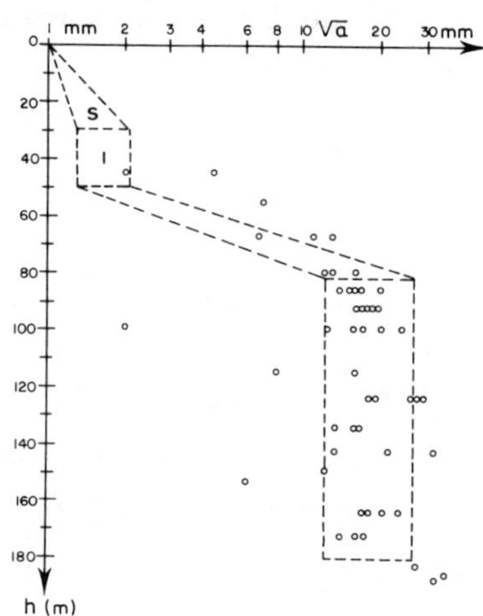


*Fig. 1.31. Average size of crystals $\sqrt{a}$ on thin horizontal sections in the Massif du Mont Blanc: (S) is in the firn and (I) in glacier ice. From Vallon et al., (1976).*

Duval (1976) has shown that syntectonic recrystallization in glaciers is at the origin of many textures with preferential orientations of the c-axis. The crystals take an orientation mainly controlled by the state of stress. Those textures formed with slow recrystallization are controlled by basal slip. Thus in simple shear, the basal planes tend to become parallel to the plane of permanent shear while in compression the normal to these planes is parallel to the direction of compression.

## 1.5.5 - FORMATION OF ICEBERGS

When a glacier front ends in the sea or in a lake, ice

floes will detach from the ice cliff. If they are large enough they will be called icebergs and the process is called the calving of icebergs. This is quite general for polar and sub-polar regions when the glaciers are close to the ocean. Iceberg production is particularly important from the Groenland indlandsis and the annual production is estimated at 240 $km^3$ (Bauer, 1955).

The mechanisms of production of icebergs from the front of a glacier moving into the sea are discussed by Lliboutry (1965) and shown on Fig. 1.32.

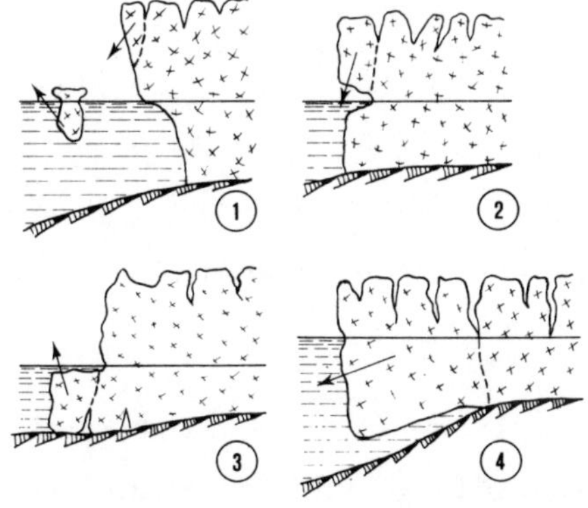


*Fig. 1.32. Modes of calving of icebergs.*

These mechanisms are:

1°) With the combined action of atmospheric ablation and melt of the ice by sea water, ice pieces fall off from the top edge of the glacier cliff;

2°) Wave action produces a local erosion at the water line that carves an overhanging ice block which falls into the sea;

3°) The Archimedean thrust on a submerged ice platform, breaks it off and makes it float;

4°) As the glacier front moves into deeper water, the vertical uplift by water increases and the ice breaks off at a distance from the edge, usually at the site of a crevasse. This mechanism is the most usual and produces the biggest icebergs.

## 1.6 - CLASSIFICATION OF ICE STRUCTURES

### 1.6.0 - INTRODUCTION

From all types of mechanisms of ice formation which have been described previously it is possible to classify the various types of structure of the solid ice cover. This will apply to river and lake ice and with some restrictions, also to sea ice. Glacier ice is not included in this classification. This classification is essentially that of Michel and Ramseier (1971).

### 1.6.1 - BASIC ICE LAYERS

When viewed parallel to the growth direction, the solid ice cover can be divided into three basic ice layers:

#### a) Primary Ice

The first type of ice of uniform structure and texture which forms on a water body is primary ice. On a calm surface the primary ice is an ice skim which grows horizontally in the supercooled layer and is a few tenths of a millimeter thick. In turbulent water it may consist of frozen frazil slush which can be very thick. If nucleation occurs by snow, the resulting congealed snow slush would also be part of the primary ice.

#### b) Secondary Ice

Secondary ice forms parallel to the heat flow which in most cases is perpendicular to the primary ice. Its structure is different from that of the primary ice and may be in the form of columnar ice, the texture of which is entirely controlled by the primary ice. It can also be in the form of frazil slush or snow slush deposited under a primary ice layer, which after some time may become an integral part of the ice cover.

c) Superimposed Ice

Superimposed ice always forms on top of the primary ice and is caused by flooding of the ice cover from any water source. If there is snow on the ice surface, snow ice will form.

Fig. 1.33 is an example of an entire river ice profile as seen under transmitted and polarized light. The superimposed ice layer is snow ice, the primary layer is frozen frazil slush, and the secondary layer is composed of columnar ice and frozen frazil slush.

## 1.6.2 - STRUCTURE AND TEXTURE OF ICE SHEETS AND ICE COVERS

The gross structure i.e., the horizontal bands, and texture of an ice sample can be observed in great detail from thin sections placed between crossed polaroid sheets. The structure, the grain size, grain shape, and global preferred crystallographic orientations of the c-axes greatly influence the mechanical properties of the ice cover and must be considered in a classification.

The grain size has been divided into five ranges:

a) fine; diameter less than 1 mm

b) medium; diameter 1-5 mm

c) large; diameter 5-20 mm

d) extra large; diameter greater than 20 mm

e) giant; grains with dimensions in meters.

Grain boundary shape and grain shape are both described in qualitative terms only. The crystallographic orientation is represented by plots on a Wulff net depicting the angular rela-

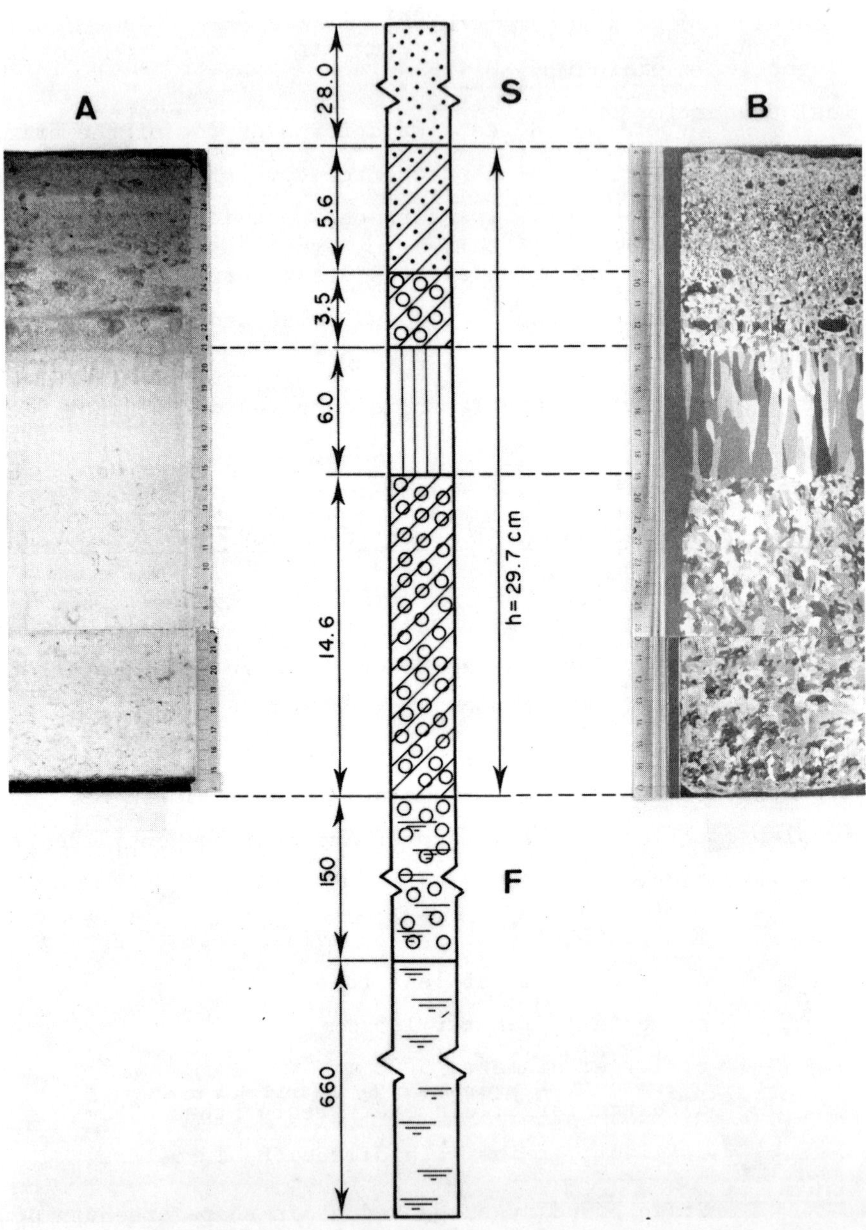


*Fig. 1.33. Profile under natural light and polarized light of the structure of an ice cover.*

tionships between the individual grains. Due to the strong influence of a preferred global c-axis orientation on the mechanical properties of the ice cover, the various distributions which have been observed are indicated in Fig. 1.34. The thin sections are mounted and analyzed with the projections of the c-axes in the upper hemisphere. The plane of the Wulff net is taken parallel to the ice cover, normal to the growth direction.

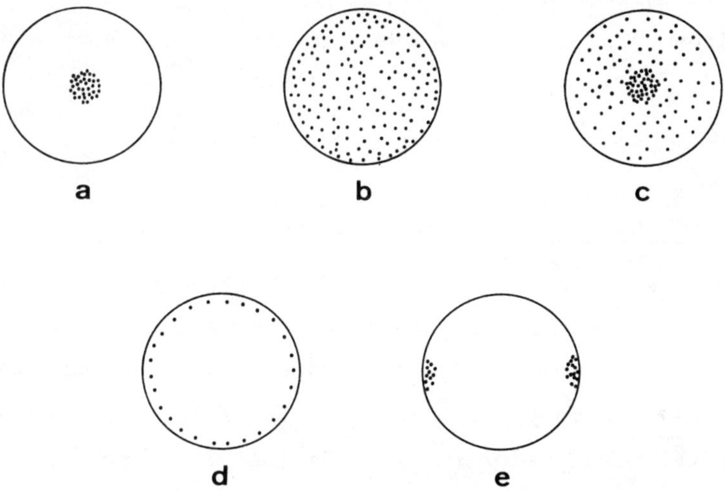


*Fig. 1.34. Various distributions of the c-axis of ice crystals that are observed in fresh-water ice: a) preferred vertical c-axis, b) random c-axis, c) preferred vertical orientation superimposed on a random orientation, d) preferred horizontal c-axis, e) preferred aligned horizontal c-axis.*

## 1.6.3 - CLASSIFICATION

The texture of the primary ice cover depends on the meteorological and hydrodynamic conditions existing at the time of formation. The secondary ice is indirectly related to these factors since its characteristics depend directly on those of the primary ice. Superimposed ice and agglomerate ice are independent of the characteristics of primary and secondary ices. They are caused by local meteorological or hydrodynamic conditions or by such conditions occurring either up or dowstream of the site of ice formation. The classification of ice describes the crystal size and shape and the crystallographic orientation. The environmental factors which cause these ice types are also indicated.

## Primary Ice

### P1 Calm Surface, Small Temperature Gradient

The grains are large to extra large with irregularly shaped boundaries. The crystallographic orientation of the c-axis is preferred vertical. This type of ice is found in reservoirs, lakes and very calm areas of rivers but has not been observe in the sea. Under ideal conditions it is possible to obtain giant sized grains which can have dimensions in meters.

Fig. 1.35. Horizontal thin section of P1 ice.

P2 Calm Surface, Large Temperature Gradient

The grain size is varied, ranging from medium to extra large. The crystallization progresses rapidly. The grain shape varies from tabular to needle-like and can be several cm in length. Dendrites are also common. The crystallographic orientation is random or vertically preferred superimposed on a random orientation. This type of ice is found in lakes, reservoirs, very calm areas of rivers and in sea ice.

*Fig. 1.36. Horizontal thin section of P2 ice.*

P3 Agitated Surface, Nucleated From Frazil

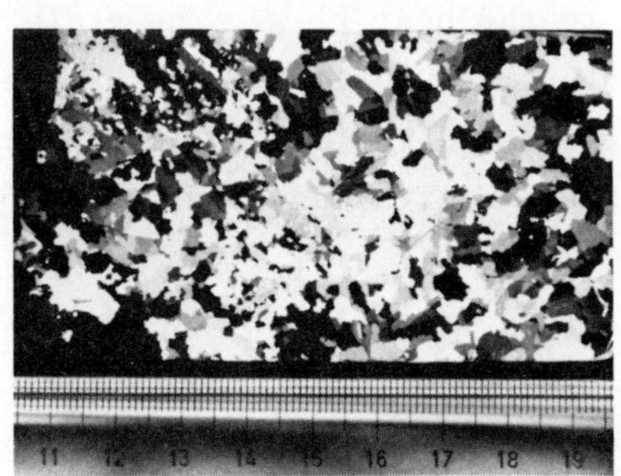


*Fig. 1.37. Horizontal thin section of P3 ice.*

The grains are fine-to-medium, the grain shape being

tabular. The crystallographic orientation is random. This type of ice can be found on lakes, reservoirs, rivers and is very general in the sea.

P4 Nucleation by Snow

The grains are fine-to-medium sized and are equiaxed. These properties depend to a large extent on the snow type. The crystallographic orientation is random. This ice can form in lakes, reservoirs and at sea.

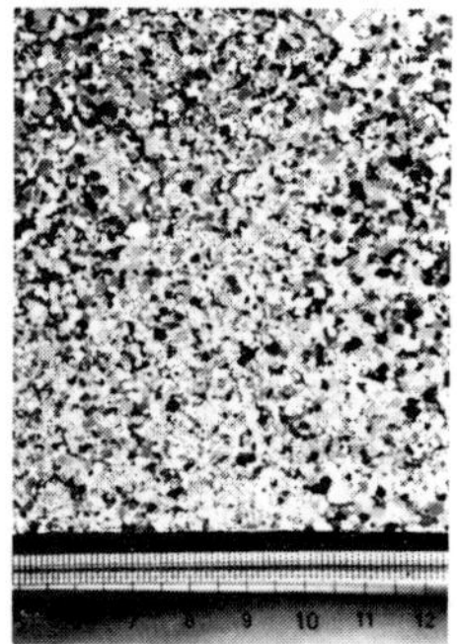

Fig. 1.38. Horizontal thin section of P4 ice.

## Secondary Ice

S1 Columnar Ice, Preferred Vertical Orientation of c-Axis

The grains form parallel to the heat flow. The grain size increases with depth, since unfavorably oriented grains are consumed by those more favorably oriented. The crystallographic orientation of the c-axis is preferred vertical and remains so for the entire columnar ice depth. The grain size is usually large to extra large; the grain shape is irregular. The long axis of the grains depends on the relative orientation of adjacent grains and can vary from equiaxed to a length equivalent to the thickness of the columnar ice layer. This type of ice forms as a result of the conditions given for primary ice types P1 or

P2; and very occasionally P3 or P4 if the ice cover is under pressure. It is found in lakes, reservoirs, and rivers with low flow velocities but not in the sea.

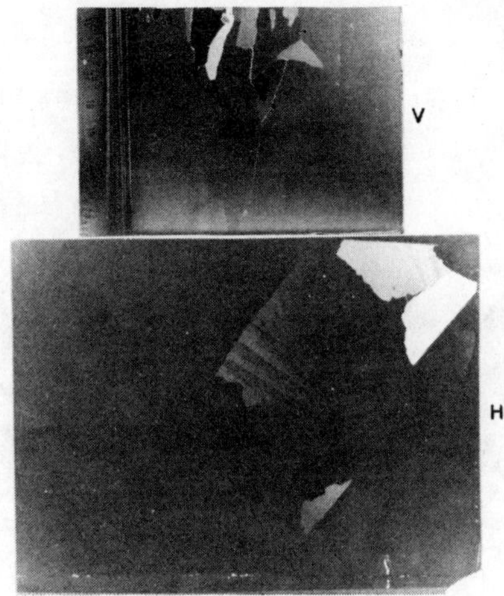


*Fig. 1.39. Vertical and horizontal thin sections of S1 ice.*

<u>S2 Columnar Ice, Preferred Horizontal Orientation of c-Axis</u>

The conditions are the same as for S1 but the primary ice has a random orientation. Grain size increases more rapidly with depth than for S1. The crystallographic orientation changes continuously with depth becoming preferred horizontal after 5 to 20 cm of growth (Shumskii, 1964). The initial grain size is the same as for the primary ice types P2, P3 and P4, increasing to large and possibly extra large at the bottom of the columnar layer, particularly in Arctic sea ice, (Fig. 1.40 and Fig. 1.41). It forms as a result of conditions given for primary ice types P2, P3 and P4.

Fig. 1.40. Vertical and horizontal thin section of S2 fresh water ice.

Fig. 1.41. Vertical and horizontal thin sections of S2 sea ice (One year Arctic ice).

S3 Columnar Ice, Preferred Aligned Horizontal Orientation of c-Axis

This type of ice forms at the bottom of thick ice sheets and covers. It is found in perennially frozen lakes, thick sea ice and Arctic ice islands. It develops from a prefered horizontal orientation to one with several of the c-axes nearly parallel. The cross-polaroid photograph for such an ice type is shown on Fig. 1.42.

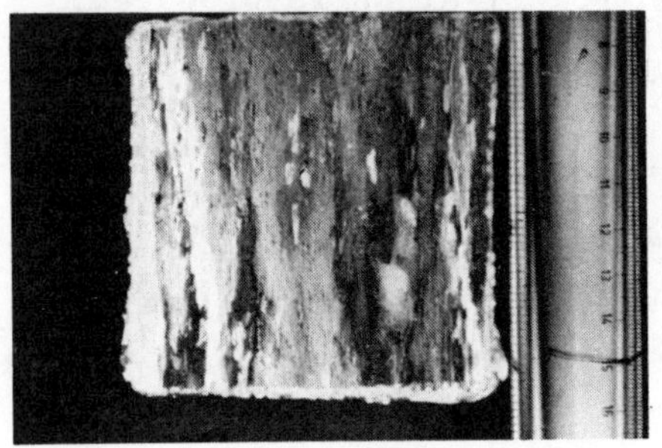

Fig. 1.42. Vertical thin section of S3 sea ice at the bottom of one year Arctic ice.

S4 Congealed Frazil Slush

Fig. 1.43. Vertical section of S4 ice.

The grains are equiaxed and tabular in shape. Their size is fine-to-medium and depends on the age of the originating frazil. The crystal boundaries are irregular and the crystallographic orientation is random. The frazil is formed in turbulent water and is swept under the secondary ice cover. An entire ice cover may consist of congealed frazil slush. It can also be found in different layers of the cover alternating with columnar ice. It is found in rivers as well as reservoirs and lakes fed by turbulent waters and at sea when frazil slush is produced.

S5 Drained Congealed Frazil Slush

The grains range from fine-to-medium size and are angular in shape. The density is low; over 0.4 g $cm^{-3}$. This ice is found where water has drained through the ice cover leaving the slush to be refrozen.

*Fig. 1.44. Appearance of S5 ice held on a rope.*

Superimposed Ice

T1 Snow Ice

The grain size ranges from fine-to-medium, the shape is

round to angular depending on the age of the snow, and the grains are equiaxed with a random orientation. The density may vary from 0.8 to 0.9 g $cm^{-3}$. This ice is formed when water saturates a snow deposit which then freezes.

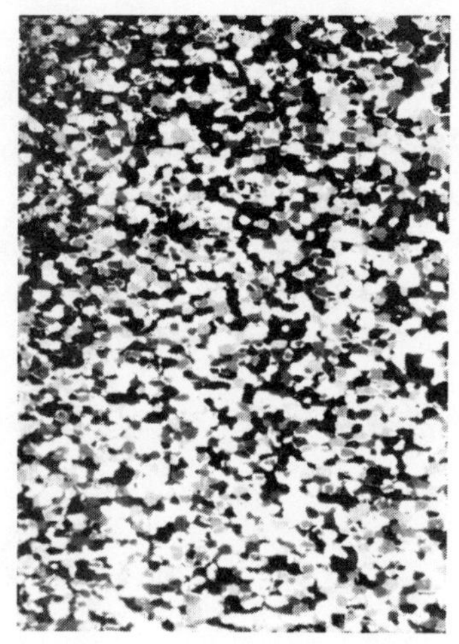

*Fig. 1.45. Section of T1 ice grown in the laboratory.*

T2 Drained Snow Ice

The grains range from fine-to-medium size and are well rounded. The layer is homogeneous and randomly oriented. The density varies approximately from 0.5 to 0.8 g $cm^{-3}$. This ice is found where water levels can vary rapidly, draining a previously saturated snow cover which then refreezes.

T3 Superimposed Layered Ice

Layers of columnar ice which have formed on top of the original primary ice.

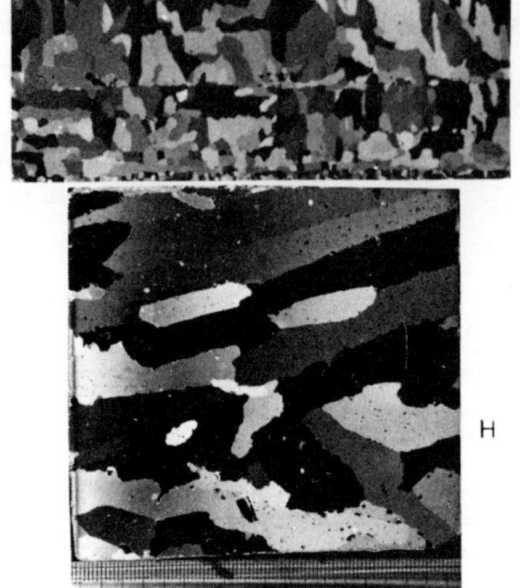


*Fig. 1.46. Vertical and horizontal sections of T3 ice with many occluded air bubbles.*

R Agglomerate Ice

This ice is an agglomeration of various ice types and forms. At any point where the process of thickening is more dynamic the horizontal layering will become indistinguishable. The grain size can vary from fine to extra large and the shape can vary from equiaxed to tabular to columnar with the crystal boundaries regular to angular in shape. The orientation can vary from random to a preferred orientation. This type of ice which is essentially an agglomeration of individual ice pieces which have refrozen, is frequently found in areas of turbulent flow or rapids in rivers and ice accumulation in the sea.

The observer may relate this ice type to particular ice forms such as rafted or ridged ice. The term agglomerate ice is purposely broad enough to permit the investigator to be more specific in describing the ice at any particular site.

*Fig. 1.47. Vertical section of R ice.*

## 1.6.4 - INCLUSIONS AND IMPURITIES

Air and brine are the predominant inclusions in ice. Organic and inorganic impurities, though they may not be present in large quantities, can significantly affect physical and mechanical properties of the ice. Some readily visible characteristics of ice due to inclusions are color, degree of opaqueness, and density. Textural changes are also present.

During solidification, air is rejected at the ice-water interface and is supplemented by gases from the biological processes taking place in a natural body of water. Air can also be incorporated from the atmosphere through cracks, drained snow ice, or an agitated water surface. If growth is slow most air will be rejected and the ice will be as transparent as glass. If the solidification proceeds rapidly air in the form of bubbles will be distributed throughout some parts of the ice cover. If air is being admitted continuously, the bubbles take on a cylindrical shape. (Peshanskii, 1969).

When ice grows in sea water there is an instability at the ice-water interface so that pure ice platelets are growing parallel to one another separated by layers of brine. The ice bridges over at various points and the brine is trapped in rows of vertical brine cells. The fraction of the volume of ice occupied by brine, the brine content, is a major factor influencing the mechanical behavior of sea ice. Because the brine has to be in thermodynamical equilibrium with the surrounding ice and because of brine drainage and loss in the ice cover, sea ice is virtually never in a perfect equilibrium state in nature. This subject was discussed in section 1.4 of this chapter.

## REFERENCES

Altberg, W.J. (1936) - "Twenty years of work in the domain of underwater ice formation" Int. Ass. Sci. Hydrology, Bull. No. 23, p. 373-407.

Assur, A. (1956) - "Airfields on floating ice sheets" U.S. Army, Snow, Ice and Permafrost Research Establishment, SIPRE Report 36.

Assur, A. (1958) - "Composition of sea ice and its tensile strength in Arctic sea ice" U.S. Nat. Acad. Sciences, N.R.C. Pub. 598, p. 106-138.

Bader, H. (1961) - "The Greenland ice sheet" Cold Regions Research and Engineering Lab., CRREL, Report 1-B2.

Barnes, W. (1929) - "The crystal structure of ice between 0°C and - 183°C" Proc. R. Soc, A215, p. 670-93.

Bauer, A. (1955) - "Le glacier de l'Ege (Greenland)" (ASI Herman no. 1225), 120 p. Rev. by W.H. Ward; J. Glac., 2, pp. 688-689.

Bigg, E.K. (1953) - "The supercooling of water" Proc. Phys. Soc. London, series B, Vol. 66, p. 688.

Bragg, W.H. (1922) - "The crystal structure of ice" Proc. Phys. Soc. 34, pp. 98-103.

Cox, G.F., Weeks, W.F. (1974) - "Salinity variation in sea ice" Journal of Glaciology, Vol. 13, No. 67, pp. 109-20.

Devik, O. (1949) - "Freezing water and supercooling" Journal of Glaciology, Vol. 1, p. 307-309

Dorsey, N.E. (1948) - "The freezing of supercooled water" Trans. Am. Phil. Soc. Vol. 38, part 3, p. 247-328.

Duval, P. (1976) - "Fluage et recristallisation des glaces polycrystallines" Thèse de Doctorat d'Etat, Université de Grenoble.

Eide, L.I., Martin, S. (1975) - "The formation of brine drainage features in young sea ice" Journal of Glaciology, Vol. 14, No. 70, pp. 137-154.

Fletcher, N.H. (1970) - "The chemical physics of ice" Cambridge University Press.

Frankenstein, G., Garner, R. (1967) - "Linear relationship of brine volume and temperature from -0.5 to -22.9$^{o}$C for sea ice" Journal of Glaciology 6, No. 48, pp. 943-945.

Geiger, R. (1965) - "The climate near the ground" Harvard University Press, Cambridge, Mass.

Ginnings, D.C., Corrucini, R.J. (1947) - "An improved ice calorimeter: the determination of its calibration factor and the density of ice at 0$^{o}$C" J. Res. Nat. Bur. Stand. 38, pp. 583-91.

Heverly, J.R. (1949) - "Supercooling and crystallization" Trans. Amer. Geo. Union, Vol. 30, no. 2, pp. 205-210.

Hobbs, P.V. (1974) - "Ice physics" Clarendon Press, London, 837p.

Katsaros, K.B., Liu, W.T. (1974) - "Supercooling and deposit of frazil ice" Proc. 'Eastern Snow Conference' Vol. 8, pp. 129-149.

Kivisild, H.R. (1959) - "Hanging ice dams" Proc. 8th Congress of the Int. Ass. for Hydraulic Research, Vol. 3, p. 1-1.

Kovacs, A., Weeks, W.F., Ackley, S., Hibler III, W.D. (1973) - "Structure of a multi-year pressure ridge" Arctic, Vol. 26, No. 1, p. 22-31.

Kumai, M. (1968) - "Hexagonal and cubic ice at low temperatures" J. Glaciol. 7, pp. 95-108.

Lewis, W.V. et al., (1960) - "Norwegian cirque glaciers" Roy. Geog. Soc. Series, No. 4, 104 p.

Lliboutry, L. (1965) - "Traité de Glaciologie" Tome I et II, Masson et Cie, Paris.

Lliboutry, L. (1976) - "Physical processes in temperature glaciers" Journal of Glaciology, Vol. 16, No. 74, pp. 151-158.

Megaw, H.D. (1934) - "Cell dimensions of ordinary and 'heavy' ice" Nature, London 134, 900-1.

Michel, B. (1965) - "Les métamorphoses du frasil en rivière" Trans. Eng. Inst. of Canada, Vol. 8, No. A-5.

Michel, B. (1965) - "Criterion for the hydrodynamic stability of the frontal edge of an ice cover" Proc. 11th Congress of the Int. Ass. for Hydraulic Research, Vol. 5, p. 1-11.

Michel, B., Ramseier, R. (1971) - "Classification of river and lake ice" Canadian Geotech. Journal, Vol. 8, No. 1, p. 38-45.

Michel, B. (1971) - "Winter regime of river and lakes" Cold Regions Science and Engineering Monograph III-Bla, U.S. Army Corps of Engineers.

Michel, B., Hanley, T.O.D. (1975) - "Mechanisms of ice growth at the ice-water interface in a laboratory tank" Proc. Seminar on the Themal Regime of River Ice. N.R.C. Canada Tech. Mem. No. 114, p. 96-104.

Michel, B. (1977) - "La température de nucléation du frasil" Rapport GCS-77-03, Dép. de Génie Civil, Université Laval.

Nye, J.F. (1951) - "The flow of glaciers and ice sheets as a problem in plasticity" Proc. Roy. Soc., 207, pp. 554-572.

Osterkamp, T.E. (1974) - "Supercooling and frazil ice formation in a small sub-arctic stream" Proc. Research Seminar Thermal Regime of River Ice, Tech. Memorandum No. 114, Natural Research Council of Canada, p. 104-109.

Pariset, E., Hausser, R. (1961) - "Formation and evolution of ice covers in rivers" Trans. Eng. Inst. of Canada, Vol. 5, No. 1, p. 40-49.

Pauling, L. (1935) - "The structure and entropy of ice and of other crystals with some randomness of atomic arrangement" J. Am. Chem. Soc. 57. 2680-4.

Peschanskii, I.S. (1969) - "Ice Science and ice technology" Institute of Hydro-Meteorology, Leningrad.

Pounder, E.R. (1965) - "The Physics of Ice" Pergamon Press, London.

Riehl, N. (1965) - "Protonic mobility and its importance for biological systems" Trans. N.Y. Acad. Sci. 27, 772-81.

Robert, S. (1978) - "Expériences sur la température de nucléation du frasil" Thèse M.Sc., Faculté des Sciences, Université Laval.

Schaefer, V.J. (1950) - "The formation of frazil and anchor ice in cold water" Am. Geo. Union, Vol. 31, No. 6, p. 885-896.

Schwarzacher, W. (1950) - "Sea ice studies in the Arctic Ocean" Journal of Geophysical Research, Vol. 64, p. 2357-2367.

Shumskii, P.A. (1964) - "Principles of Structural Glaciology" Dover Publications Inc., New York.

Schnell, R.C., Vali, G. (1972) - "Atmospheric ice nuclei from decomposing vegetation" Nature, Vol. 236, No. 5343, p. 163-165.

Transehe, N.A. (1928) - "The ice cover of the Arctic Sea with a genetic classification of sea ice" Publ. of Polar Research, A.G.S., Special Publ., No. 7, pp. 91-123.

Uzuner, M.S., Kennedy, J.F. (1972) - "Stability of floating ice blocks" Journ. of the Hydraulics Div., ASCE, Vol. 93, No. 93, No. H12, Proc. Paper 9118, pp. 2117-2133.

Vallon, M., Petit, J.R., Fabre, B. (1976) - "Study of an ice core to the bedrock in the accumulation zone of an alpine glacier" Journal of Glaciology, Vol. 17, No. 75, pp. 13.28.

Weeks, W., Assur, A. (1967) - "The mechanical properties of sea ice" Cold Regions Science and Engineering Monograph 11-C3, U.S. Army, 80 p.

Whalley, E. (1969) - "Structural problems of ice" In Physics of Ice, pp. 19-43.

# CHAPTER 2

# MECHANICAL PROPERTIES OF ICE

## 2.0 - INTRODUCTION

Before any discussion can be made on the strength of ice, it is most important to distinguish between two different meanings for this property of ice which correspond to completely different modes of failure of the material. One is the fracture strength which occurs under brittle conditions when a brittle crack is formed that propagates across an ice sample for a complete fracture. The other is the yield strength which occurs under ductile conditions such that the ice does not fail at this stress level but continues to deform with corresponding lower stresses. This process may, or may not, lead to complete failure of an ice sample, but the yield strength corresponds to the maximum stress that the ice can resist during its ductile mode of deformation.

## 2.1 - BRITTLE BEHAVIOR

### 2.1.0 - INTRODUCTION

The strength of polycrystalline ice depends on a number of factors, the most important being the temperature, the internal structure and the conditions of loading (Michel, 1970). Because of these factors, there is a wide variation in ice strength measurements quoted in the literature for thousands of tests on ice made during the last hundred years. For example, the crushing strength has been found to vary from $4 \times 10^5$ to $130 \times 10^5$ Pa and the flexural strength from $3 \times 10^5$ to $30 \times 10^5$ Pa. Unfortunately, in practically all cases, the results are given without adequate information on the structure of the ice and the conditions of loading, so that they cannot be used to develop general relationships for the strength of ice.

## 2.1.1 - BRITTLE BEHAVIOR OF MONOCRYSTALS

For very short periods of time, the deformation of monocrystals is proportionnal to the load and fully recovered upon removal of the latter. This is true however only if the dislocations do not have the time to move along the basal planes and this happens only for very high loading rates up to the speed of propagation of elastic waves, which is close to 3900 $ms^{-1}$ for cold ice.

### 2.1.1.1 - ELASTIC MODULII

When the ice is deformed elastically, it follows Hooke's law and the strain $\varepsilon$ is proportionnal to the applied stress $\sigma$, so that:

$$\varepsilon = S\sigma \tag{2-1}$$

where S is the compliance constant of the ice. This equation may also be written:

$$\sigma = C\varepsilon \tag{2-2}$$

where C is the stiffness constant of the ice.

When a tensile or compressive force is applied to a body, the stress is equal to the force acting on a unit area of the body normal to the direction of the force, the longitudinal strain is the fractional increase in the length of the body in the same direction and the stiffness constant is referred to as the Young's modulus E. The ratio of the longitudinal strain $\varepsilon_x$ along the direction of the force to the strain $\varepsilon_y$ produced in a direction perpendicular to the force is called the Poisson's ratio $\nu$, so that:

$$\nu = \varepsilon_y/\varepsilon_x \tag{2-3}$$

If a pure compressionnal stress is applied to a body, its volume will change. In this case, the stiffness constant which is the ratio of the applied pressure to the fractionnal change in the volume of the body, is called the bulk modulus K. The reciprocal of the bulk modulus is the compressibility. Under a shear stress, a body will suffer shear strain, the stiffness constant is then referred to as the rigidity modulus (or shear modulus) G. For an isotropic body there are two independant elastic parameters, and it may be shown that:

$$G = E/2\ (1 + \nu) \tag{2-4}$$

and

$$K = E/3\ (1 - 2\nu) \tag{2-5}$$

For a non isotropic crystal, the elastic behavior is described by the more general relationships (Nye, 1957).

$$\varepsilon_i = S_{ij}\ \sigma_j \tag{2-6}$$

$$\sigma_i = C_{ij}\ \varepsilon_j \tag{2-7}$$

where $S_{ij}$ and $C_{ij}$ are matrix coefficients representing compliance and stiffness constants respectively and i and j take integral values from 1 to 6 inclusive. For hexagonal crystals, and for ice, there are only five non-zero independant elastic constants, namely $S_{11}$, $S_{12}$, $S_{13}$, $S_{33}$, $S_{44}$ and $S_{66}$ with the corresponding $C_{ij}$'s.

These notations can be taken as follows in relation to the orientation of the main optical crystal axis:

- $S_{11}$ gives the extension perpendicular to the c-axis due to a longitudinal tensile stress also perpendicular to the c-axis, in the same direction.

- $S_{12}$ gives the extension perpendicular to the c-axis due to a longitudinal stress also perpendicular to the c-axis but also perpendicular to the direction the extension is measured.

- $S_{13}$ gives the extension perpendicular to the c-axis due to a longitudinal stress along the c-axis and is equal to the extension along the c-axis due to a tensile stress perpendicular to it.

- $S_{33}$ gives the extension along the c-axis due to a tensile stress along it.

- $S_{44}$ gives the shear strain on a plane containing the c-axis due to a shear stress in the same plane.

- $S_{66}$ gives the shear strain on a plane perpendicular from the c-axis due to a shear stress in the same plane. It is derived from the other constants.

The elastic constants for single ice crystals have been determined by many authors, in function of the temperature. Frequently cited results are those of Dantl (1968) which were obtained with an ultrasonic method. They are given below:

$C_{11} = 1.2904\ (1 - 1.489 \times 10^{-3}\ \theta - 1.85 \times 10^{-6}\ \theta^2) \times 10^{10}\ \text{N m}^{-2} \pm 0.3\%$

$C_{12} = 0.6487\ (1 - 2.072 \times 10^{-3}\ \theta - 3.62 \times 10^{-6}\ \theta^2) \times 10^{10}\ \text{N m}^{-2} \pm 2\%$

$C_{13} = 0.5622\ (1 - 1.874 \times 10^{-3}\ \theta) \times 10^{10}\ \text{N m}^{-2} \pm 7\%$

$C_{33} = 1.4075\ (1 - 1.629 \times 10^{-3}\ \theta - 2.93 \times 10^{-6}\ \theta^2) \times 10^{10}\ \text{N m}^{-2} \pm 0.4\%$

$C_{44} = 0.2819\ (1 - 1.601 \times 10^{-3}\ \theta - 3.62 \times 10^{-6}\ \theta^2) \times 10^{10}\ \text{N m}^{-2} \pm 0.7\%$

$C_{66} = 0.3708\ (1 - 0.90 \times 10^{-3}\ \theta) \times 10^{10}\ \text{N m}^{-2} \pm 3\%$

From the above, the values of $S_{ij}$ can be deduced:

$S_{11} = 1.040\ (1 + 1.070 \times 10^{-3}\ \theta + 1.87 \times 10^{-6}\ \theta^2) \times 10^{10}\ \text{m}^2\ \text{N}^{-1} \pm 1\%$

$S_{12} = -\ 0.442\ (1 + 0.463 \times 10^{-3}\ \theta - 2.06 \times 10^{-6}\ \theta^2) \times 10^{10}\ \text{m}^2\ \text{N}^{-1} \pm 6\%$

$S_{13} = -\ 0.189\ (1 + 1.209 \times 10^{-3}\ \theta + 6.15 \times 10^{-6}\ \theta^2) \times 10^{10}\ \text{m}^2\ \text{N}^{-1} \pm 20\%$

$S_{33} = 0.848\ (1 + 1.405 \times 10^{-3}\ \theta + 4.66 \times 10^{-6}\ \theta^2) \times 10^{10}\ \text{m}^2\ \text{N}^{-1} \pm 1\%$

$S_{44} = 3.342\ (1 + 1.505 \times 10^{-3}\ \theta + 4.04 \times 10^{-6}\ \theta^2) \times 10^{10}\ \text{m}^2\ \text{N}^{-1} \pm 1\%$

$S_{66} = 2.964\ (1 + 0.89 \times 10^{-3}\ \theta + .70 \times 10^{-6}\ \theta^2) \times 10^{10}\ \text{m}^2\ \text{N}^{-1} \pm 7\%$

From these values the compressibility can also be calculated; it is:

$K = 1.194\ (1 + 1.653 \times 10^{-3}\ \theta + 3.12 \times 10^{-6}\ \theta^2)\ 10^{10}\ \text{m}^2\ \text{N}^{-1} \pm 15\%$

The values of the dynamic modulii, for single ice crystals, vary with the direction of loading. For any direction the compliance constants can be obtained from an equation by Fletcher (1970).

$$S_\alpha = S_{33} \cos^4 \alpha + S_{11} \sin^4 \alpha + (S_{44} + 2\ S_{13}) \sin^2 \alpha \cos^2 \alpha \quad (2\text{-}8)$$

where the S's are the compliance constants and $\alpha$ is the angle between the direction of loading and the crystal c-axis.

The Young's modulus of ice is the reciprocal to the corresponding compliance constant. It varies then by about 30% depending upon $\alpha$ and its minimum value is smaller than either that parallel or perpendicular to the c-axis. This is shown on Fig. 2.1.

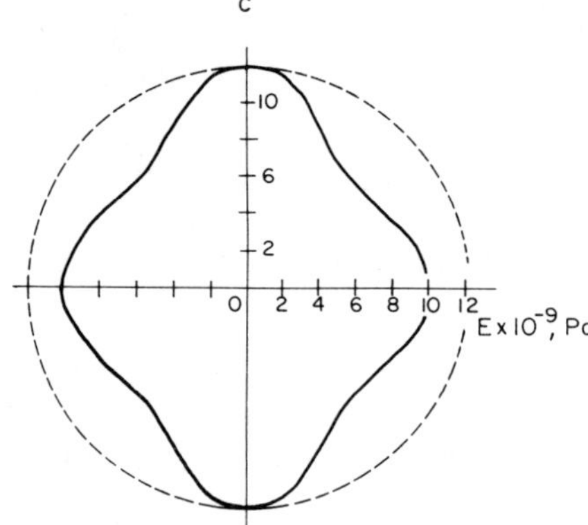


*Fig. 2.1. Young's modulus of an ice monocrystal $E = 1/S_{ij}$ as a function of stress orientation. From Fletcher (1970).*

## 2.1.1.2 - FRACTURE AND CRACK PROPAGATION UNDER TENSION

A purely brittle fracture will occur in monocrystalline ice when the concentrated elastic energy in one particular region will be released to initiate and propagate a crack through the ice.

Pseudo-monocrystalline $S_1$ ice is a very similar material because it is made up of vertical columnar crystals which all have a vertical c-axis. The basal plane of all crystals is thus co-planar, and the ice acts much like a monocrystal except that there are distinct changes in the orientation of the a-axis of the basal plane, at the frontier of each grain.

When a crystal of $S_1$ ice is deformed, the energy required to deform it in a direction is proportional to the value of

$\varepsilon^2/S$ in this direction. For a given strain $\varepsilon$ and with the numerical values of $S_{44}$ and $S_{66}$, this shows that there is about 10% more shear strain energy available to fail ice in the basal plane, under pure elastic conditions, than normal to it.

The non-isotropic behaviour of ice explains its fracture. The experimental evidence (Carter and Michel, 1971) shows that, for all cases of uniaxial tension or compression, the surface of fracture is unique and always along the basal plane of $S_1$ ice, whatever be its orientation, except at the extremes when the basal plane is parallel or normal to the direction of loading. The aspect of fracture for various inclinations of the basal plane is shown in Fig. 2.2.

There is always some structural stress risers, present in natural ice, that can start the nucleation of a crack. The usual one is the air bubble that will set up a concentrated stress field. But, the ice fails at stress levels much lower than those given by these air inclusions. There is considerable experimental evidence that in polycrystalline ice, micro-cracks when formed, extend to the full size of each grain (Gold 1960, Michel *et al.*, 1978). This is why Michel has proposed (1978) that the critical condition of formation of a crack is the release of the elastic energy of a grain of ice to form a crack extending throughout the basal plane of this individual grain. This is the basis of the theory which is used here.

Let us consider a crystal of $S_1$ ice (A), as shown on Fig. 2.3 which is surrounded by the lateral frontiers of the neighbouring crystals (B). In the columnar crystal itself, we will consider that the shear stress is unchanged in the outer regions (C) after the formation of a central crack. Grain (A) is acted upon by a main shear stress $\tau$ in the direction of the basal

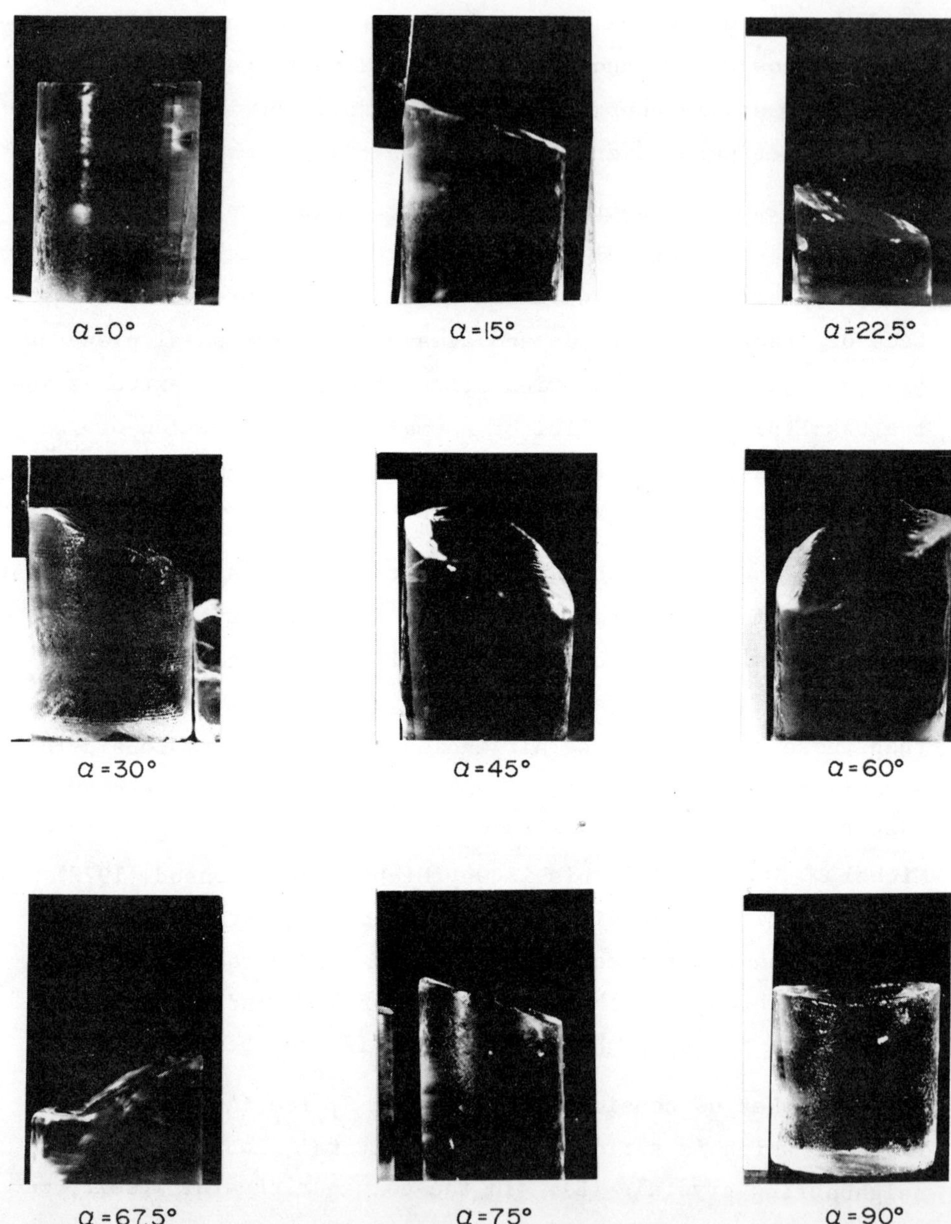


*Fig. 2.2. Fracture surfaces of $S_1$ monocrystals in function of angle α of basal plane with the normal to the loading direction.*

plane and a normal tensile stress σ. The crystal can be considered a section of a cylinder whose diameter is d and height H. This latter is, indeed, choosen as the equivalent height of ice where the shear strains are relaxed after the formation of the crack.

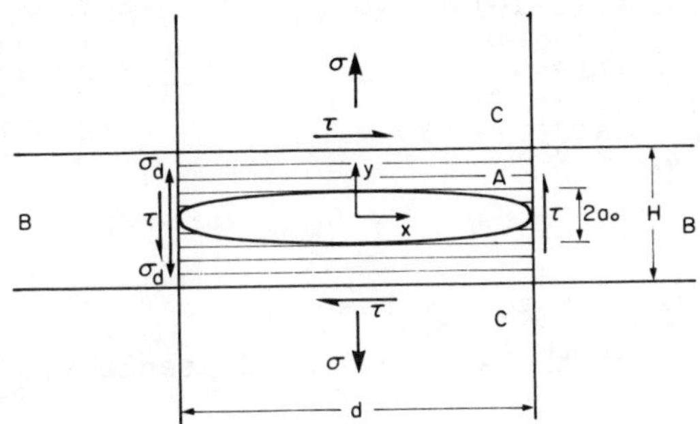


*Fig. 2.3. Sketch of crack nucleation in crystal of $S_1$ ice (A) surrounded by adjoining crystals (B) and undeformed ice (C).*

The elastic energy which is made available from the crystal for the formation of a crack is:

$$W_e = \int_o^V \int_o^\tau \tau \, d\gamma \, dV + \int_o^V \int_o^\sigma \sigma d\varepsilon \, dV \tag{2-9}$$

where

γ - shear strain caused by the shear stress τ

ε - tensile strain caused by the tensile stress σ

V - volume of the crystal

For a monocrystal under elastic conditions:

$$\begin{aligned} d\gamma &= S_{66} \, d\tau = d\tau/G_{66} \\ d\varepsilon &= S_{33} \, d\sigma = d\tau/E_{33} \end{aligned} \tag{2-10}$$

The first term of Eq. (2-9) is easy to compute from the definition of the equivalent thickness H of the crystal.

$$W_\tau = \frac{\pi \, d^2 \, H \, \tau^2}{8 \, G_{66}} \tag{2-11}$$

The second term in Eq. (2-9) can be evaluated from the instantaneous energy change in the normal component of the stress to form a crack of oblate spheroid form in the basal plane on the crystal. The diameter is d and the minor semi-axis is $a_o$.

The instantaneous energy change in the volume of the crack is given from Eqs (2-9) and (2-10) by:

$$W_\sigma = 2\int_o^{a_o} \frac{\sigma^2}{2\ E_{33}}\ s\ dy \tag{2-12}$$

where

y - distance perpendicular to the basal plane

s - area of a section of the spheroid at distance y.

For this geometry, the integral of (2-12) gives:

$$W_\sigma = \frac{\pi\ d^2\ a_o\ \sigma^2}{6\ E_{33}} \tag{2-13}$$

The surface energy needed for the creation of the two surfaces of a thin crack is:

$$W_s = \frac{\pi\ d^2}{2}\ \sigma_{cv} \tag{2-14}$$

where

$\sigma_{cv}$ = 0.109 $Jm^{-2}$, is the surface energy of ice in contact with its saturated vapor (Ketcham and Hobbs, 1969).

The condition for crack formation is then:

$$W_\tau + W_\sigma \geq W_s \tag{2-15}$$

With Eqs (2-11), (2-13) and (2-14) this gives:

$$\frac{\tau^2}{G_{66}} + \frac{4}{3}\left(\frac{a_o}{H}\right)\frac{\sigma^2}{E_{33}} \geq \frac{4\sigma_{cv}}{H} \tag{2-16}$$

Let us assume that, normally, $a_o < H$. We also have, $E_{33} > G_{66}$. The second term on the left is then usually negligible before the first, except when $\tau$ is very small or $\sigma$ is large. The usual condition of nucleation of a crack is then:

$$\tau \geq 2\sqrt{\frac{\sigma_{cv} G_{66}}{H}} \tag{2-17}$$

Once a crack has nucleated, there is a large increase in tensile stresses at the tips of the crack. For this crack configuration, Sadowsky and Sternberg (1949) have obtained the stress concentration at the tip of the crack, $\sigma_d$ in the y direction in function of stress at infinity $\sigma$. For a ratio $^{a}o/d$ = 0.05, this value is:

$$\sigma_d = 13.5\ \sigma \tag{2-18}$$

The crack will propagate, in the Griffith's sense, when the tensile strain energy is related to the energy needed to form a surface:

$$\frac{\sigma_d^{\ 2}}{2\ E_{33}}\ dV > 2\ \sigma_{cv}\ ds \tag{2-19}$$

ds - unit area parallel to the basal plane.

With Eq. (2-18) this gives for a fictitious crystal of height H:

$$90.6\ \frac{\sigma^2\ H}{E_{33}} > 2\ \sigma_{cv} \tag{2-20}$$

This condition is realized as long as there is a small tensile stress $\sigma$ normal to the basal plane. This is also the condition necessary for a crack to form in the basal plane, when the load is not either normal or parallel to this plane. Thus in tensile testing, a crack will propagate and fail the sample as soon as it is formed except under these extreme conditions.

For tests in uniaxial tension, the values of $\tau$ and $\sigma$ in a basal plane making an angle $\alpha$ with the normal to the loading direction, are given, from simple stress analysis, by:

$$\tau = \sigma_{\alpha} (\sin 2\alpha)/2 \tag{2-21}$$

$$\sigma = \sigma_{\alpha} (1 + \cos 2\alpha)/2 \tag{2-22}$$

where the subscript $\alpha$ denotes the uniaxial strength in tension of an ice sample where the basal plane has an orientation $\alpha$. The condition of fracture given by Eq. (2-17) becomes in Eq. (2-21):

$$\sigma_{\alpha} = 4 \operatorname{cosec} 2\alpha \sqrt{\frac{\sigma_{cv} G_{66}}{H}} \tag{2-23}$$

The lowest value of $\sigma_{\alpha}$ is for $\alpha = 45^{\circ}$. It gives:

$$\sigma_{45} = 4\sqrt{\frac{\sigma_{cv} G_{66}}{H}} \tag{2-24}$$

The tensile strength of $S_1$ ice at $-10^{\circ}C$ has been measured by Carter and Michel (1971), in function of the angle $\alpha$ and the results are given in Fig. 2.4. For $\alpha = 45^{\circ}$ the average value is $\sigma_{45} = 22 \times 10^5$ Pa. The value of $G_{66}$ can be obtained from the value of $S_{66}$:

$$G_{66} = 1/S_{66} = 3.374 (1 - 0.89 \times 10^{-3} \theta) \times 10^9 \text{ Pa} \tag{2-25}$$

With this value, the thickness of the relaxed elastic band H for $S_1$ ice can be computed and it is found to be H = 1.2 mm.

The value of $\sigma_{\alpha}$ is related to $\sigma_{45}$ by Eqs (2-21) and (2-24):

$$\sigma_{\alpha} = \sigma_{45} \operatorname{cosec} 2\alpha \tag{2-26}$$

This relationship is traced as a full line in Fig. 2.4. It fits perfectly well the data for all $\sigma$'s except $\alpha < 15^{\circ}$ and

$\alpha > 75^{\circ}$. For $\alpha = 0$ and $\alpha = 90^{\circ}$, Eq. (2-26) gives a $\sigma_{\alpha}$ value of infinity. The experimental result show essentially that the brittle fracture is associated with the maximum shear stress in the basal plane. For the extreme conditions of loading, there are mechanisms of fracture in non-basal planes, that limit the fracture strength to the experimental values shown in Fig. 2.4.

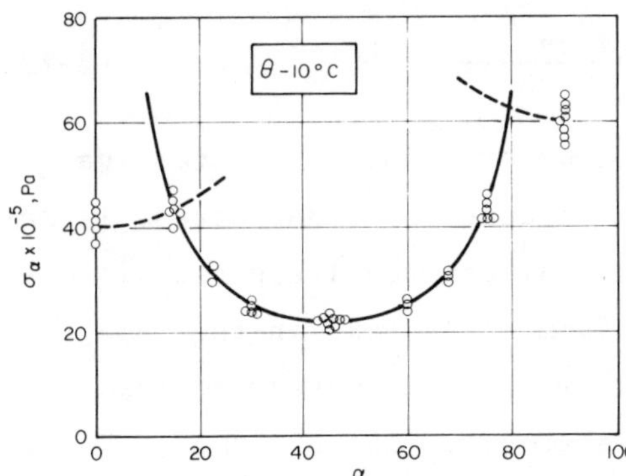


*Fig. 2.4. Variation of tensile strength of S1 ice in function of angle α of basal plane with the normal to the loading direction. (Carter and Michel, 1971).*

## 2.1.1.3 - FRACTURE UNDER COMPRESSION

Under compressive stresses, the major difference is that the formation of a crack will not release any stored elastic energy by compression, as it does in tension. In fact, the compressive stress, normal to the crack, will be transmitted integrally from one side to the other. It will even introduce an additionnal resistance to cracking in the form of the cohesive force resisting the formation of the free surfaces of the crack. Because of this, only the stored elastic energy caused by shear, is available in the equivalent crystal, for the formation of the crack. Eq. (2-15) (2-15) then reduces to:

$$W_{\tau} \geq W_{s} \qquad (2\text{-}27)$$

Furthermore, the total shear stress $\tau'$ that will produce cracking will be given by:

$$\tau' = \tau^* + C_i \tag{2-28}$$

where $\tau^*$ is the effective shear stress that causes fracture by the release of the elastic energy and $C_i$ the cohesive strength of the ice. From Eq. (2-17):

$$\tau^* = 2\sqrt{\frac{\sigma_{cv} G_{66}}{H}} \tag{2-29}$$

Once a crack has formed under a normal compressive stress, it is inherently stable, because the shear stress is reduced to $C_i$ at its tip and the normal compressive stress is unchanged. However, further away from the crack, in the same plane, the effective shear stress will have increased because of this local reduction in shear resistance. In this manner, separated new cracks would successively appear but because of the fast increase in available energy, the ice fails brittely by coalescence of these cracks, in the same basal plane.

In uniaxial compression, Eq. (2-21) becomes with Eq. (2-28):

$$\sigma_\alpha' = 2\ \tau'/\sin 2\ \alpha \tag{2-30}$$

The experimental results for the strength of $S_1$ ice in uniaxial compression at $-10^\circ$ are given in Fig. 2.5 (Carter and Michel, 1971), for various angles $\alpha$ of the basal plane. It can be seen again that they fit perfectly well Eq. (2-30), traced in full line. The value of $\tau^*$ can be obtained from the tensile tests and the value of $C_i$ computed from $\tau'_{45}$ in compression. They are: $\tau^* = 11 \times 10^5$ Pa and $C_i = 50 \times 10^5$ Pa for $S_1$ ice at $-10^\circ$C.

As for the cases of tension when $\alpha < 15^{\circ}$ and $\alpha > 75^{\circ}$, the uniaxial crushing strength does not correspond to a fracture in the basal plane and is limited to the values obtained experimentally.

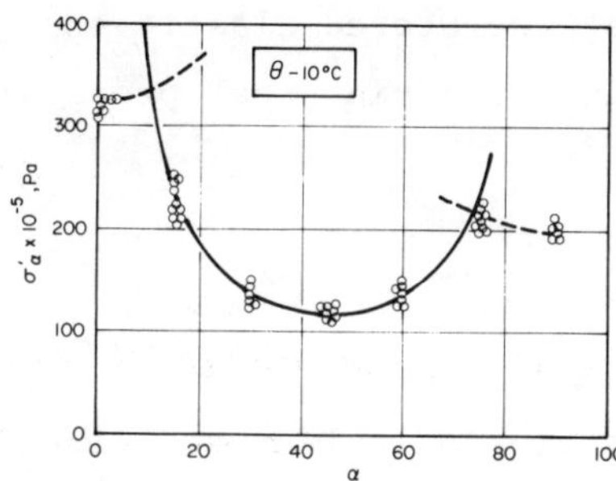


*Fig. 2.5. Variation of crushing strength of $S_1$ ice in function of angle α of basal plane with the normal to the loading direction. (Carter and Michel, 1971).*

## 2.1.2 - BRITTLE BEHAVIOUR OF POLYCRYSTALLINE ICE

### 2.1.2.0 - INTRODUCTION

The brittle behaviour of polycrystalline ice is characterized by elastic deformation followed by a sudden fracture. Over a given loading rate, the fracture strength depends only on temperature and ice type. We have found this threshold to be close to $10^{-4}$ $s^{-1}$ for tensile fractures and around $10^{-2}$ $s^{-1}$ for fractures in compression. It depends a little on the ice structure and on temperature.

The elastic deformations are caused primarily by changes in intermolecular distances under the applied stresses, and under these circumstances, there is no permanent plastic deformation which would have been induced by the movement of dislocations.

As for $S_1$ ice, the brittle strength of polycrystalline ice has been determined from the study of the elastic stability of an individual grain. The analysis is based on the assumption that when the stress level is adequate, the stored elastic energy is released to produce a micro-crack in the grain, that propagates and fails the ice sample.

### 2.1.2.1 - ELASTIC MODULII

The litterature on the properties of polycrystalline ice give values for the elastic modulii that vary within a wide range, usually by a factor of 1 to 5, and this is confusing. The highest values are obtained when loads are applied very quickly, at the speed of elastic waves, and the lowest ones are obtained under static conditions or with soft loading machines due to their own elastic constants.

Under static conditions, it takes a certain time to apply a load to ice and to remove it. During that time, even if it is very short, polycrystalline ice will show considerable delayed elasticity and this added with the instantaneous deformation will give a much lower apparent elastic modulus. It is not the dynamic elastic modulii.

The effect of delayed elasticity can be taken into account by modifying the value of the instantaneous Young's modulus E by an equivalent secant modulus E* obtained by dividing the stress at fracture by the corresponding deformation. Because the fracture stress level is the same for polycrystalline ice for a large range of loading rates, the secant modulus will vary and attain its maximum value when the rate is equal to that of the celerity of the elastic wave in the ice. This is called the dynamic modulus of polycrystalline ice.

The values of the dynamic modulii for single ice crystals under various directions of loading are given in Section 2.1.1.1. The elastic parameters for polycrystalline ice can be obtained by noting that in a polycrystalline sample, under a given stress $\sigma$, the compliance constant will be the average of the sum of the compliance constant for each crystal. For any direction the compliance constant can be obtained from Eq. (2-8).

For the case of columnar $S_1$ ice with an aligned c-axis orientation, the Young's modulus parallel to the basal plane is given by:

$$E_{11} = 1/S_{11} \tag{2-31}$$

and perpendicular to this plane:

$$E_{33} = 1/S_{33} \tag{2-32}$$

For columnar $S_2$ fresh water ice, which is columnar ice where all the c-axis are randomly oriented in the horizontal plane only, the resulting Young's modulus is, horizontally, i.e. perpendicular to the long axis of the columns.

$$E_{s_2} = 1/S_{s_2} \tag{2-33}$$

$$S_{s_2} = \frac{2}{\pi}\int_0^{\pi/2} S_\alpha \, d\alpha \tag{2-34}$$

For granular snow ice $T_1$, either fresh water or sea ice, all the c-axis are randomly oriented in space. For an equal surface distribution over a half sphere, we then have:

$$E_{t_1} = 1/S_{t_1} \tag{2-35}$$

$$S_{t_1} = \frac{2}{\pi}\int_o^{\pi/2} S_\alpha \sin\alpha d\alpha \qquad (2\text{-}36)$$

The results of the computations of these elastic modulii are given in Table 2.1 for the various ice types.

| ICE TYPES | | | E in $10^9$ Pa |
|---|---|---|---|
| $S_2$ | - | horizontal | $9.27 [1 - 1.36 \times 10^{-3}\theta]$ |
| | | vertical | $9.62 [1 - 1.07 \times 10^{-3}\theta]$ |
| $S_1$ | - | horizontal | $9.62 [1 - 1.07 \times 10^{-3}\theta]$ |
| | | vertical | $11.79 [1 - 1.40 \times 10^{-3}\theta]$ |
| $T_1$-$S_4$ | - | agglomerate any direction | $8.93 [1 - 1.28 \times 10^{-3}\theta]$ |

Table 2.1 - Dynamic Young's modulus of polycrystalline ice

Measured values of the Young's modulus, rigidity modulus G, bulk modulus K, and Poisson's ratio $\nu$ for polycrystalline ice at $\theta = -5^{o}C$ are given in Table 2.2. Gold's measurements were made perpendicular to the long direction of grains of $S_2$ ice. Not much is known on the structure of the ice for the other measurements. Grossly, they seem to confirm the computed values given in Table 2.1.

| E | G | $\nu$ | AUTHOR |
|---|---|---|---|
| x $10^9$ Pa | x $10^9$ Pa | | |
| 9.17 | 3.36* | 0.36* | Ewing *et al.*, (1934) |
| 9.80 | 3.68* | 0.33 | Northwood (1947) |
| 9.94 | 3.80 | 0.31 | Gold (1958) |

Table 2.2 - Measured elastic parameters for polycrystalline ice. (The asterix indicates a computed value)

The effect of the temperature on the dynamic modulus of ice is not very important as can be ascertained in Table 2.1. However, the porosity of the ice has a major influence. Nakaya (1959) found that E would decrease by as much as 20% if the density of glacier ice (with bubbles) would decrease from 0.915 to 0.903 gr $cm^{-3}$.

In general, it appears that the elastic modulus can be related linearly with the porosity by an expression of the form:

$$E/E_o = 1 - e/e_o \tag{2-37}$$

where $E_o$ is the dynamic modulus for pure ice given in Table 2.1, e is the porosity of the ice $e_o$ a porosity of reference for which the ice is becoming a firn. This porosity is defined here as the ratio of the volume of cavities to the total volume of ice. This expression appears to be quite general in the sense that the pores might be filled either with air, water or brine and the results may take the same form.

Fig. 2.6 shows the variation of Young's modulus in function of the porosity. Data points are from Langleben and Pounder (1963) for Arctic ice of different conditions where the porosity e is the brine volume $\nu$. The dotted line was proposed by Nakaya (1959) for ice having air bubbles. The full line is Eq. (2-37), with a best fit value of $e_o = 0.285$. This value corresponds to dense snow (density 0.65 gr $cm^{-3}$) for which the Young's modulus is of the order of $1.3 \times 10^9$ Pa, i.e. eight times less than that of pure ice.

A large number of measurements have been made on the static modulus of elasticity of polycrystalline ice. Values range from $0.2 \times 10^9$ Pa to that of the dynamic modulus. These measurements do not give usually the loading time so they cannot be interpreted to include the amount of delayed elastic strains.

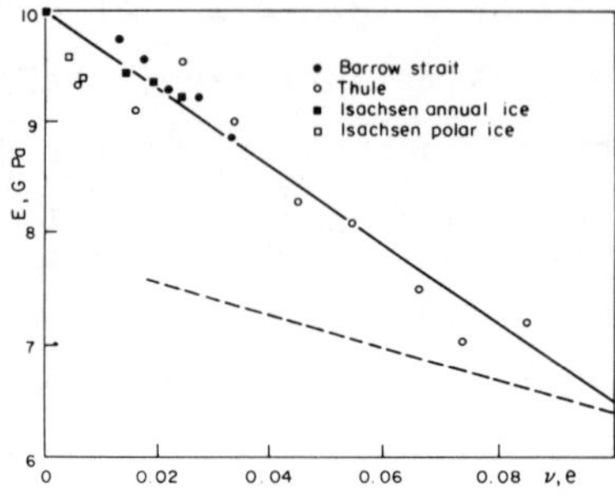


*Fig. 2.6. Young's modulus is function of brine volume ν or ice porosity e. The full line is for arctic sea ice (Langleben et al., 1963). The dash line is for porous ice (Nakaya, 1959).*

## 2.1.1.2 - BRITTLE CRACK FORMATION AND PROPAGATION

For pseudo monocrystalline $S_1$ ice, cracks are formed in the basal plane only for directions of loading such that $15^o < \alpha < 75^o$. The fracture strength of ice is then given by the release of the elastic energy of a grain to form a crack along the basal plane. We will apply this condition to the fracture of polycrystalline ice. (Michel, 1978).

Let us consider in Fig. 2.7 the two basic types of crystals that are found in polycrystalline ice. One type can be idealized by a sphere and the other by a cylinder. Each crystal is surrounded by other crystals with different crystallographic orientations. For $S_1$ ice we have shown that the crystal orientation which requires less energy for fracture is the orientation where the basal plane coincides with the main shear direction. For polycrystalline ice, we assume that the less resistant crystal will also be the one which makes an angle $\alpha = 0$ with the main shear direction. We will also consider, as for the grain of $S_1$ ice, that the formation of a crack will release all the strain energy in this crystal, the outer ice crystals of different orientation being considered as rigid bodies. This energy will be transformed into the surface energy required to form the crack surfaces all across the crystal. These conditions are given by Eq. (2-15):

$$W_\tau + W_\sigma \geq W_s \tag{2-15}$$

In the case of polycrystalline ice where all the elastic energy in shear and tension is released from the crystal

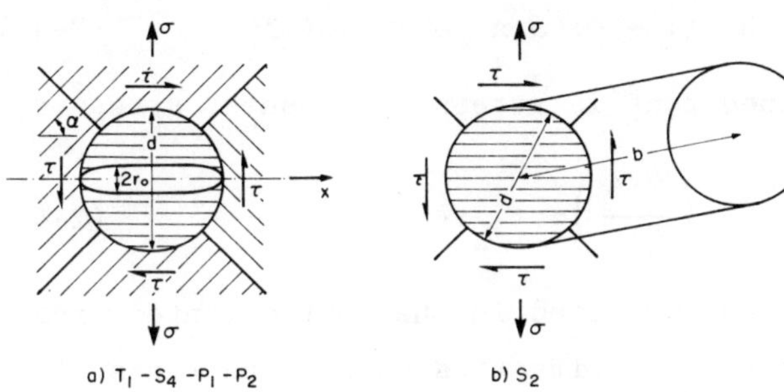


*Fig. 2.7. Sketch of crack formation for two basic types of grains of polycrystalline ice.*

$$W_\tau = \frac{\tau^2}{2\ G_{66}} \int dV = \frac{\tau^2\ V}{2\ G_{66}} \tag{2-36}$$

$$W_\sigma = \frac{\sigma^2\ V}{2\ E_{33}} \tag{2-37}$$

the surface energy needed to form a crack is given by Eq. (2-14),

$$W_s = 2\ \sigma_{cv}\ a \tag{2-38}$$

where a is the area of the crack in the basal plane and V the volume of the crystal. These equations lead to:

$$\frac{\tau^2}{2\ G_{66}} + \frac{\sigma^2}{2\ H_{33}} = \frac{3\ \sigma_{cv}}{d^*} \tag{2-39}$$

where d* is the diameter of an equivalent spherical crystal. In the case of columnar ice (cylinder) it is:

$$d^* = 3\ \pi\ d/8 \tag{2-40}$$

Under a pure shear stress, the crack will nucleate when:

$$\tau^* = \sqrt{\frac{6\ \sigma_{cv}\ G_{66}}{d^*}} \tag{2-41}$$

where $\tau^*$ is the shear stress needed to nucleate a crack in a given crystal.

With the numerical values of the parameters:

$$\tau^* = 4.7 \times 10^4\ (1 - 0.45 \times 10^{-3}\ \theta) d^{*-1/2} \tag{2-42}$$

Under a combined tensile stress $\sigma$ and shear stress $\tau$, we will have:

$$\tau^2 + \left(\frac{G_{66}}{E_{33}}\right)\sigma^2 = \tau^{*2} \tag{2-43}$$

Once a crack has nucleated in the best oriented crystal, it will act as a stress riser at its tips under conditions of normal tension. A demonstration similar to that used for $S_1$ ice will show that the crack is unstable and will propagate catastrophically through the whole sample.

### 2.1.2.3 - TENSILE STRENGTH

When an ice sample is tested in pure tension with a stress $\sigma_o$, the shear stress $\tau$ and tensile stress $\sigma$ on any plane making an angle $\beta$ with the load are given by Eqs (2-21) and (2-22):

$$\tau = \sigma_o(\sin 2\ \beta)/2 \tag{2-44}$$

$$\sigma = \sigma_o(1 + \cos 2\ \beta)/2 \tag{2-45}$$

The strength in tension under brittle conditions is given by Eq. (2-43). With Eqs (2-44) and (2-45) this gives:

$$\frac{\sigma_o^2}{4}\left[\sin^2 2\beta + \left(\frac{G_{66}}{E_{33}}\right)(1 + \cos\ 2\beta)^2\right] = \tau^{*2} \tag{2-46}$$

The left hand side of Eq. (2-46) can be optimized to give its highest value. It is maximum when:

$$\cos 2\beta \ \max = G_{66}/(E_{33} - G_{66}) \qquad (2\text{-}47)$$

Which gives:

$$\frac{\sigma_o^{\ 2}}{4}\left(1 - \frac{G_{66}}{E_{33}}\right) = \tau^{*2} \qquad (2\text{-}48)$$

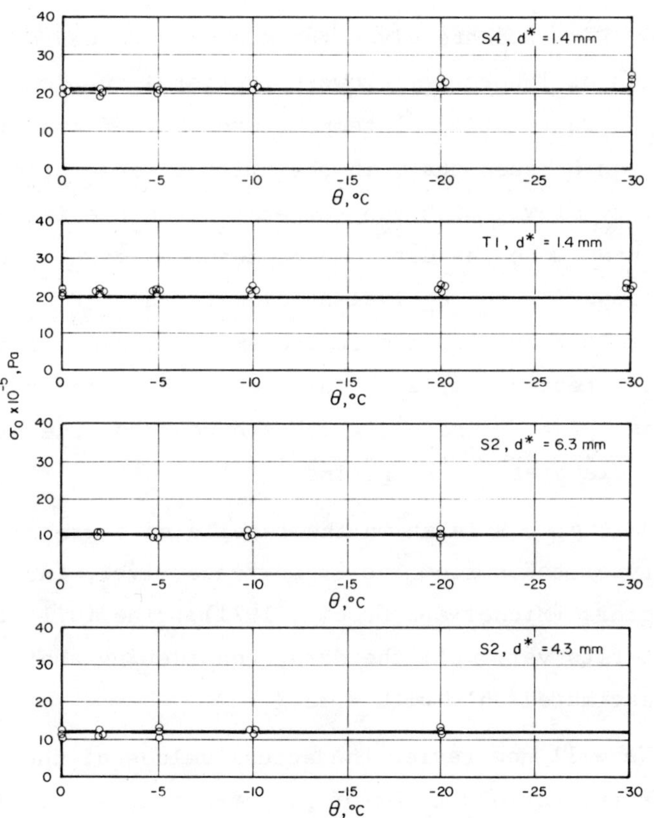


*Fig. 2.8. Tensile strength of various types of polycrystalline ice in function of temperature.*

The ice sample will start to fail in a plane making an

angle of $33^0$ with the load and the tensile strength of the ice is:

$$\sigma_o = 2\ \tau^* \sqrt{1 - \frac{G_{66}}{E_{33}}} \qquad (2\text{-}49)$$

With the numerical values of the parameters and $\tau^*$ obtained from Eq. (2-41),

$$\sigma_o = 7.94 \times 10^4 \sqrt{(1 - 0.9 \times 10^{-3}\ \theta)/d^*} \qquad (2\text{-}50)$$

On Fig. 2.8 are given the measured values of the tensile strength of laboratory grown polycrystalline ice of type $S_4$, $T_1$ and $S_2$ in function of temperature (Carter and Michel, 1971). It can be seen that, in the brittle range, the strength decreases very slowly with the temperature as predicted by Eq. (2-50). Furthermore the data can be adjusted exactly to Eq. (2-50) by computing the equivalent diameter $d^*$. For $S_2$ and $S_4$ ice this diameter corresponds to the smallest dimension of the grains in the tested samples. This can be expected for elongated grains. For $T_1$ grains, which are rounded, the diameter is the average diameter of the grains.

On Fig. 2.9 is shown the tensile strength of ice of different types taken from the St-Lawrence river, in function of grain diameters (Michel and Drouin, 1971). The full line is Eq. (2-50). It fits very well the data, considering that the grain sizes are estimated only within ±20%.

We will now review the actual values of the fracture strength of the various types of polycrystalline ice, in tension.

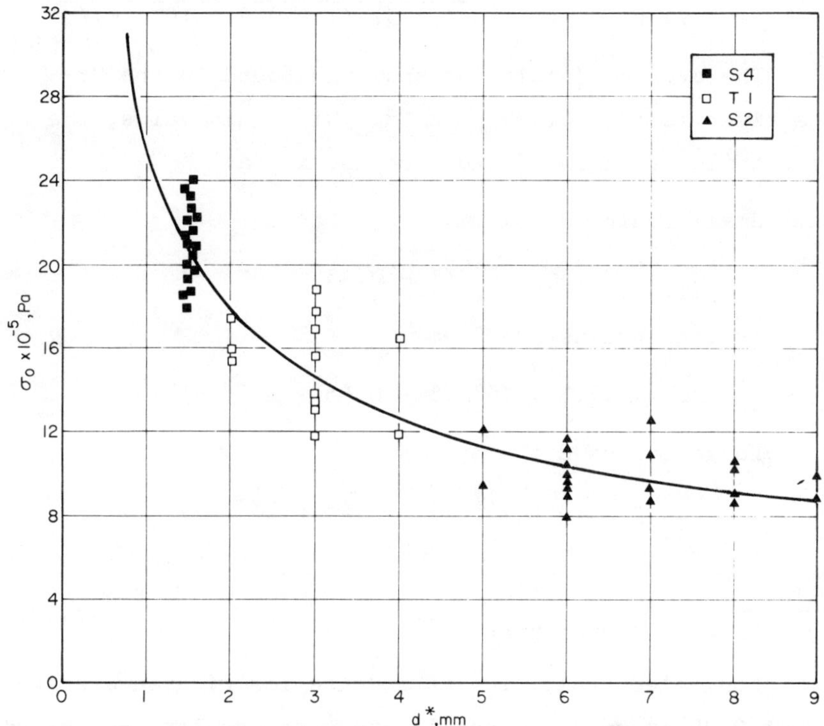


*Fig. 2.9. Tensile strength of natural and laboratory ice in function of grain size.*

Snow ice-$T_1$

The tensile strength of this ice can be obtained from Eq. (2-50) by adding one parameter to take into account the porosity of snow ice e:

$$e = 1 - \rho''/\rho' \qquad (2\text{-}51)$$

where $\rho''$ and $\rho'$ are the respective densities of snow ice and pure ice, $\rho' = 0.916$ gr cm$^{-3}$. This dependence can be written by introducing the effect of the air content on the elastic modulus of the ice. We get an expression of the type:

$$\sigma_{T_1} = 7.94 \times 10^4 \sqrt{(1 - e/.285)\ (1 - 0.9 \times 10^{-3}\ \theta)/d^*} \tag{2-52}$$

The average density of snow ice found in the St-Lawrence river is $\rho'' = 0.875$ gr cm$^{-3}$. With this value, the typical tensile strength of snow ice is, in Pa:

Small grains d = 1 mm

$$\sigma_{T_1} = 24.5 \times 10^5\ (1 - 0.45 \times 10^{-3}\ \theta)$$

Average grains d = 2 mm

$$\sigma_{T_1} = 17.3 \times 10^5\ (1 - 0.45 \times 10^{-3}\ \theta)$$

Large grains d = 4 mm

$$\sigma_{T_1} = 12.3 \times 10^5\ (1 - 0.45 \times 10^{-3}\ \theta)$$

Frazil ice-$S_4$

Congealed frazil ice is a dense ice with an average density of 0.90 gr cm$^{-3}$. Crystals are very angular and elongated so that the equivalent diameter is that of the smallest cross-section. For St-Lawrence river $S_4$ ice (Michel and Drouin, 1971), tests made on 17 samples where the smallest average diameter was 2 mm gave the following results in Pa:

$\sigma_{S_4}$ max - $22.3 \times 10^5$

$\sigma_{S_4}$ average - $18.0 \times 10^5$

$\sigma_{S_4}$ min - $15.0 \times 10^5$

Eq. (2-50) with d* = 2 mm gives $\sigma_{S_4} = 17.7 \times 10^5$ Pa. Four tests were also made with large size ice crystals, whose average size was more than 5 mm. They gave an average strength of $\sigma_{S_4} = 13.5 \times 10^5$ Pa. Eq. (2-50) gives a corresponding value of d* = 3.5 mm. This confirms the elongated character of these grains which was observed in thin section studies. The smallest diameter should be used in that case.

## Columnar ice-$S_2$

This is also a very dense ice with an average density of 0.910 gr $cm^{-3}$ in the St-Lawrence river. One major characteristics of the vertical columns of this ice is that, after a certain growth, all the c-axis of the grains take a preferred horizontal orientation. Another characteristic is the growth with depth in diameter of the columns. Starting from diameters of a few mm at the top, the columns attain a diameter of 1 cm at a depth of the order of 30 cm. From thereon, the growth is disorderly and we once observed (Michel and Ramseier, 1969) an average crystal size of 3 cm at a depth of 60 cm. Because of large crystal sizes, the bottom of $S_2$ ice cannot be tested with small laboratory samples and its strength can be obtained only from in-situ beam tests. For this type of ice, Eq. (2-40) shows that the equivalent diameter should be taken as:

$$d^* = 3\pi d/8 \qquad (2\text{-}40)$$

Tests on 14 samples of St-Lawrence river $S_2$ ice with grains ranging from 5 to 20 mm gave a strength of $11.0 \times 10^5$ Pa (Michel and Drouin, 1971). The tensile strength of $S_2$ ice with large crystals at the bottom can be obtained by breaking upwards in-situ cantilever beams. The results of 18 tests made by Drouin and Michel (1972) on the St-Lawrence river gave the following value:

$$\sigma_{S_2} = (5.0 \pm 0.5) \times 10^5 \text{ Pa}$$

With this strength, Eq. (2-50) gives a grain diameter of 2.5 cm which seems appropriate from measurements of grain sizes under similar conditions. Summarizing, typical tensile strengths of $S_2$ ice could be taken as follows, in Pa:

Upper layer, d* = 5 mm

$$\sigma_{S_2} = 11.2 \times 10^5 \ (1 - 0.45 \times 10^{-3} \ \theta)$$

Middle layer, d* = 1 cm

$$\sigma_{S_2} = 8.0 \times 10^5 \ (1 - 0.45 \times 10^{-3} \ \theta)$$

Lower layer, d* = 2.5 cm

$$\sigma_{S_2} = 5.0 \times 10^5 \ (1 - 0.45 \times 10^{-3} \ \theta)$$

Columnar-$S_2$ sea ice

For sea ice, the strength is considerably reduced by the presence of air and brine pockets. Contrary to snow ice where the cavities are more or less uniformly distributed, the cavities in sea ice are concentrated on lines in the basal planes. The reduction in cross-sectional area along the basal plane is thus much more important than would be suggested only by the porosity because it depends more on a surface effect than a volume effect. The reduction in strength has been studied by Weeks and Assur (1969) and they proposed a formula of the type:

$$\frac{\sigma_{S_2 s}}{\sigma_o} = 1 - \sqrt{\nu/\nu_o} \qquad (2\text{-}53)$$

where $\sigma_o$ is the tensile strength of fresh water ice, $\nu$ is the brine volume and $\nu_o$ is a brine volume of reference with an approximate value of 0.25. There is a large dispersion of experimental results for the tensile strength interpreted with Eq. (2-53) when compare to other types of polycrystalline ice. This is due in part to the difficulty in measuring precisely the total volume of air and brine in the ice. We also think that another cause of dispersion is the ignorance that is made of grain size. In fact the tensile strength of sea ice $S_{2_s}$ should be written with Eqs (2-50) and (2-53) and by dropping the very small direct influence of temperature in the formula:

$$\sigma_{S_2 s} = 7.94 \times 10^4 \ d^{*-1/2} \ (1 - \sqrt{\nu/0.25}) \qquad (2\text{-}54)$$

for $\nu \leq 0.11$

For brine volumes higher than $\nu = 0.16$, observations show that large interconnecting drains are formed within the ice, which drains the brine quickly without any further reduction in strength. Experimental results on the flexural strength of sea ice in function of brine volume are shown on Fig. 2.10 with Eq. (2-54) where d* is computed to be 1.1 cm.

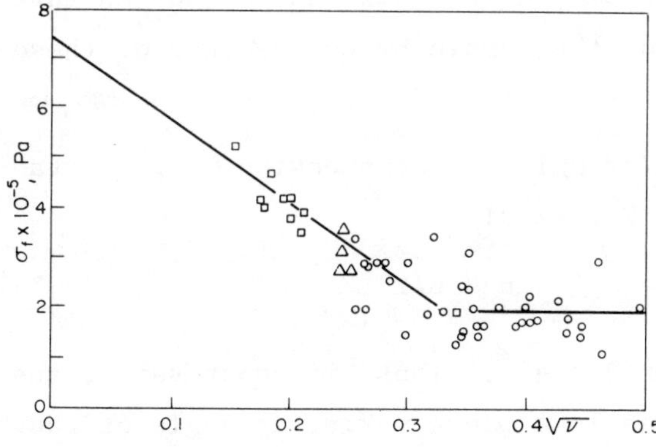


*Fig. 2.10. Flexural strength of sea ice. (o Weeks and Anderson, 1958; □ Brown, 1963; Δ Butkovich, 1956). From Weeks and Assur, 1967).*

## 2.1.2.4 - CRUSHING STRENGTH

As for the case of pseudo-monocrystalline ice, it is

obvious that the formation of a crack under a normal compressive stress will not release the compressive strain energy, but only the shear strain energy. Considering only the shear stresses, the condition necessary for crack formation will be the same as for pseudo-monocrystalline ice. It is given by Eq. (2-28):

$$\tau' = \tau^* + C_i \tag{2-28}$$

where $\tau'$ is again the total stress needed for fracture and $C_i$ the cohesive strength of polycrystalline ice. The value of $\tau^*$ will thus be the same as for tension, when all the shear strain energy of a crystal is made available to form the crack. This is given by Eq. (2-41):

$$\tau^* = \sqrt{\frac{6\sigma_{cv} G_{66}}{d^*}} \tag{2-41}$$

Once a crack is formed in the weakest crystal, it is inherently stable. A reasoning similar to that used for pseudo-monocrystalline ice shows, however, that other cracks will immediately follow in similar placed crystals and in the fracture plane leading to collapse of the sample by coalescence of these cracks.

Under conditions of uniaxial compression $\sigma_o'$, the tangential stress is given by Eq. (2-21):

$$\tau' = \sigma_o' \sin 2\beta/2 \tag{2-55}$$

It is maximum for $\beta = 45^o$. Thus, in compression, the cracks will align themselves in a plane making an angle of roughly $45^o$ with the load. The brittle strength in compression of polycrystalline ice is thus from Eqs (2-29) and (2-55):

$$\sigma_o' = 2\,\tau' = 2\left(C_i + \sqrt{\frac{6\,\sigma_{cv} G_{66}}{d^*}}\right) \tag{2-56}$$

This equation has the same form as the Petch's equation used to obtain the yield strength for ductile materials, except that it is derived for brittle conditions and not because of a dislocation pile-up mechanism.

The brittle strength in compression of polycrystalline ice was obtained by Carter and Michel (1971). The results could be interpreted by substracting twice the value of the pure shear strength $\tau^*$ to obtain the value of the cohesive strength $C_i$. This value is given on Fig. 2.11 from that data, in function of the temperature. It can be seen that the cohesive strength is almost entirely independant of the type of polycrystalline ice but depends only on the ice temperature. It is also most interesting to note that its value is close to zero at melting point and that it increases rapidly with the temperature.

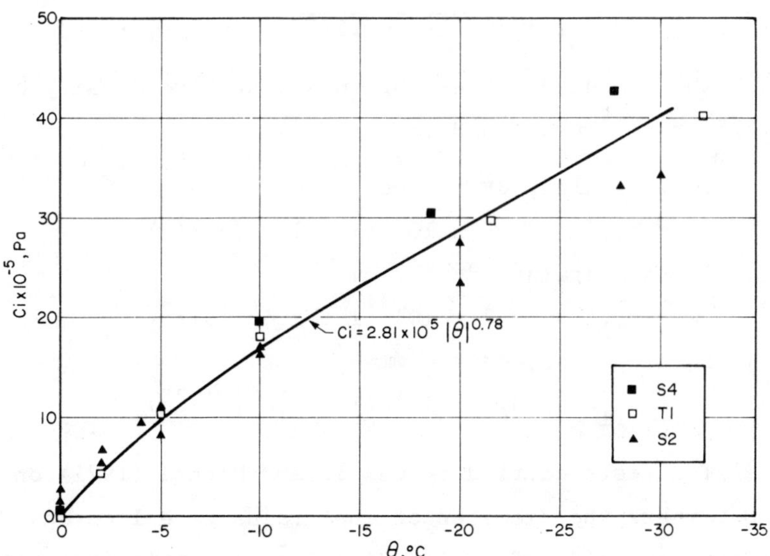


*Fig. 2.11. Cohesive strength of various types of polycrystalline ice in function of temperature. Each point represents the average between 5 to 10 tests.*

A good fit function, going through zero, is:

$$C_i = 2.81 \times 10^5 \; |\theta|^{0.78} \text{ in Pa} \qquad (2\text{-}57)$$

The crushing strength of any type of polycrystalline ice, in the brittle range, is then given by Eqs (2-56) and (2-57) in Pa.

$$\sigma_o' = 9.4 \times 10^4 \; (d^{*-1\;2} + 3 \; |\theta|^{0.78}) \qquad (2\text{-}58)$$

With this formula it is possible to predict the crushing strength of the various types of polycrystalline ice and compare it with the experimental data.

Snow ice - T1

As for tension, a correction has to be introduced for the variation of the elastic modulus with the porosity. This gives with Eq. (2-58) and as in Eq. (2-52) for the porosity term:

$$\sigma_{T1}' = 9.4 \times 10^4 [\sqrt{(1 - e/.285)/d^*} + 3 \; |\theta|^{0.78} ] \qquad (2\text{-}59)$$

The computed values for an average ice density of $\rho' = 0.875$ gr $cm^{-3}$ are, in Pa:

Small grains, $d^* = 1$ mm

$$\sigma_{T1}' = 27.9 \times 10^5 \; (1 + 1.0 \; |\theta|^{.78})$$

Average grains, $d^* = 2$ mm

$$\sigma_{T1}' = 19.7 \times 10^5 \; (1 + 1.42 \; |\theta|^{.78})$$

Large grains, $d^* = 4$ mm

$$\sigma_{T1}' = 13.9 \times 10^5 \; (1 + 2.0 \; |\theta|^{.78})$$

Ten tests carried by Carter and Michel (1971) on snow ice at $0^oC$ where the diameter of the grains were 1 mm$<d^*<$ 2 mm, gave an average value of $\sigma_{T1}' = 24 \times 10^5$ Pa. This corresponds well to the computed values.

Columnar - $S_2$ ice

For columnar ice tested in the horizontal direction, Eq. (2-58) gives in Pa:

Upper layer, d* = 5mm

$$\sigma_{S_2}' = 13.3 \times 10^5 \ (1 + 2.1 \ |\theta|^{.78})$$

Middle layer, d* = 1 cm

$$\sigma_{S_2}' = 9.4 \times 10^5 \ (1 + 3.0 \ |\theta|^{.78})$$

Lower layer, d* = 2.5 cm

$$\sigma_{S_2}' = 5.4 \times 10^5 \ (1 + 5.2 \ |\theta|^{.78})$$

Eight tests were done at $0^{\circ}$C on laboratory grown $S_2$ ice taken 10 cm from the surface, that gave an average value of $15 \times 10^5$ Pa. The average size of the crystals was 8 mm. One explanation of the difference might be that for $S_2$ ice the controlling size is the smallest dimension in the cross-section of the grains, probably because of the substructures within the largest grains.

Frazil ice - $S_4$

This is a dense ice, for which Eq. (2-58) gives in Pa:

Fine size, d* = 1 mm

$$\sigma_{S_4}' = 29.7 \times 10^5 \ (1 + 0.95 \ |\theta|^{.78})$$

Large size, d* = 4 mm

$$\sigma_{S_4}' = 14.8 \times 10^5 \ (1 + 1.9 \ |\theta|^{.78}).$$

Six tests in compression on $S_4$ ice at $0^{\circ}$C gave an average value of $24.5 \times 10^5$ Pa. This corresponds to a crystal size of d* = 1.4 mm which was the observed average size of the grains.

## 2.1.2.5 - ICE STRENGTH UNDER A COMBINED STATE OF STRESS

The Mohr's envelope of the fracture strength, in the brittle range, for all types of polycrystalline ice is shown in Fig. 2.12.

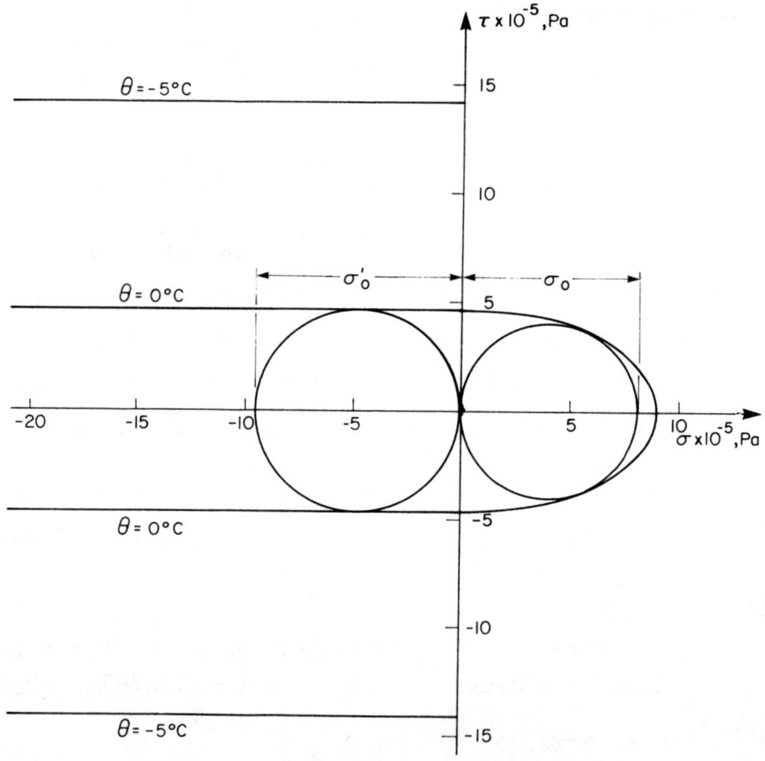


*Fig. 2.12. Mohr's failure envelope for fracture of $S_2$ ice, d* = 1 cm.*

In the tensile region, the criterion of fracture is that of minimum strain energy on the basal plane of the best oriented crystal. This law is given by Eq. (2-43).

$$\tau^2 + \mu\,\sigma^2 = \tau^{*2} \qquad (2\text{-}60)$$

where $\mu = G_{66}/E_{33}$

and $$\tau^* = \sqrt{6\,\sigma_{cv}\,G_{66}/d^*}$$

The failure line is an ellipse with major axis $\tau = \tau^*$ when $\sigma = 0$ and $\sigma_o = \tau^*/\sqrt{\mu}$ when $\tau = 0$. For any inscribed circle within the envelope, the coordinates of the principal stresses and the orientation of the failure plane, are:

$$\sigma_1 = \sigma_3\,(1 - 2\,\mu) + 2\sqrt{(1-\mu)\,(\tau^{*2} - \mu\,\sigma_3{}^2)} \qquad (2\text{-}61)$$

$$\cos 2\,\beta = \frac{\mu}{1-\mu}\left(\frac{\sigma_1 + \sigma_3}{\sigma_1 - \sigma_3}\right) \qquad (2\text{-}62)$$

In particular, for uniaxial tension:

$$\sigma_1 = 2\,\tau^*\,\sqrt{1-\mu} \qquad (2\text{-}63)$$

$$\cos 2\,\beta = \mu/(1-\mu) \qquad (2\text{-}64)$$

In the region of normal compressive stresses, the criterion of failure is the Tresca's criterion of maximum shear stress, independently of the value of the normal compressive stress. The value of this stress is given by Eq. (2-28).

$$\tau = \tau^* + C_i\,(\theta) \qquad (2\text{-}65)$$

where $C_i$, the cohesive strength, is a function of the temperature and equal to zero for $\theta = 0^\circ C$. The representative lines on the Mohr's circle envelope are given by:

$$\sigma_1{}' - \sigma_3{}' = 2\,(\tau^* + C_i) \qquad (2\text{-}66)$$

For $\theta = 0$, the lines are continuous with the fracture line in tension. For $\theta < 0$ these lines are discontinuous at the $\sigma$ origin and farther apart with decreasing values of $\theta$.

For $\sigma_3{}' = 0$, the crushing strength in uniaxial compression is given by:

$$\sigma_o' = 2\ (\tau^* + C_i) \tag{2-67}$$

At $0^oC$, the ratio of the uniaxial crushing strength to the uniaxial tensile strength is:

$$\sigma_o'/\sigma_o = 1/\sqrt{1 - \mu} = 1.18 \tag{2-68}$$

## 2.2 - DUCTILE BEHAVIOR

### 2.2.0 - INTRODUCTION

Many experiments have been made on monocrystals of ice which show that they behave more or less like a pack of cards which deforms very easily when a shear stress is applied parallel to the basal plane but has a very high resistance to deformation when the stress is applied perpendicularly to this plane. In this later direction, it would deform only elastically and break but the stress needed for fracture is then extremely higher than the stress required to make the ice glide on the basal plane.

The viscous properties of ice along the basal plane are remarkable. Despite the great forces holding the atoms together, a suitably oriented crystal can be stretched plastically, at room temperature, many times its initial length by a stress so small that the crystal can hardly support its own weight. In fact with a shear stress as low as $2 \times 10^4$ Pa it is possible to deform indefinitely the ice as long as the stress remains on this basal plane.

This facility of crystals to deform in this particular direction is caused by the defects in the atomic structure, called dislocations, that can make whole slabs of crystals glide past one another in a slip or climb process.

Polycrystalline ice is a most interesting crystalline material to study because each of its crystal is strongly anisotropic, its bulk is made up of a large variety of grain forms and arrangements and finally, it can be observed at a human scale because of the moderate dimensions of its grains and their transparency. It possesses a unique combination of strength and ductility, because of the unsymmetrical behavior under load of each of its crystals.

## 2.2.1 - MOVEMENT OF DISLOCATIONS IN ICE

All crystalline materials possess defects in their lattices. These imperfectious may occur in many forms varying from point defects, to line defects and stacking faults.

The simplest and most easily understood defects are those which are localized to the region of a single lattice site, called point defects. One of the most important is the simple vacancy or absence of an atom at a lattice site. Because of this the crystal is weakened. In fact, if sufficient vacancies are generated in the crystal structure, it looses its long range order and symmetry.

When the dimensions of an imperfection become large by atomic dimensions but are still small when compared with visible macrocrystal the imperfection is called a dislocation. Dislocations are linear arrays of atoms where the coordination characteristics differ systematically from the ideal. Depending upon the type of coordination mismatch which occurs at a dislocation line, we may find an edge, a screw or a mixed dislocation. Many direct observations have been made, using techniques of high resolution electron microscopy, that show clearly the edge and the screw dislocations in ice crystals (Hayes and Webb, 1965; Higashi, 1968).

The edge dislocation is a configuration of continuous rows of atoms, each with one less atom coordinating it than it would require for the ideal structure of the crystal. When a row of insufficiently coordinated atoms is viewed at its end, it appears as the terminus of an extra plane of atoms extending only part way into the crystal structure. This is represented on Fig. 2.13. The atoms immediately above the slip plane are in a state of compression, while those immediately below are under tension. Thus, the edge dislocation is surrounded by a stress field, roughly inversely proportional in magnitude to the distance from the dislocation.

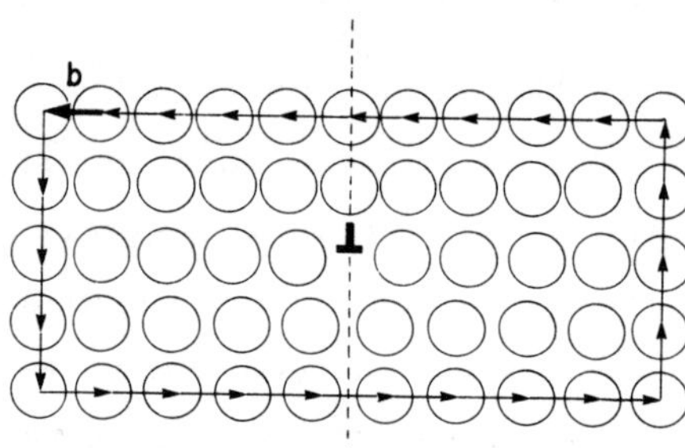


*Fig. 2.13. Schematic illustration of an edge dislocation, showing Burgers vector b. Note that b is normal to the plane containing the dislocation, shown in dashed line.*

The screw dislocation is a configuration in which the line of atoms defining the dislocation has the proper number of coordinating neighbours, but the coordination polyhedron is distorted. This in viewed on Fig. 2.14 where the screw dislocation can be visualized as originating from the partial slipping of a section of a crystal plane, with the slip displacement terminating internally to the crystal and on the dislocation line which will lie in the slip plane.

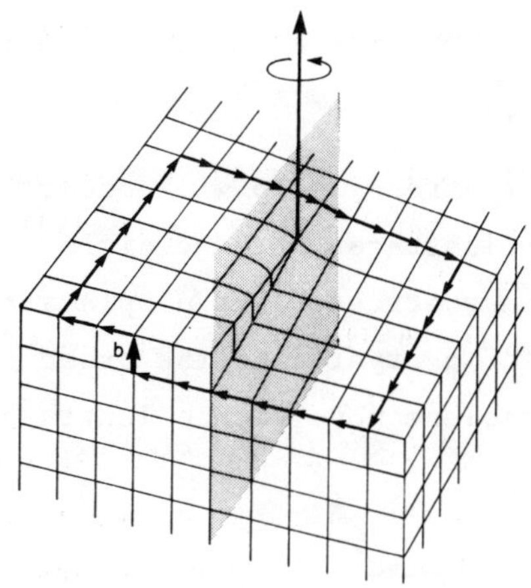


*Fig. 2.14. Schematic illustration of a screw dislocation showing Burgers vector b. Note that b is parallel to the plane in grey, containing the screw dislocation.*

In general the dislocation line through a crystal will be curved, and will be a combination of the two pure types of dislocations. This is called a mixed dislocation. One particular type is the dislocation loop where an edge dislocation is located between two screw dislocations.

The Burgers circuit, or the closed path through a crystal lattice, is used to differentiate between types of dislocations. This circuit is composed of equal number of lattice parameter translations taken sequentially in positive and negative directional senses, as shown both on Figs 2.13 and 2.14. If the Burgers circuit encloses only perfect crystalline material, it will close on itself. If the loop encloses a dislocation, however, the circuit will fail to close in the manner characteristic of that particular dislocation. It is evident, from the figures, that the closure failure for a dislocation is described by a vector with a magnitude equal to the unit lattice parameter normal to the plane of the edge dislocation. The closure vector, called the Burgers vector, and denoted by b, is characteristic of a given dislocation and is constant for the entire region of the crystal enclosed by a dislocation loop. For ice, the most common

Burgers vector has been found to be (Glen, 1975):

$$b = a/3 \simeq 1.5\text{Å} \qquad (2\text{-}69)$$

Under the action of a shear stress, slip will occur in a crystalline material. Fig. 2.15 illustrates an edge dislocation moving under the action of a shear stress. The displacement of such a dislocation from one side of a crystal to the other will make the whole crystallographic plane move with a length b.

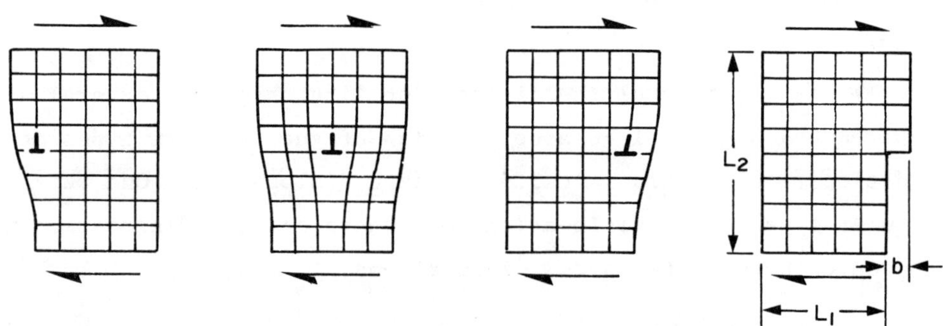


*Fig. 2.15. Displacement of an edge dislocation under a shear stress which produces the crystal slip.*

Let us call n the density of mobile dislocations, i.e. their number per unit area of the section and v their velocity of displacement. In a time $L_1/v$ one dislocation will completely cross the crystal and produce a shear strain $b/L_2$. The total rate of plastic shear $\gamma$ is thus given by (Orowan, 1934):

$$\dot{\gamma} = \frac{b}{L_2} \cdot \frac{v}{L_1} \cdot n\, L_1 L_2 = bn\, v \qquad (2\text{-}70)$$

In the creep of ice it is found in general, that the mechanism of dislocation movement is one of climb instead of slip. The edge dislocation has then to move in a direction perpendicular to the slip plane, in the climb process. The only way that climb can occur is by mass transport, by diffusion, and it is therefore a thermally activated process. It is also a very slow process compare to that of slip, i.e. that the velocity of displacement of the dislocations is low compare to metals. The usual way for climb to occur is by a vacancy diffusing to the dislocation and the extra atom moving into the vacant lattice site, as shown on Fig. 2.16. It is also possible, but not energitically favorable, for the atom to break loose from the extra half plane and become an interstitial atom. To produce negative climb, atoms must be added to the extra half plane of atoms. This can occur by atoms from the surrounding lattice joining the extra half plane, which create vacancies, or, less probably, by an interstitial atom diffusing to the dislocation.

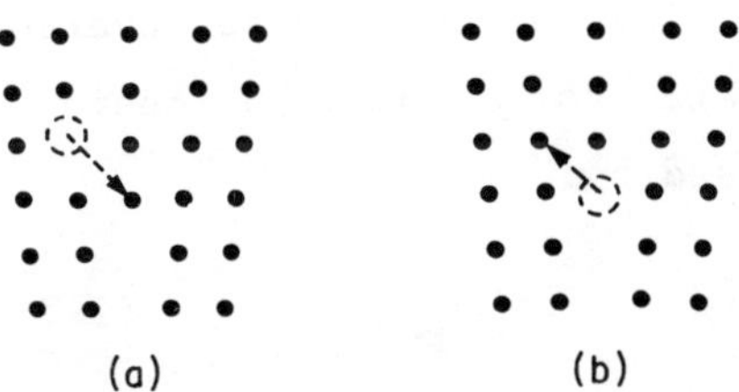


*Fig. 2.16. a) Diffusion of vacancy to edge dislocation; b) dislocation climbs up one lattice spacing.*

Broadly, the process of climb can also be described by Orowan's Eq. (2-70), but where the velocity v is much reduced, as observed for ice. This velocity, computed from the results of Jones and Glen (1969) and Wakahama (1962) is computed to be:

$$v_o \simeq 10^{-4} \text{ cm s}^{-1} \tag{2-71}$$

for a shear stress of $10^5$ Pa at $\theta = 0^oC$.

## 2.2.2 - BASIC LAW OF CREEP OF AN ICE CRYSTAL

Ice is thus a material similar to a metal at high temperatures and the controlling creep mechanism is dislocation climb.

Most of the theories based on dislocation climb make use of the following relationships (Poirier, 1976):

$$n \propto \frac{M}{d^2} \qquad (2\text{-}72)$$

where:

M - number of sources of emission of dislocations (Frank Read or other types) in the basic network or grain.

d - characteristic dimension of this basic element.

The stress field around a dislocation is also inversely proportionnal to the distance. Related to the basic dimension of the network, this is usually expressed by:

$$\tau \propto \frac{\mu b}{d} \qquad (2\text{-}73)$$

where $\mu$ is the equivalent viscosity, in the stress field.

Finally, Friedel (1956), has shown that in the climb mechanism the velocity of climb is related to the stress field by:

$$v \propto D\tau \qquad (2\text{-}74)$$

where D is the coefficient of diffusion for creep. With Arrhenius equation, it can be expressed by:

$$D = D_o \exp [- Qc/R\theta^*] \qquad (2\text{-}75)$$

where:

Qc - activation energy for creep, cal/mole $^oK$

R - universal gas constant, 1.986 cal/mole

$\theta^*$ - in temperature in $^{\circ}$K

Combining Eqs (2-74) and (2-75) we can write:

$$v = v_o \left(\frac{\tau}{\tau_o}\right) \exp\left[-\frac{Qc}{R\theta^*} + \frac{Qc}{273R}\right] \tag{2-76}$$

Thus $v_o$ is the velocity of dislocations under a stress $\tau_o$, at a temperature of $\theta = 0^{\circ}$C.

Combining Eqs (2-72) and (2-73) we also can write:

$$n \propto M\,\tau^2 \tag{2-77}$$

It has been observed both in ice (Fukuda and Higashi, 1969) and other crystalline materials that the number of mobile dislocations increases during the creep process under a constant stress. We will admit, empirically, that this increase, in function of the deformation, has the form (Michel, 1978):

$$n = n_o\,(1 + \alpha\,\gamma^m)\,\left(\frac{\tau}{\tau_o}\right)^2 \tag{2-78}$$

where:

$n_o$ - initial number of mobile dislocations

$\alpha$ - coefficient of multiplication of dislocations

m - characteristic power of growth of mobile dislocations.

Combining Eqs (2-76) and (2-78) in Orowan's Eq. (2-70) we get:

$$\dot{\gamma} = bn_o\,v_o\,(1 + \alpha\,\gamma^m)\,\left(\frac{\tau}{\tau_o}\right)^3 \exp\left[Q_c\theta/273R\theta^*\right] \tag{2-79}$$

This is the basic law of creep, with dislocation characteristics, inside an ice crystal. The dependance to the third power of the stress, is usually considered as a classical mode of dislocation climb (Friedel 1956, Nabarro 1967, Weetman 1972).

However, it is more convenient in experimentation to use a value of n as suggested by Glen (1953) for ice, which might be a little different than three. With the value of b and $v_o$ from Eqs (2-69) and (2-71), we obtain:

$$\dot{\gamma} = 1.5 \times 10^{-12}\ n_o\ (1 + \alpha\ \gamma^m)\ (\frac{\tau}{\tau_o})^n\ \exp\ [Q_c\theta/273R\theta^*] \quad (2\text{-}80)$$

This is the general form of the creep law along the basal plane of an ice crystal that will be used in this text.

It is also observed, in the creep of ice, that after a certain deformation $\gamma_t$, the number of dislocations stops to increase and becomes a constant. The total number of mobile dislocations (in $cm^{-2}$) is then equal to:

$$n_t = n_o\ (1 + \alpha\ \gamma_t{}^n) \quad (2\text{-}81)$$

And steady state creep is given by:

$$\dot{\gamma} = 1.5 \times 10^{-12}\ n_t\ (\frac{\tau}{\tau_o})^n\ \exp\ [Q_c\theta/273R\theta^*] \quad (2\text{-}82)$$

The corresponding formula under a normal stress $\sigma$ and a longitudinal deformation $\varepsilon$, is:

$$\dot{\varepsilon} = 1.5 \times 10^{-12}\ n_t\ (\frac{\sigma}{\sigma_o})^n\ \exp\ [Q_c\theta/273R\theta^*] \quad (2\text{-}83)$$

where $\sigma_o$ must be taken equal to $2\tau_o$.

With Eqs (2-80) and (2-82) or (2-83) we have computed the initial and total number of dislocations with n taken equal to 3 and $\tau_o = 10^5$ Pa, $\theta = 0^oC$, from creep curves obtained by various authors whose results are discussed in section 2.2.5. The results are shown in Table 2.3.

| ICE TYPE | AUTHORS | $n_o$ ($cm^{-2}$) | $n_t$ ($cm^{-2}$) |
|---|---|---|---|
| Monocrystal | Steineman 1958 | $2.5 \times 10^5$<br>$1.9 \times 10^5$<br>$4.3 \times 10^4$ | $4.7 \times 10^7$<br>$4.7 \times 10^7$<br>$1.0 \times 10^7$ |
| T1 | Brill and Camp 1961 | $4 \times 10^2$<br>$1.4 \times 10^2$ | $8.0 \times 10^3$<br>$6.0 \times 10^3$ |
| T1 | Mellor and Testa 1969 | $2.8 \times 10^2$ | $2.8 \times 10^4$ |
| Glacier ice | Duval 1976 | $2.6 \times 10^2$ | $2.9 \times 10^3$<br>$3.4 \times 10^3$ |
| S2 | Paradis 1978 | $2.2 \times 10^3$ | $2.4 \times 10^4$<br>$4.6 \times 10^4$ |

Table 2.3 - Initial and total number of mobile dislocations in different types of ice.

## 2.2.3 - CLASSIFICATION OF CREEP PROCESSES AND TYPICAL CREEP CURVES

A few investigators have observed the physical processes in the deformation under load of polycrystalline ice (Steineman, 1958; Rigsby, 1960; Wakahama, 1964; Gold, 1960, 70).

They found that various mechanisms would intervene at one time or another in the creep process. In addition to elastic deformation and plastic deformation by climb of dislocation they observed grain boundary slip, cavity formation at the grain boundaries, formation of low-angle boundaries, polygonization and recrystallization. Recrystallization was noticed mainly when the ice deformation became large. It either took the form of grain boundary migration, i.e. some grains growing at the expense of others, or the formation of completely new crystals.

Gold (1960, 1970) was the first to observe extensively the formation of microcracks in ice. These cracks, which formed within a crystal of the aggregate, are either parallel or perpendicular to the basal plane of the grain. He found that the cracks would start to appear only at a certain stress level in compression. They would then increase, during transitory creep, but come to a stop if the stress did not go over a certain stress limit. At still higher stresses, the process continued and creep accelerated leading to complete fracture of the ice. Gold also observed cracks forming at grain boundaries that were related to the process of grain boundary slip.

Because of the complexity of ice deformation in the ductile range, it seems appropriate to make a distinction between the physical processes that are involved and classify the types of creep curves (Michel, 1978).

Let us first define and symbolize the various types of deformations that are involved in the creep processes:

$\varepsilon$ - pure elastic deformation

$\nu$ - pure plastic deformation

$\alpha$ - deformation caused by an increase in the number of mobile dislocations in the ice and corresponding to a decreased viscosity of the ice. It is in itself a strain softening process.

$\kappa$ - deformation accompanied by micro-cracking activity in the ice which also decreases the apparent viscosity of the ice.

$\delta$ - deformation with decreased viscosity caused by syntectonic recrystallization (Lliboutry, 1964).

With these symbols for the different types of deformation, we can now describe the various creep curves of polycrystalline ice. We have divided these curves according to the classical phases of primary (transitory), secondary (permanent) and tertiary (accelerated) creep; even if these phases do not necessarily occur in that order in a natural loading process. The greek letters are used here only in a symbolic manner to define the type of creep. The first symbol in each group represents the dominant type of deformation. The various creep processes are illustrated on Fig. 2.17, γ being the shear strain and t the time.

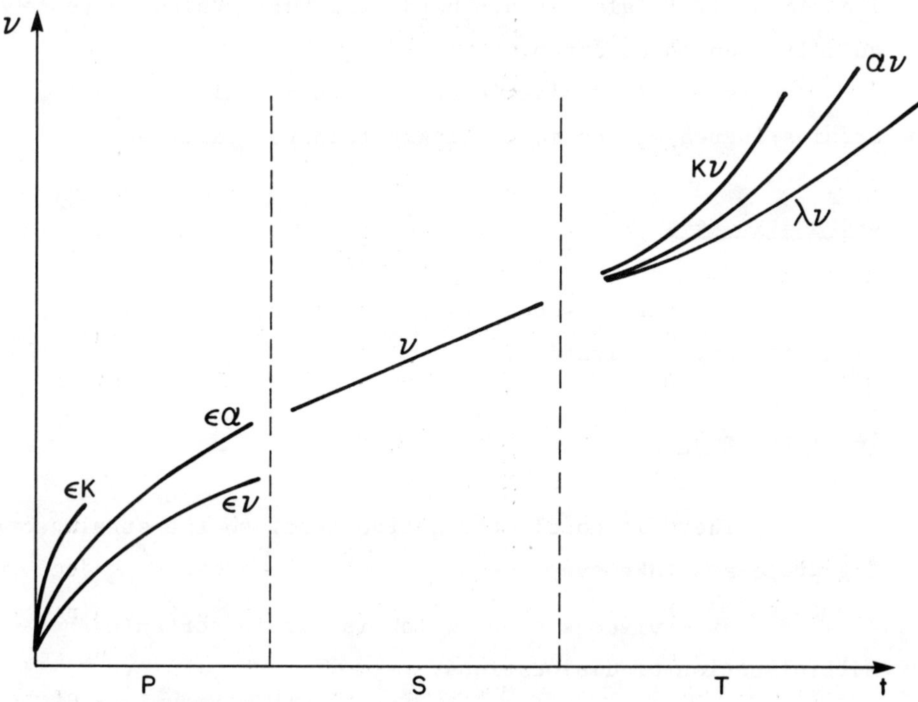


*Fig. 2.17. Form of creep curves in polycrystalline ice during the phases of primary (P), secondary or permanent (S) and tertiary (T) creep. The shear strain γ is given in function of time t for the various creep processes.*

Types of primary creep

Primary creep is strain hardening ($d\gamma/dt$ decreases with time) because of the dominance of the retarded elastic effect.

$\varepsilon\nu$ - classical visco-elastic creep which is strain hardening.

$\varepsilon\alpha$ - visco-elastic creep with an increase in mobile dislocations; it is also strain hardening but strains increase more quickly than in $\varepsilon\nu$ creep.

$\varepsilon\kappa$ - visco-elastic creep with cracking activity. Strains increase even more quickly than for $\varepsilon\alpha$ creep.

Secondary creep

$\nu$ - pure viscous creep at a constant speed, with or without recrystallization.

Tertiary creep

There is no elastic action left, so the strain softening processes take over.

$\alpha\nu$ - viscous creep which is strain softening because of multiplication of dislocations.

$\kappa\nu$ - accelerated creep caused by increased crack formation leading to fracture of the sample.

$\delta\nu$ - accelerated creep caused by syntectonic recrystallization, leading to a second stage of permanent creep.

## 2.2.3.1 - TYPICAL CREEP CURVES UNDER A CONSTANT SHEAR STRESS

With this classification of creep curves, it is possible to discuss the typical creep curves of polycrystalline ice loaded in shear at a constant stress, but under a condition of normal compression. This discussion is illustrated in Fig. 2.18 for increasing stresses $\tau_3 > \tau_2 > \tau_1$.

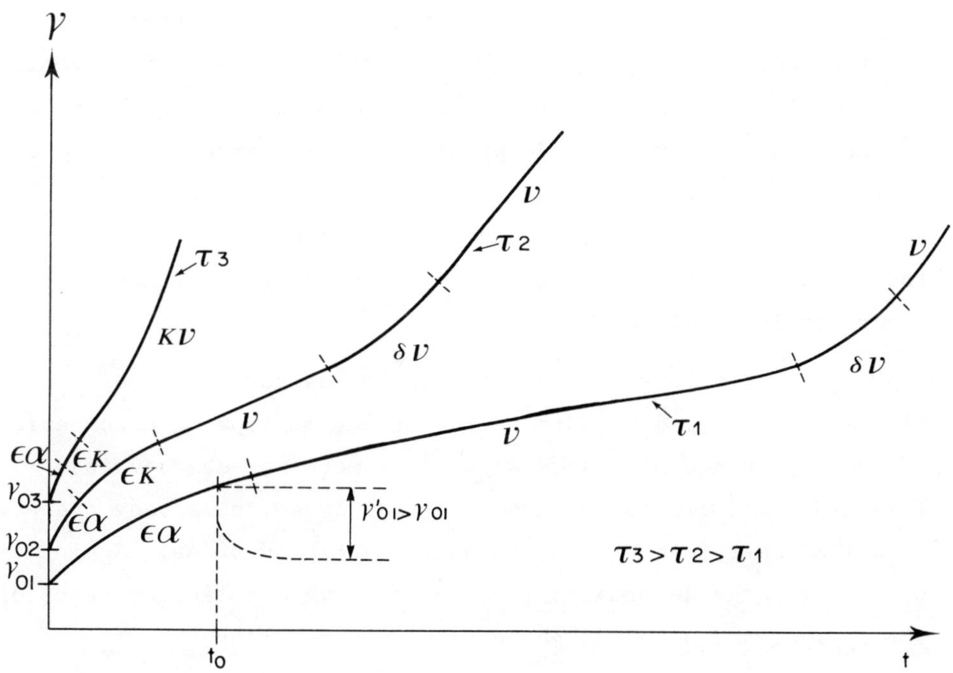


*Fig. 2.18. Typical creep curve under a constant shear stress with increasing values* $\tau_3 > \tau_2 > \tau_1$.

For a low shear stress like $\tau_1$, there is a small instantaneous elastic deformation $\gamma_{01}$, followed by transitory creep which is strain hardening of the $\varepsilon\alpha$ type, then followed by steady creep $\nu$ at a given strain rate $\gamma_o$. If the load is kept for a very long time and the deformation attains many percents of the original length (3 to 8%), the creep rate starts to accelerate and syntectonic recrystallization takes place, with type $\delta\nu$ creep, until a new steady state is attained which is around one order of magnitude faster than the first part of secondary creep. This is particularly true for glacier ice as measured by Duval (1976), under hydrostatic stresses.

If the load is removed at time $t_o$, the strain will relax as shown in Fig. 2.18. This elastic after-working consists of an instantaneous relaxation plus a time dependent relaxation, so that the total relaxed deformation $\gamma'_{01}$ is larger than the original elastic deformation $\gamma_{01}$. The delayed elasticity can be many times greater than the instantaneous elasticity and this explains the low values of the Young's modulus found by experimentation under static conditions.

For a larger stress $\tau_2$, the transitory creep of type $\varepsilon\alpha$ will be followed by a transitory creep of type $\varepsilon\kappa$ accompanied by the appearance of cracks within the more stressed grains. This type of creep softens the ice rapidly so the steady creep $\nu$ is quickly attained when the cracking activity stops. Again with very large deformation, we may observe accelerated creep of the $\delta\nu$ type and a second phase of permanent creep.

Finally, if the stress is still increased to $\tau_3$, the cracking activity will increase considerably after the initial deformation. The first part of the curve may still be of the $\varepsilon\alpha$ type followed by a very short phase of $\varepsilon\kappa$ creep and, then, finally, accelerated creep of the $\kappa\nu$ type will set in so that the increased cracking activity will lead to failure of the ice.

If the creep curves had been obtained under a normal tensile stress, they will all stop at the beginning of the κ deformation process. The appearance of the first transcrystalline crack leads immediately to complete fracture of the sample (Carter and Michel, 1971).

### 2.2.3.2 - TYPICAL CREEP CURVES UNDER A CONSTANT SHEAR STRAIN

For most problems in ice engineering, the conditions of loading are better represented by tests where the loading is made at a constant rate of strain. Although such tests are usually made in simple compression, we are representing here the effect of the resolved shear stress. Typical curves of deformation under those conditions are shown on Fig. 2.19 for loading under a constant shear stress with a normal compressive stress.

For a small shear strain rate $\dot{\gamma}_1$, the stress increases with time at a decreasing rate corresponding to εα type creep. The stress then attains a maximum which corresponds to the yield strength of ice, without cracking, and may also be considered equivalent to permanent creep of the ν type. For long periods of deformation, recrystallization will occur with δν creep and the stress will finally settle at a much lower level of permanent creep with recrystallization.

For a larger strain rate $\dot{\gamma}_2$, the transitory εα creep will be followed by transitory εκ when cracking will start. Yield may then occur when the number of cracks attains a maximum and cracking activity ceases. There will then be αν creep with a decrease in stress caused by the continued increase in the number of mobile dislocations. Finally, the stress will settle at a lower constant level, corresponding to another stage of permanent creep of the $\nu$ type. For much larger deformations we may also move into $\delta\nu$ creep.

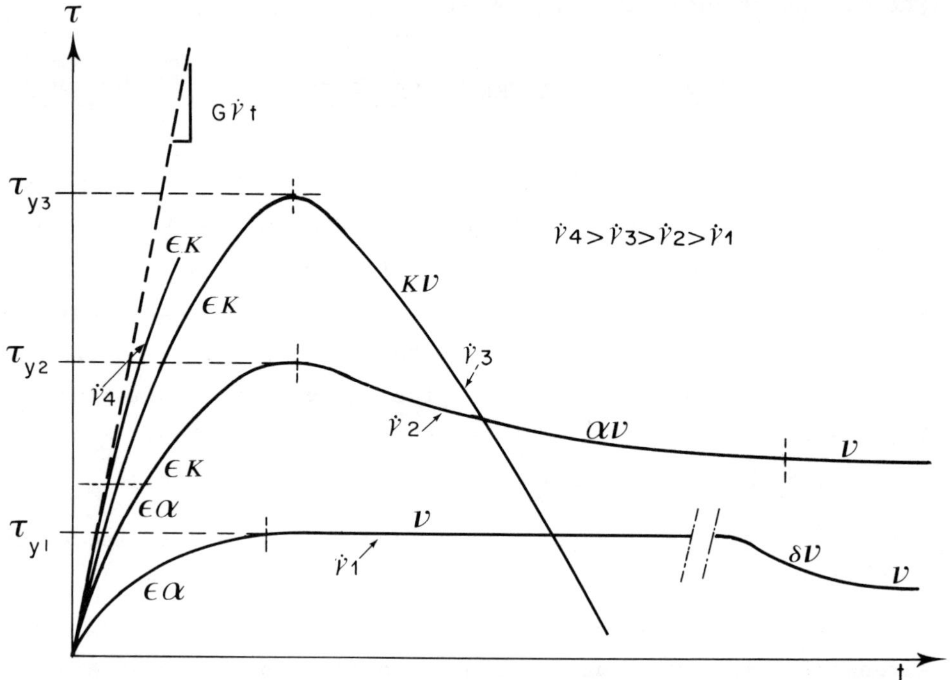


*Fig. 2.19. Typical creep curves under constant strain rates for increasing values $\dot{\gamma}_4 > \dot{\gamma}_3 > \dot{\gamma}_2 > \dot{\gamma}_1$. The corresponding shear stress is given in function of time with indications of the creep types in various parts of the curves. The dash line shows the pure elastic limit.*

For still a larger strain rate $\dot{\gamma}_3$, the maximum yield strength will occur after $\varepsilon\alpha$ and $\varepsilon\kappa$ creep, during the cracking activity. It will be followed by decreasing stresses with the continued increase in cracking activity and finally $\kappa\nu$ creep will lead to failure of the ice sample.

Finally, for a still higher strain rate $\dot{\gamma}_4 > \dot{\gamma}_3$ the cracks will multiply in a catastrophic manner, after the first one has appear, and the ice will fail in a brittle manner, usually at a lower stress level than that attained previously, when the ice was yielding.

Under conditions of a normal tensile stress, the first crack is catastrophic and fails the sample as for the case of loading under a constant stress.

## 2.2.4 - MECHANICAL MODEL OF CREEP IN POLYCRYSTALLINE ICE

### 2.2.4.1 - BASIC MODEL

Let us consider a two dimensional simplified model of polycrystalline ice as shown in Fig. 2.20 where there are j rows of crystals, each containing i rystals (Michel, 1978). The whole block is subjected to a shear stress $\tau$ and an hydrostatic stress $\sigma$. Each crystal in this group is acted upon by a shear stress $\tau_{ij}$ in one direction and $\tau_{ji}$ in the other which are producing the total shear deformations $\gamma_{ij}$ and $\gamma_{ji}$.

Because of the extreme difficulty in moving dislocations and producing plastic flow in directions other than that of the basal plane we consider, as for the case of monocrystals, that each crystals slips only along its basal plane which has an orientation $\alpha_{ij}$ with the j shear direction.

It is obvious that in such a non-uniform model, some grain boundaries will be subjected to intense stressing at the beginning of loading and local boundary slip will happen to accomodate the total deformation of the ice aggregate. Even some small boundary cracks might appear. The worst oriented crystals will be subjected to a moment of rotation and will initially rotate in order to reduce the angle of their basal plane to that favorizing an easier flow direction. We will thus consider the rotation as the first mechanism to intervene in the ductile behavior of ice; reducing the angle of some key placed crystals to the critical angle of equilibrium $\alpha_m$ where the orthogonal stresses on each crystal become equal; $\tau_{ij} = \tau_{ji}$.

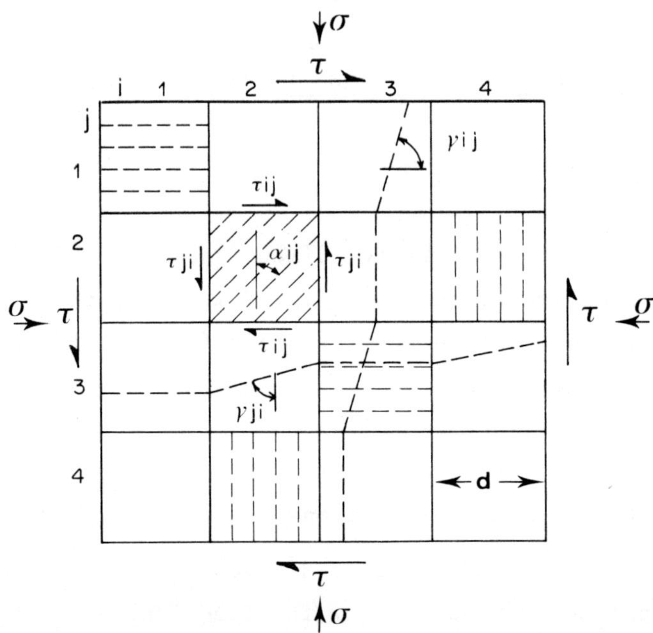


*Fig. 2.20. Basic model of polycrystalline ice containing 4 crystals in the horizontal rows i and the same in the vertical rows j. It is acted upon by a shear stress τ and an hydrostatic compressive stress σ. Dashed lines indicate the orientation of basal planes in each crystal and also the line of shear deformation.*

Once a first rearrangement of crystals has taken place

by preliminary rotation, experiments show that preferential slip bands form, where crystals in a row deform considerably under plastic flow, while the neighbouring rows deform little and only in an elastic manner. The deformation pattern is shown with pointed lines in Fig. 2.20. The major hypothesis of this mechanical model is that the bulk of each crystal, outside a thin boundary zone, is deformed in a manner that is compatible with that of its neighbours of the same slipping band.

Thus, on any horizontal or vertical row:

$$\gamma_{ij} = \gamma_j = \text{constant} \qquad \gamma_{ji} = \gamma_i = \text{constant} \tag{2-84}$$

where the $\gamma_i$ are the deformations at $90^\circ$ with those of the main considered shear plane $\gamma_j$.

Condition (2-84), obviously produces shear stresses which are different from one crystal to the other, on the same slip band.

We have to consider that the bulk of each crystal deforms like a monocrystal where there is an addition of elastic and plastic deformation as in a Maxwell rheological model. Taking into account the values of the plastic deformations in function of the angles $\alpha_{ij}$ for each grain, but only in the slip direction, we get:

$$\dot{\gamma}_j = \dot{\gamma}_{j\varepsilon} + \dot{\gamma}_{j\nu} \tag{2-85}$$

where the subscripts $\varepsilon$ and $\nu$ refer respectively to the elastic and viscous component of the deformation. Thus:

$$\dot{\gamma}_{j\varepsilon} = \frac{1}{G_{ij}} \frac{d\tau_{ij}}{dt} \tag{2-86}$$

and the viscous component is given by an expression, also based on the movement and multiplication of mobile dislocations given by Eq. (2-80) and resolved on the basal plane of each crystal.

$$\dot{\gamma}_{j\nu} = C\,(\tau_{ij}\ \cos 2\alpha_{ij})^n\ \sin \alpha_{ij} \tag{2-87}$$

in which the $\cos 2\alpha_{ij}$ resolves the stress along the basal plane and $\sin \alpha_{ij}$ resolves the strain along the main shear direction.

For the multiplication of mobile dislocations we use basic Eq. (2-80):

$$C = Co\ (1 + \alpha\gamma^m) \tag{2-88}$$

where Co is a constant for a given ice sample at a given temperature:

$$Co = \frac{b\ n_o\ v_o}{\tau_o^{\ n}}\ \exp\ \{Qc\theta/273R\theta^*\} \tag{2-89}$$

in which:

$v_o$ - velocity of reference of mobile dislocations at $0^oC$.

$n_o$ - initial number of mobile dislocations.

$\alpha$ - coefficient of multiplication of mobile dislocations.

n - Glen's exponent.

m - characteristic power of growth of mobile dislocations.

$\tau_o$ - shear stress of reference, taken as $10^5$ Pa.

The shear modulus of ice depends little on crystal orientation and its variability is not significant in viscous flow so that $G_{ij} \simeq G$. In any one row, if we assume identical dislocation characteristics, we then have:

$$\dot{\gamma}_{j\nu} = \dot{\gamma}_{ij\nu} = C\ (\tau_{ij})^n\ CR\ (\alpha_{ij}) \tag{2-90}$$

The creep function CR $(\alpha_{ij})$ compares the creep rate of each crystal for a given shear stress, in function of its orientation $\alpha_{ij}$.

$$CR\ (\alpha_{ij}) = (\cos\ 2\alpha_{ij})^n \sin\ \alpha_{ij} \tag{2-91}$$

This function is shown on Fig. 2.21. It is theoretically equal to zero for $\alpha_{ij} = \pi/2$ and $\alpha_{ij} = \pi/4$. On the same Fig. is also shown the curve CR $(\alpha_m)$ where the angles are classed irrespective of their actual values, but in function of the resistance to viscous flow of a number of grains in the polycrystalline aggregate.

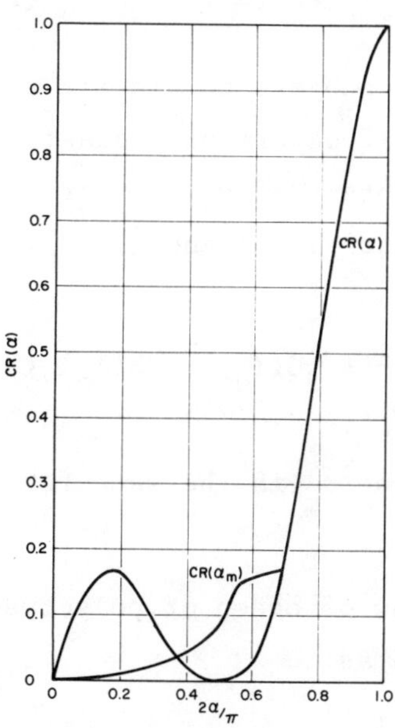


*Fig. 2.21. Creep function CR (α) in relation to the angle of the corresponding crystal α. Also shown is the classified value CR $(\alpha_m)$ for increasing values of the function.*

Because many crystals are blocking viscous flow in one direction it can be excepted, and it is in fact observed, that slip bands will form in prefered directions whereas other bands will be rigid. This is why, in this proposed model, we may expect that in a randomly oriented ice sample, half of the slip bands will form in one direction and the other half in the other.

Certain rows, alternating with slip bands, will deform only in an elastic manner. For these rows j + 1, the strain rate is:

$$\dot{\gamma}_{j+1} = \frac{1}{G}\frac{d\tau}{dt} \qquad (2\text{-}92)$$

Once a slip band is mobilized to flow in one direction, the deformation is controlled by that of the worst oriented crystal which will have pre-rotated to a critical angle $\alpha_m$. From Eqs (2-86), (2-87) and (2-91):

$$\dot{\gamma}_j = \frac{1}{G}\frac{d\tau_{j\alpha m}}{dt} + C\ (\tau_{j\alpha m})^n\ CR\ (\alpha_m) \qquad (2\text{-}93)$$

As only half of the rows are subject to plastic flow, as shown in Fig. 2.20 and the other half deform only elastically under the average stress $\tau$, then the total deformation of the sample $\dot{\gamma}$ is given by the addition of Eqs (2-92) and (2-93):

$$\dot{\gamma}_j = \frac{1}{\ell}\sum_{1}^{\ell}\dot{\gamma}_j = \frac{1}{2G}\frac{d\tau}{dt} + \frac{1}{2G}\frac{d\tau_{\alpha m}}{dt} + \frac{1}{2}C(\tau_{\alpha m})^n\ CR(\alpha_m) \qquad (2\text{-}94)$$

where the critical angle of pre-rotation $\alpha_m$ is the same for each slip band; $\gamma_{jam} = \gamma_{\alpha m}$.

Eq. (2-94) is the general law of creep of polycrystalline ice. It can also be expressed as:

$$\dot{\gamma} = \frac{1}{G}\frac{d\tau}{dt} + \frac{1}{2G}\frac{d\ (\tau_{\alpha m} - \tau)}{dt} + \frac{1}{2}C\ (\tau_{\alpha m})^n\ CR\ (\alpha_m) \qquad (2\text{-}95)$$

The first term represents pure elastic deformation, the second term, delayed elasticity and the third term pure viscous flow. The rheological representation of this mechanical model is shown in Fig. 2.22. It is made of a serie of a spring representing the pure elastic bands with a network of parallel elements each combining a spring with a non-linear dashpot and representing a crystal in the slip bands.

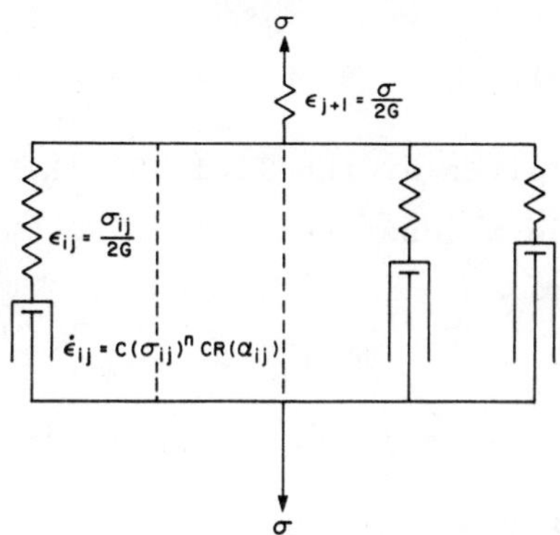


*Fig. 2.22. Sketch of rheological model of creep of ice under a tensile stress σ.*

The general law of creep of polycrystalline ice applies also to monocrystalline ice if:

$$\tau_{\alpha m} = \tau \qquad (2\text{-}96)$$

which means that all the crystals have the same orientation α. Eq. (2-95) then reduces only to two terms:

$$\dot{\gamma} = \frac{1}{G}\frac{d\tau}{dt} + \frac{1}{2} C\ (\tau)^n\ CR\ (\alpha) \qquad (2\text{-}97)$$

## 2.2.5 - CREEP OF ICE UNDER A CONSTANT LOAD

We will analyse the different types of creep with displacement of mobile dislocations as defined in section 2.2.3.

### 2.2.5.1 - EQUATIONS OF εα AND ν CREEP UNDER A CONSTANT STRESS

If creep occurs under a constant shear stress $\tau_o$, we can separate the components of the deformation:

$$\gamma = \gamma_{\varepsilon} + \gamma_{\varepsilon\alpha} + \gamma_{\nu} \tag{2-98}$$

where the instantaneous elastic deformation is the first term of Eq. (2-95):

$$\gamma_{\varepsilon} = \tau_o/G \tag{2-99}$$

The permanent creep is given at the limit by the third term of Eq. (2-95), when C becomes a constant:

$$\dot{\gamma}_{\nu} = \frac{1}{2} C(\tau_{\alpha m})^n \text{ CR } (\alpha_m) \tag{2-100}$$

and the transitory creep of types $\varepsilon\alpha$ is given by the second term

$$\dot{\gamma}_{\varepsilon\alpha} = \frac{1}{2G} \frac{d\ (\tau_{\alpha m} - \tau_o)}{dt} \tag{2-101}$$

which integrated, also gives:

$$\tau_o = \tau_{\alpha m} - 2G\ \gamma_{\varepsilon\alpha} \tag{2-102}$$

On a slip band, during all phases of creep the strain rate is the same from one crystal to the other. Thus, even during transitory creep, the creep rate can be obtained from that of the crystal which makes an angle $\alpha = 90^{\circ}$ with the direction of slip, then:

$$\dot{\gamma}_{\varepsilon\alpha j} = C\ (\tau_{90})^n \tag{2-103}$$

where $\tau_{90}$ is the stress on that particular crystal.

The total strain rate of the sample, will be given by substracting half of the bands which have deformed only in an elastic manner:

$$\dot{\gamma}_{\varepsilon\alpha} = \frac{C}{2} (\tau_{90})^n \tag{2-104}$$

We need only relate the value of the stress on this particular crystal $\tau_{90}$ to the average stress on the row $\tau_o$.

If we admit that the stress distribution during transitory creep is very close to that of permanent creep, we then have:

$$\dot{\gamma}_{ij\nu} = \dot{\gamma}_{j\nu} = 2\dot{\gamma}_{\nu} = C\ (\tau_{ij})^{n}\ CR\ (\alpha_{ij}) \quad (2\text{-}105)$$

Isolating the $\tau_{ij}$, we have:

$$\tau_{ij} = \left(\frac{2\dot{\gamma}\nu}{C}\right)^{1/n} / CR\ (\alpha_{ij})^{1/n} \quad (2\text{-}106)$$

If the distribution of $\alpha$'s is uniform from $\alpha_m$ to $\alpha = \pi/2$ and if the CR $(\alpha_{ij})$ is equal to CR $(\alpha_m)$ from $\alpha = o$ to $\alpha = \alpha_m$, the average stress on a slip band is given by:

$$\tau_o = \frac{1}{\ell}\sum_{1}^{\ell} \tau_{ij} = \left(\frac{2\dot{\gamma}_{\nu}}{C}\right)^{1/n} \Big/ \Omega' \quad (2\text{-}107)$$

where:

$$\frac{1}{\Omega'} = \frac{2}{\pi}\left[\int_{o}^{\pi/2-\alpha m} \frac{d\alpha}{CR\ (\alpha)^{1/n}} + \frac{\alpha m}{CR\ (\alpha m)^{1/n}}\right] \quad (2\text{-}108)$$

Because in the same row:

$$\tau_{90} = \left(2\ \frac{\dot{\gamma}_{\nu}}{C}\right)^{1/n} \quad (2\text{-}109)$$

The value of $\Omega'$ can also be expressed with Eqs (2-107) and (2-109):

$$\Omega' = \tau_{90}/\tau_{o} \quad (2\text{-}110)$$

The function $\Omega'$ has been integrated numerically and its value is given in Fig. 2.23. Also given in this figure is the value of $\Omega$ which relates the stress on the worst oriented crystal to the average stress:

$$\Omega = \tau_{o}/\tau_{\alpha m} = CR\ (\alpha m)^{1/n}/\Omega' \quad (2\text{-}111)$$

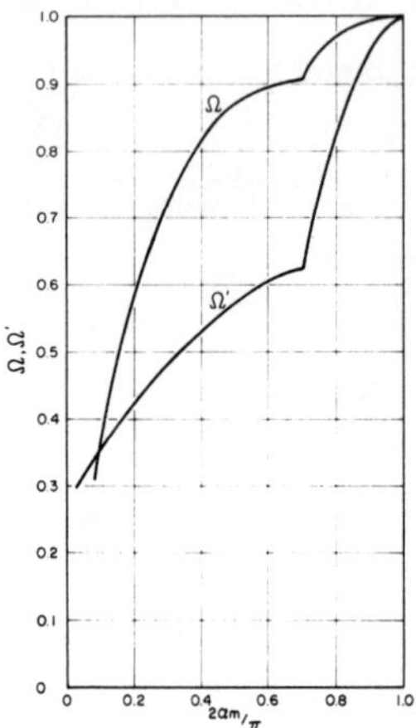


*Fig. 2.23. Values of the function Ω and Ω' in function of the worst angle of rotation of the head crystal $\alpha_m$.*

The creep function can be expressed by Eq. (2-104) with Eq. (2-110):

$$\dot{\gamma}_{\varepsilon\alpha} = \frac{C}{2} (\Omega' \tau_o)^n \qquad (2\text{-}112)$$

The component of the elastic deformation in the rheological model can be expressed with the value of $\tau_o$ in Eq. (2-102). This gives in Eq. (2-112):

$$\dot{\gamma}_{\varepsilon\alpha} = \frac{C\, \Omega'^n}{2} (\tau_{\alpha m} - 2G\gamma_{\varepsilon\alpha})^n \qquad (2\text{-}113)$$

With $\tau_{\alpha m}$ in Eq. (2-111):

$$\dot{\gamma}_{\varepsilon\alpha} = \frac{C(\Omega')^n}{2} (\frac{\tau o}{\Omega} - 2G\gamma_{\varepsilon\alpha})^n \qquad (2\text{-}114)$$

The value of C in Eq. (2-114) is not a constant, but depends on the number of mobile dislocations which is a function of the plastic deformation $\gamma_{\varepsilon\alpha}$. With Eq. (2-89) it can be represented by:

$$C = Co\ (1 + \alpha\ \gamma_{\varepsilon\alpha}^m) \qquad (2\text{-}115)$$

When $\gamma_{\varepsilon\alpha} \geq \gamma_{\varepsilon\alpha o}$, then C stays a constant. Permanent creep is given by the last term of Eq. (2-95):

$$\dot{\gamma}_{\nu} = \frac{1}{2} \text{Co } (1 + \alpha\, \gamma_{\varepsilon\alpha o}^{\,m})\ (\tau_{\alpha m})^{n}\ \text{CR } (\alpha_{m}) \qquad (2\text{-}116)$$

Or with Eqs (2-111) and (2-115):

$$\dot{\gamma}_{\nu} = \tfrac{1}{2} \text{ Co } (1 + \alpha\, \gamma_{\varepsilon\alpha o}^{\,m} (\Omega')^{n} (\tau_{o})^{n} \qquad (2\text{-}117)$$

The transitory creep of $\varepsilon\alpha$ type is obtained from Eqs (2-114) and (2-115):

$$\dot{\gamma}_{\varepsilon\alpha} = \tfrac{1}{2} \text{ Co } (1 + \alpha\, \gamma_{\varepsilon\alpha}^{\,m})\ (\Omega')^{n}\ (\frac{\tau o}{\Omega} - 2G\ \gamma_{\varepsilon\alpha})^{n} \qquad (2\text{-}118)$$

If we define the delayed elastic deformation $\gamma_{\varepsilon r}$, as:

$$\gamma_{\varepsilon r} = \tau_{o}/2\ \Omega G \qquad (2\text{-}119)$$

and an equivalent rate of steady creep for $\varepsilon\alpha$ type, $\dot{\gamma}_{\varepsilon r}$, as:

$$\dot{\gamma}_{\varepsilon r} = \tfrac{1}{2} \text{ Co } [\frac{\Omega'}{\Omega}\ \tau_{o}]^{n} \qquad (2\text{-}120)$$

With the new reduced variables:

$$\overline{\gamma}_{\varepsilon\alpha} = \gamma_{\varepsilon\alpha}/\gamma_{\varepsilon r}$$

$$\overline{t} = \dot{\gamma}_{\varepsilon r} t/\gamma_{\varepsilon r}$$

$$\overline{\alpha} = \alpha\gamma_{\varepsilon r}^{\,m}$$

The basic Eq. (2-118) of transitory creep of the $\varepsilon\alpha$ type becomes:

$$\frac{d\ \overline{\gamma_{\varepsilon\alpha}}}{d\overline{t}} = (1 + \overline{\alpha}\ \overline{\gamma}_{\varepsilon\alpha}^{\,m})\ (1 - \overline{\gamma}_{\varepsilon\alpha})^{n} \qquad (2\text{-}122)$$

and the permanent creep $\dot{\gamma}_{\nu}$ rate is, with this notation, from Eqs (2-119) and (2-120):

$$\dot{\gamma}_{\nu} = \Omega^{n}\ (1 + \overline{\alpha}\ \overline{\gamma}_{\varepsilon\alpha o}^{\,m})\ \dot{\gamma}_{\varepsilon r} \qquad (2\text{-}123)$$

It is found experimentally, that the value of m that best fits the creep curves for many types of polycrystalline ice is around the value of m = 0.5. Eq. (2-122) has been integrated for n = 3, m = 0.5 and values of $\bar{\alpha}$ of 0 to 100. The results are shown in Fig. 2.24. It can be seen that, for high values or $\bar{\alpha}$, that is for crystals with a relatively small initial number $n_o$ of mobile dislocations compare to the one which sets in when permanent creep occurs, that there is practically no transitory creep, but the delayed elasticity is transformed immediately into an added elasticity, that is much larger than the real instantaneous deformation. Indeed, transitory creep of the $\varepsilon\alpha$ type stops after a certain relative deformation $\gamma_{\varepsilon\alpha o}$ when all the dislocations become mobile, and is followed by permanent creep.

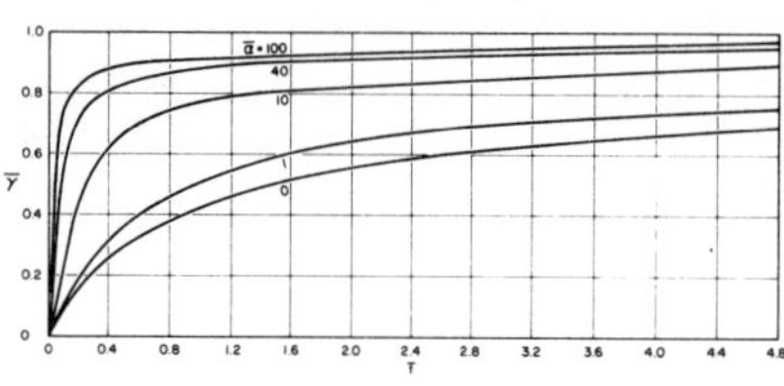


*Fig. 2.24. Curves of $\varepsilon\alpha$ creep in non-dimensionnal shear strain $\bar{\gamma}$ versus $\bar{t}$. They are computed for values of n = 3, m = 0.5 and various values of the parameter $\bar{\alpha}$.*

## 2.2.5.2 - EQUATIONS OF $\alpha\nu$ CREEP UNDER A CONSTANT STRESS

When transitory creep of the $\varepsilon\alpha$ type is completed, it may be followed by accelerated creep of the $\alpha\nu$ type if the number of mobile dislocations is still climbing. This would also be the case when there are very few crystals in an ice sample, with c-axis aligned in the same direction, as in a monocrystal, so that the initial transitory creep would be replaced by accelerated creep of the $\alpha\nu$ type. In this type of creep, the transitory deformation $\gamma_{\varepsilon\alpha}$ does not appear in Eq. (2-98), so the deformation is reduced only to $\gamma_\varepsilon$, the instantaneous elastic deformation and the quasi-permanent deformation $\gamma_\nu$. This last term has however to include the increasing number of mobile dislocations, so it is from Eq. (2-100):

$$\dot{\gamma}_{\alpha\nu} = \tfrac{1}{2}\, C\, (\tau_{\alpha m})^n\, CR\, (\alpha_m) \qquad (2\text{-}124)$$

with Eq. (2-111):

$$\tau_o = \Omega\tau_{\alpha m} = CR\, (\alpha m)^{1/n}\, \tau_{\alpha m}/\Omega' \qquad (2\text{-}125)$$

Eqs (2-111) and (2-115) in Eq. (2-125), with appropriate notation instead of the deformation $\gamma_{\varepsilon\alpha}$, give:

$$\dot{\gamma}_{\alpha\nu} = \tfrac{1}{2}\, Co\, (1 + \alpha\, \gamma^m_{\alpha\nu})\, (\Omega')^n\, (\tau_o)^n \qquad (2\text{-}126)$$

The permanent creep at the end of $\alpha\nu$ creep, is given when $\gamma_{\alpha\nu} = \gamma_{\alpha\nu o}$. Defining an equivalent rate of steady creep $\dot{\gamma}_\nu$ given by:

$$\dot{\gamma}_\nu = \tfrac{1}{2}\, Co\, (\Omega')^n\, (\tau_o)^n \qquad (2\text{-}127)$$

With the reduced variables:

$$\begin{aligned} \overline{\gamma}_{\alpha\nu} &= \gamma_{\alpha\nu}/\gamma_{\alpha\nu o} \\ \overline{\alpha} &= \alpha\gamma^m_{\alpha\nu o} \\ \overline{t} &= \dot{\gamma}_\nu\, t/\gamma_{\alpha\nu o} \end{aligned} \qquad (2\text{-}128)$$

The equation of $\alpha\nu$ creep, from Eq. (2-126) is:

$$\frac{d\,\overline{\gamma_{\alpha\nu}}}{d\overline{t}} = (1 + \overline{\alpha}\,\overline{\gamma}^m_{\alpha\nu}) \qquad (2\text{-}129)$$

The permanent state of creep is given by:

$$\overline{\gamma}_{\nu o} = (1 + \overline{\alpha}\,\overline{\gamma}^m_{\alpha\nu o})\, t + B \qquad (2\text{-}130)$$

where B is a constant.

This is the equation of accelerated $\alpha\nu$ creep which shows the acceleration of the deformation with time, up to a value $\overline{\gamma}_{\alpha\nu} = \overline{\gamma}_{\alpha\nu o}$ when all the dislocations are in movement and permanent creep follows.

### 2.2.5.3 - CREEP CURVES FOR MONOCRYSTALS

The equation for creep under a constant load of monocrystal is given by Eq. (2-97). If the shear stress is applied parallel to the basal plane, $\alpha = 90^{o}$ and CR ($\alpha$) = 2. The equation then reduces with Eq. (2-115) to one particular type of $\alpha\nu$ creep:

$$\dot{\gamma}_{\alpha\nu} = Co\ (1 + \alpha\ \gamma_{\alpha\nu}^{m})\ \tau o^{n} \qquad (2\text{-}131)$$

The creep of monocrystals of ice has been partly explained by many authors (Higashi *et al.*, 1964; Ready and Kingery, 1964; Krausz, 1968; Jones and Glen, 1969) from a model proposed by Johnston (1962) to explain the ductile behavior of monocrystals of LiF. This model gives an expression similar to Eq. (2-88) in this particular case, when m = 1. Jones and Glen (1969) have used Johnston's equation to check the experimental curve of a monocrystal oriented at $45^{o}$ with the load and they found a fairly good fit. They got a $n_o$ value of $5 \times 10^5\ cm^{-2}$, very similar to that given on Table 2.3, and a velocity of dislocation, when reduced to $\tau o = 10^5$ Pa and $\theta = 0^{o}C$, of $5.5 \times 10^{-5} cm\ s^{-1}$.

Some of the earlier results were obtained by Steineman (1958) for monocrystals sheared along the basal plane at $-2.3^{o}C$. The data points for three curves are shown in Fig. 2.25 for very low shear stresses.

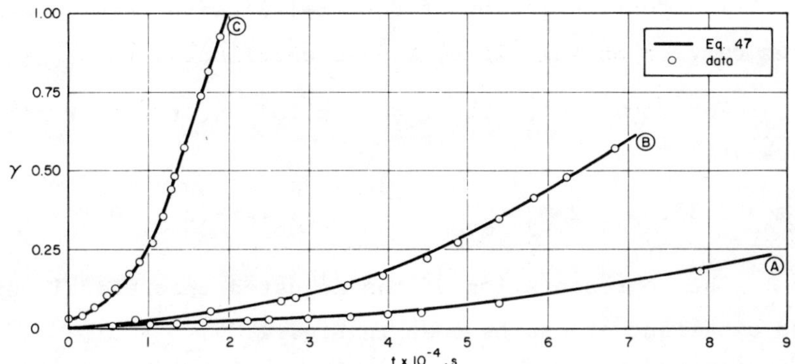


*Fig. 2.25. Measured (circles) and computed creep curves (full lines) of monocrystals under a constant shear stress. The measurements are by Steineman (1958). The adjusted creep curves are made using Eq. (2-132).*

Eq. (2-131) or (2-129) of αν creep has been used to represent these curves:

$$\frac{d\bar{\gamma}}{d\bar{\tau}}\,\alpha\nu = (1 + \bar{\alpha}\,\bar{\gamma}^{m}_{\alpha\nu}) \qquad (2\text{-}132)$$

In all cases an excellent fit has been obtained and the resulting continuous curves are shown on Fig. 2.25.

For curve (A), with $\tau = 4.7 \times 10^4$ Pa, the best fit parameters are:

$$m = 0.33,\ \alpha = 186,\ \gamma_{\alpha\nu o} = 0.23,\ \dot{\gamma}_{\nu o} = 5.84 \times 10^{-6}\ s^{-1}$$

where $\gamma_{\alpha\nu o}$ is the deformation at the beginning of permanent creep and $\dot{\gamma}_{\nu o}$ is the rate of permanent creep.

For curve (B), with $\tau = 6.5 \times 10^4$ Pa, the best fit parameters are:

$$m = 0.33,\ \alpha = 242,\ \gamma_{\alpha\nu o} = 0.25,\ \dot{\gamma}_{\nu o} = 1.55 \times 10^{-5}\ s^{-1}$$

The representation of Eq. (2-132) with this data is shown on curve B in full line, and is excellent.

For curve (C), with $\tau = 19 \times 10^4$ Pa, the best fit parameters are:

$$m = 0.33, \ \alpha = 230, \ \gamma_{\alpha\nu o} = 0.37, \ \dot{\gamma}_{\nu o} = 8.2 \times 10^{-5} \ s^{-1}$$

The representation of the adjusted theoretical curve is also excellent as can be seen in Fig. 2.25.

In all cases the number of mobile dislocation as been multiplied by a value between 187 and 242 from the beginning of deformation to the onset of permanent creep. The Johnston's model would have been inadequate to represent these creep curves; the value of m for best fit being 0.33 instead of one.

### 2.2.5.4 - CREEP CURVES FOR POLYCRYSTALLINE ICE

Most of the testing in the litterature is for uniaxial tension or compression. The relationship between the maximum shear strain rate $\dot{\gamma}$ and the uniaxial strain rate $\varepsilon$ is given, for an incompressible medium, by Lliboutry (1964):

$$\dot{\gamma} = \sqrt{3}\,\dot{\varepsilon} \qquad (2\text{-}133)$$

The uniaxial stress $\sigma$ is also related to the maximum shear stress, which occurs at $45^{o}$ with the loading direction, by:

$$\tau = \sigma/2 \qquad (2\text{-}134)$$

Thus, the results of tests in uniaxial compression or tension differ only by a constant from creep curves caused by maximum shear on slip bands. These tests can then be used directly to verify the proposed creep laws.

Early results on creep in tension under a constant load of polycrystalline ice were obtained by Brill and Camp (1961). They worked with fine grain snow-ice of type T1 with grain diameters between 1 and 2 mm and some of their more interesting results were obtained for two tests where the loads were removed after a certain time and the deformation recovery-curve was measured. The data point are shown on Fig. 2.26.

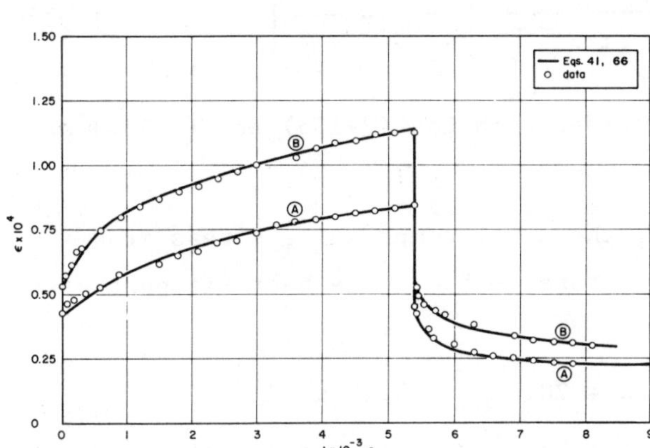


*Fig. 2.26. Measured (circles) and computed creep curves (full lines) of T1 snow ice under a constant load in tension, and after load removal. The measurements are by Brill and Camp (1961). The adjusted creep curves are with Eqs (2-135) and (2-137).*

For the loading part the primary creep curve is given by Eq. (2-122) under longitudinal deformation:

$$\frac{d\,\overline{\varepsilon_{\varepsilon\alpha}}}{d\bar{t}} = (1 + \bar{\alpha}\,\bar{\varepsilon}_{\varepsilon\alpha})^{m}\,(1 - \bar{\varepsilon}_{\varepsilon\alpha})^{n} \qquad (2\text{-}135)$$

When the load is removed, there is an instantaneous elastic recovery equal to $\varepsilon_{\varepsilon o}$ and a visco-elastic recovery given by Eq. (2-135) but where the value of 1 representing the constant load, is removed:

$$\frac{d\,\overline{\varepsilon_{\varepsilon\alpha}}}{d\bar{t}} = -\left(1 + \bar{\alpha}\,\bar{\varepsilon}_s^{-m}\right) \varepsilon_{\varepsilon\alpha}^{\;n} \qquad (2\text{-}136)$$

The value of $\varepsilon_s$ is the one which occurs at the time of removal of the load, when the number of dislocations is blocked. With a n value of 3, the time recovery curve is given by the integration of Eq. (2-136).

$$\bar{t} = \frac{1}{2\left(1 + \bar{\alpha}\bar{\varepsilon}_s^{-m}\right)} \left[\frac{1}{\varepsilon_{\varepsilon\alpha}^{\;2}} - \frac{1}{\varepsilon_s^{\;2}}\right] \qquad (2\text{-}137)$$

Two adjusted curves with Eqs (2-135) and (2-137) are shown in Fig. 2.26.

For curve (A), the axial compressive stress is $\sigma = 2.45 \times 10^5$ Pa, the temperature, $-15^{\circ}$C. The best fit parameters for primary creep are:

$$m = 0.5,\ \alpha = 20,\ \varepsilon_{\varepsilon o} = 0.40 \times 10^{-4}$$

$$\varepsilon_{\varepsilon r} = 0.32 \times 10^{-4},\ \dot{\varepsilon}_o = 3.57 \times 10^{-9}\ s^{-1}$$

During the time-recovery creep, the full elastic deformation $\varepsilon_{\varepsilon o}$ is recovered. Because of a different crystal orientation during unloading it can be expected that the total time dependant elastic recovery will be different than that obtained under loading conditions. It is found to be $\varepsilon_{\varepsilon r} = .20 \times 10^{-4}$. The difference between computed values gives a total permanent deformation of $.22 \times 10^{-5}$. The full line on the figure represents Eqs (2-135) and (2-137) with these values.

The same computations were carried for curve (B) in the figure where the load is very close to be the same $\sigma = 2.43 \times 10^5$ Pa but the temperature has been increased to $-10^{\circ}$C. The best fit values of the parameters are:

$$m = 0.5,\ \alpha = 40,\ \dot{\varepsilon}_o = 4.76 \times 10^{-9}\ s^{-1}$$

$$\varepsilon_{\varepsilon o} = .53 \times 10^{-4},\ \varepsilon_{\varepsilon r} = .42 \times 10^{-4},\ \varepsilon'_{\varepsilon r} = .34 \times 10^{-4}$$

The permanent set is $.26 \times 10^{-4}$.

Classical results of primary creep of fine-grained polycrystalline ice where crystal orientation was initially random have been obtained by Mellor and Testa (1969). The constant axial compressive stress was low, $\sigma = .43 \times 10^5$ Pa and test temperature $-2.06^oC$. Two set of data points are shown in Fig. 2.27.

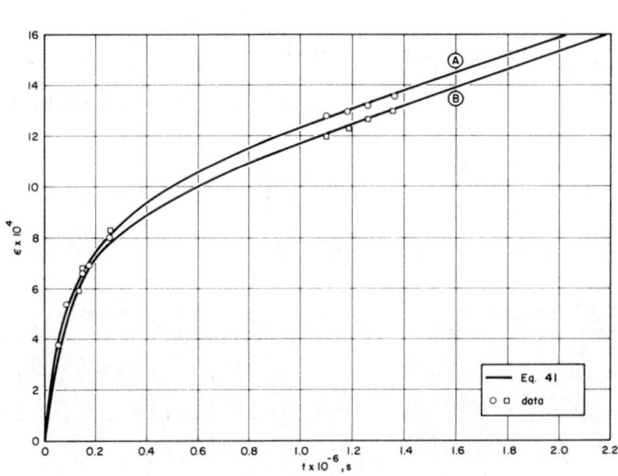


*Fig. 2.27. Measured (circles) and computed creep curves (full lines) of fine-graines polycrystalline ice under compression. The measurements are by Mellor and Testa (1969). The adjusted curves are obtained with Eq. (2-135).*

Curve (A) is Eq. (2-135) with the best fit parameters:

$$m = 0.5,\ \alpha = 100,\ \dot{\varepsilon}_o = 3.33 \times 10^{-10} s^{-1},\ \varepsilon_{\varepsilon r} = 1.03 \times 10^{-3}$$

Curve (B) is the same representation with the best fit parameters:

$$m = 0.5,\ \alpha = 100,\ \dot{\varepsilon}_o = 3.33 \times 10^{-10}\ s^{-1},\ \varepsilon_{\varepsilon r} = 0.96 \times 10^{-3}$$

Thus both curves differ only by the value of retarded elasticity. This may be expected even if the samples were choosen side by side in the same ice piece.

An interesting application of the present theory is for the case of load reversal. This was done for glacier ice by Duval (1976). This is for large grain polycrystalline ice at $-10^oC$, $\tau = 2.5 \times 10^5$ Pa in torsion-compression. The experimental data is shown in Fig. 2.28.

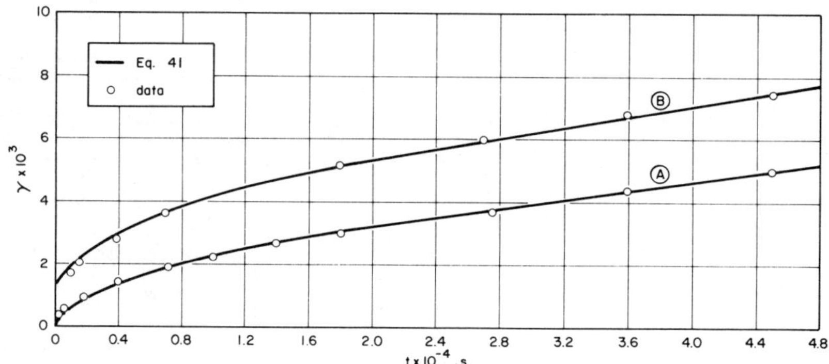


*Fig. 2.28. Measured (circles) and computed creep curves (full lines) of glacier ice under torsion and with load reversal. The measurements are by Duval (1976). The adjusted curves are obtained with Eqs (2-122) and (2-138).*

Curve (A) is the creep curve for the first load. The full line is Eq. (2-122) for classical primary creep with the best fit parameters.

$$m = 0.33,\ \alpha = 10,\ \dot{\gamma}_o = 6.1 \times 10^{-8}\ s^{-1}$$

$$\gamma_{\varepsilon o} = 0.2 \times 10^{-3},\ \gamma_{\varepsilon r} = 2.6 \times 10^{-3}$$

Curve (B) is the creep curve with Eq. (2-122) but where all the dislocations are in movement at 10.3 times the original amount which occured at the end of creep with first load, without any further increase in number. Thus Eq. (2-122) becomes:

$$\frac{d\,\overline{\gamma_{\varepsilon\alpha}}}{d\overline{t}} = 10.3\ (1 - \overline{\gamma}_{\varepsilon\alpha})^3 \qquad (2\text{-}138)$$

Because of the change in orientation there is a change in the geometry of the grains relative to the load. The best fit parameters for Eq. (2-138) become:

$$m = 0.33,\ \alpha = 10,\ \dot{\gamma}_o = 7.22 \times 10^{-2}\ s^{-1}$$

$$\gamma_{\varepsilon o} = 1.3 \times 10^{-3},\ \gamma_{\varepsilon r} = 3.6 \times 10^{-3}\ s^{-1}$$

Finally, the present creep formulas can represent even abnormal creep curves. This is often the case when testing is done on very small samples of polycrystalline ice where only a few grains will be present in slip bands. On Fig. 2.29 are shown three tests on $S_2$ ice with samples 2.5 cm in diameter, tested in tension at $-10^{\circ}C$ (Paradis, 1978). The ice was taken from the same batch.

Sample A shows classical primary-secondary creep. The best fit parameters with Eq. (2-135) are:

$$m = 2.0,\ \alpha = 10,\ \dot{\varepsilon}_o = 2.58 \times 10^{-7}\ s^{-1}$$

$$\varepsilon_{\varepsilon o} = 1 \times 10^{-4},\ \varepsilon_{\varepsilon r} = 4.5 \times 10^{-4},\ \sigma = 6.24 \times 10^5\ Pa$$

Sample B, does not show any primary creep, but only an elastic deformation followed by secondary creep. The best fit parameters are:

$$\varepsilon_{\varepsilon o} = 1.3 \times 10^{-4},\ \dot{\varepsilon}_o = 7.5 \times 10^{-7}\ s^{-1},\ \sigma = 7.22 \times 10^5\ Pa$$

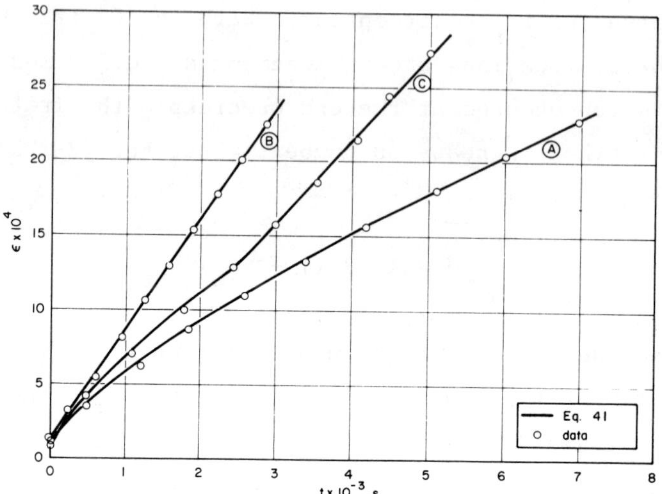


*Fig. 2.29. Measured (circles) and computed creep curves (full lines) of $S_2$ columnar ice under tension. The measurements are by Paradis (1978). The adjusted curves are obtained with Eq. (2-135).*

The number of mobile dislocations must increase very rapidly in this sample and the initial deformation should included both elastic and retarded elastic deformation.

Finally sample C shows a bizarre behavior. After a normal strain-hardening primary creep phase and the beginning of permanent creep, the sample turns soft and the strains pass suddently to a higher permanent creep rate. This may also be explained by the creep model. When all the dislocations are mobile, the rotation of hard controlling crystals bring in place softer controlling ones (with more mobile dislocations). The best fit parameters with Eq. (2-136) are:

<u>First phase</u> $\varepsilon < 13.3 \times 10^{-4}$

$$m = 2, \ \alpha = 10, \ \dot{\varepsilon}_o = 3.57 \times 10^{-7} \ s^{-1}$$

$$\varepsilon_{\varepsilon o} = 0, \; \varepsilon_{\varepsilon r} = 4 \times 10^{-4}, \; \sigma = 6.24 \times 10^{5} \text{ Pa}$$

<u>Second phase</u> $\varepsilon > 13.3 \times 10^{-4}$

$$\dot{\varepsilon}_o = 2.58 \times 10^{-7} \text{ s}^{-1}$$

More complex behaviors could also be explained with the creep model which is presented in this text.

## 2.2.6 - CREEP OF ICE AT CONSTANT SPEED

This type of creep is much more representative of actual conditions of ice loading on engineering structures than creep obtained under a constant load.

### 2.2.6.1 - EQUATIONS OF $\varepsilon\alpha$ AND $\alpha\nu$ CREEP UNDER A CONSTANT STRAIN RATE

For creep under a constant strain rate, the deformation is still controlled by that of the more stressed crystal in a slip band. In basic Eqs (2-94) or (2-95), we then have:

$$\dot{\gamma} = \dot{\gamma}_o = \text{constant} \tag{2-139}$$

Assuming the same stress distribution on each of the crystal as for permanent flow, Eq. (2-111) also applies with:

$$\tau = \Omega \, \tau_{\alpha m} = \tau(\alpha m) \; CR \; (\alpha m)^{1/n}/\Omega' \tag{2-140}$$

Basic Eq. (2-94) then becomes with (2-139) and (2-140):

$$\dot{\gamma}_o = \frac{1}{2G} \left(1 + \frac{1}{\Omega}\right) \frac{d\tau}{dt} + \tfrac{1}{2} \, C \, (\Omega')^n \, (\tau)^n \tag{2-141}$$

With the Eq. (2-115):

$$C = C_o \, (1 + \alpha\gamma^m) \tag{2-142}$$

If we define an apparent yield stress $\tau_o$, as:

$$\dot{\gamma}_o = \tfrac{1}{2} C_o (\Omega')^n (\tau_o)^n \qquad (2\text{-}143)$$

the total instantaneous and delayed elasticity, $\gamma_{\varepsilon t}$, for this stress is:

$$\gamma_{\varepsilon t} = \frac{\tau_o}{2G} [1 + 1/\Omega] \qquad (2\text{-}144)$$

With the reduced variables:

$$\begin{aligned} \bar{\gamma} &= \gamma/\gamma_{\varepsilon t} = \dot{\gamma}_o t/\gamma_{\varepsilon t} \\ \bar{\alpha} &= \alpha \gamma_{\varepsilon t}^{m} \\ \bar{\tau} &= \tau/\tau_o \end{aligned} \qquad (2\text{-}145)$$

basic Eq. (2-141) becomes with Eqs (2-143) and (2-145):

$$\frac{d\bar{\tau}}{d\bar{\gamma}} = 1 - (1 + \bar{\alpha}\,\bar{\gamma}^{m})\,\bar{\tau}^{n} \qquad (2\text{-}146)$$

This is the complete equation of $\varepsilon\alpha$, $\nu$ and $\alpha\nu$ creep under a constant shear strain rate. It has been integrated numerically with $n = 3$, $m = 0.5$, for values of $\bar{\alpha}$ of 1, 10 and 40 and the results are shown in Fig. 2.30. The corresponding curves show a first part of $\varepsilon\alpha$ creep when the stress increases with the deformation, a maximum is attained for the yield strength of ice and then there is a reduction in stress with deformation after yield point, corresponding to $\alpha\nu$ creep.

The yield strength $\bar{\tau}_y$ of the ice will be given the first time $\frac{d\bar{\tau}}{d\bar{\gamma}} = 0$, for a given deformation $\bar{\gamma}_y$. With Eq. (2-146) this corresponds to the conditions:

$$\bar{\tau}_y = \frac{1}{(1 + \bar{\alpha}\,\bar{\gamma}_y^{m})^{1/n}} \qquad (2\text{-}147)$$

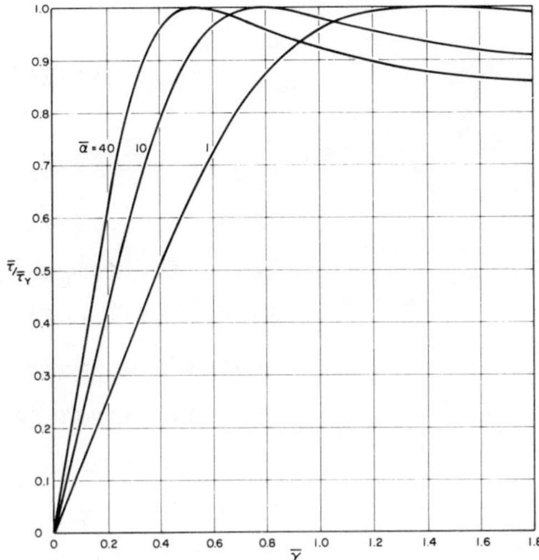


*Fig. 2.30. Curves of εα and αν creep under a constant strain rate. They are given in non-dimensionnal parameters $\bar{\tau}/\bar{\tau}_y$ and strain $\bar{\gamma}$ and computed for values of n = 3, m = 0.5 and three values of $\bar{\alpha}$.*

Where all the dislocations are mobile, i.e. $\gamma = \gamma_{\nu o}$, a new steady ν type creep is attained for a given residual stress $\bar{\tau}_\nu$:

$$\bar{\tau}_\nu = 1/(1 + \bar{\alpha}\,\bar{\gamma}_{\nu o}^{\,m})^{1/n} \qquad (2\text{-}148)$$

It is also interesting to note, on these curves, that the largest is the value of $\bar{\alpha}$, the more the ice behaves like an ideal elasto-plastic body. But the steepest is also the reduction in stress after yield point. This behavior might be called an elasto-ductile deformation process.

## 2.2.6.2 - CREEP CURVES FOR MONOCRYSTALS

The experiments done by Higashi *et al.*,(1964) are very useful to discuss the role of the various parameters in the creep behavior of monocrystals. Figs 2.31, 2.32, 2.33 show respectively the effects of temperature, strain rates and of a succession of unloading and reloading.

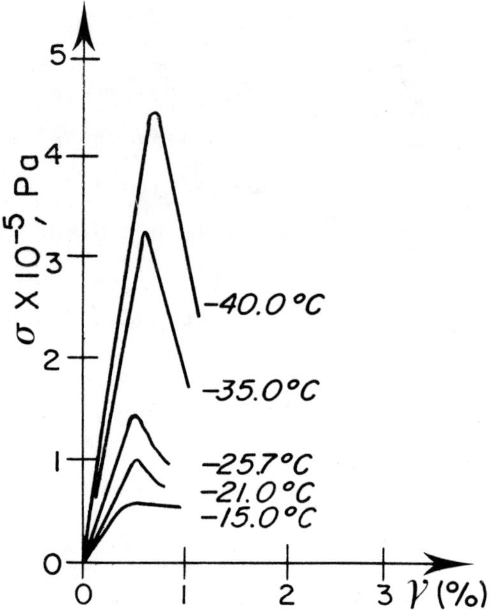


*Fig. 2.31. Stress-strain curves at different temperatures for single crystals deformed by basal shear at a constant strain rate of 1..3 × $10^{-1}$ $s^{-1}$ (Higashi et al, 1964).*

These effects could be predicted both with the Johnston's model or with the present theory by Eq. (2-97). A very interesting effect is that of reloading after the yield point has been reached. With repetition of straining and unloading, the stress at yield point gradually decreases. Because the number of mobile dislocations is increased after each test, the equation with a higher initial $n_o$ shows a reduction in yield strength. However past a certain threshold, the total number of dislocations stays a constant, so the yield strength does not decrease anymore as shown both by theory and experiments.

The activation energy found by Higashi *et al.*, was 15.9 Kcal/mole a little higher than the activation energy for molecular self-diffusion in ice which is close to 13.1 Kcal/mole (Kuroiwa, 1964).

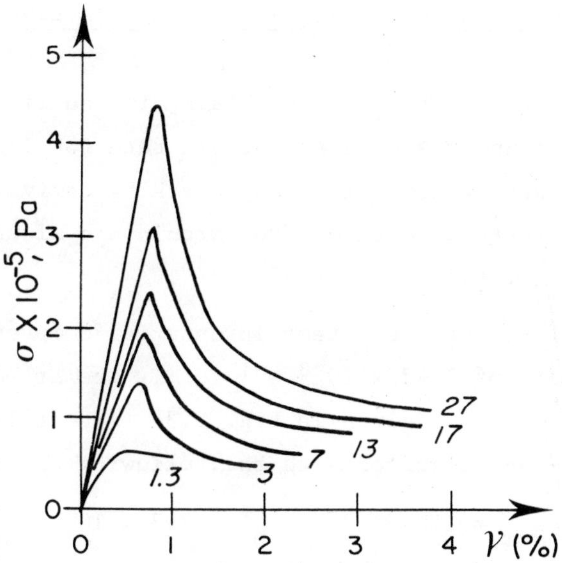


*Fig. 2.32. Stress-strain curves at -15°C for single crystals of ice deformed by basal shear at constant strain rates shown as parameters in units of $10^{-1}$ $s^{-1}$. (Higashi et al. 1964)*

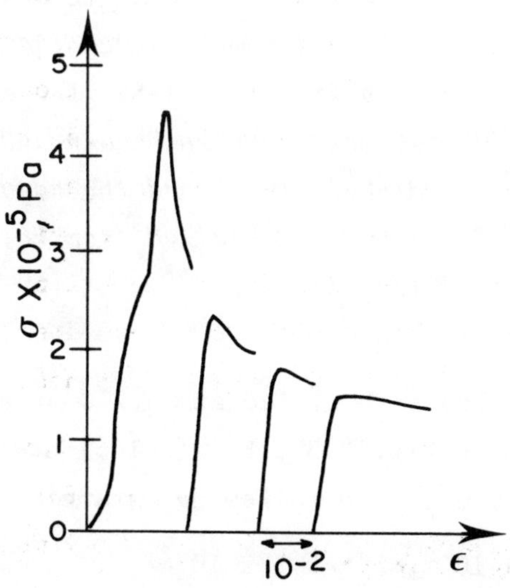


*Fig. 2.33. Stress-strain curve for easy glide with repetition of straining and unloading. Temperature $\theta_i$ = -16°C, strain rate 2.66 × $10^{-1}$ $s^{-1}$. (Higashi et al. 1964).*

## 2.2.6.3 - CREEP CURVES FOR POLYCRYSTALLINE ICE

It is very difficult to obtain, in the litterature, results of creep under a constant strain rate for low enough rates so that there would not be any cracking activity in the ice. Two such tests were taken from Drouin and Michel (1972) and were used to verify Eq. (2-146).

The first of these test in uniaxial compression was carried at a rate of $9.42 \times 10^{-8}\ s^{-1}$ for $T_1$ ice at $-4.2^{\circ}C$. The experimental points are given in Fig. 2.34. The full line represents Eq. (2-146) adjusted with the following parameters:

$$\varepsilon_{\varepsilon t} = 1.8 \times 10^{-3},\ \varepsilon_{\nu o} = 3.7 \times 10^{-3}$$

$$m = 0.5,\ \alpha = 5.9,\ \sigma_o = 20 \times 10^5\ Pa$$

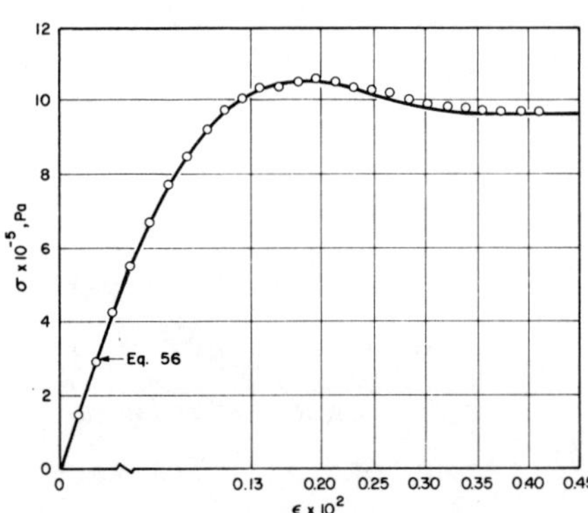


*Fig. 2.34. Measured (circles) and computed creep curves (full lines) for $T_1$ snow ice under a constant compressive strain rate of $9.42 \times 10^{-8}\ s^{-1}$. The measurements are by Drouin and Michel (1971) and the adjusted curve is obtained from Eq. (2-146).*

The second test, also with $T_1$ ice at $-4.2^{\circ}C$ at a rate of $2.46 \times 10^{-8}\ s^{-1}$ is shown in Fig. 2.35, Eq. (2-146) has been fitted with the data points, with the following parameters:

$$\varepsilon_{\varepsilon t} = 1.8 \times 10^{-3},\ \varepsilon_{\nu o} = 5 \times 10^{-3}$$

$$m = 0.5,\ \alpha = 4,\ \sigma_o = 13.74 \times 10^5\ Pa$$

It can be seen that the theory gives also very good fits for these types of tests.

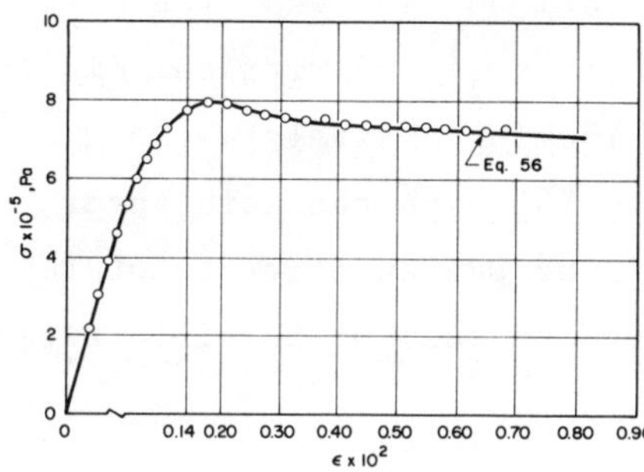


*Fig. 2.35. Measured (circles) and computed creep curves (full lines) for T1 snow ice under a constant compressive strain rate of $2.46 \times 10^{-8}$ $s^{-1}$. Measurements by Drouin and Michel (1971); adjusted curve with Eq. (2-146).*

## 2.2.7 - RECRYSTALLIZATION AND GLACIER FLOW

At the beginning of the creep process, we have seen that crystal rotation is responsible for boundary crack formation and boundary slip, accompanied by a decrease of the maximum angle of the more stressed crystals to $\alpha_m$.

Once this has happened, there is still a local stress concentration, much smaller and with more subtle effects which is continuously applied to a number of crystals and will favorize, either the nucleation and growth of new crystals from boundaries changes, or the growth of the best oriented crystals at the expense of the more stressed ones. This is syntectonic recrystallization as described by Steinemann (1958) and Lliboutry (1964).

We have observed that the phenomenon of recrystallization begins to accelerated at strains of the order of 3 to 8%. Although we are not proposing here a law for accelerated creep of the $\delta\nu$ type, it is interesting to see that the proposed basic mechanical model is entirely compatible with the flow model of recrystallized ice, as would happen in a glacier, for example.

The basic arrangement of recrystallized ice, compatible with the conditions given for the mechanical model, is shown on Fig. 2.36. In this model, the dislocations flow and pass over the boundaries of the crystals in horizontal and vertical directions which join each other at interlocking crystals where the basal plane makes an angle of 15° and 75° with the main shear plane. The flow model is in a continued permanent state of recrystallization and is the one giving the maximum possible rate of creep.

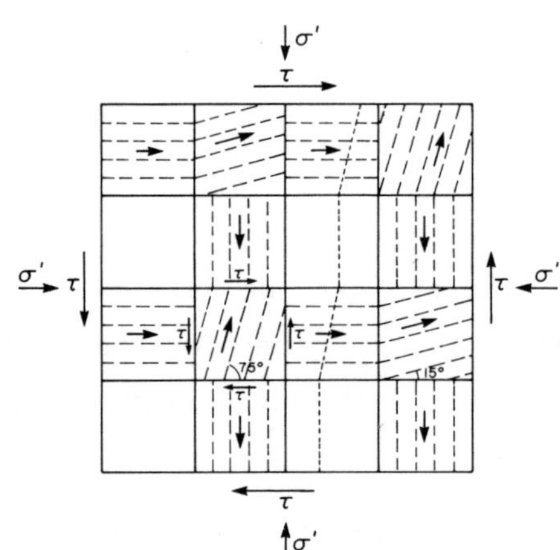


*Fig. 2.36. Ad hoc model of recrystallized glacier ice under a permanent state of creep. Dashed lines indicate the orientation of basal planes.*

There are thus horizontal and vertical slip bands alternating with rigid bands that have deformed only elastically. Under conditions of constant shear strain, all the stresses are balanced, those producing plastic flow act in either one or the other direction, the orthogonal stresses deform elastically the crystals.

This model shows four definite prefered orientations of the basal plane and one which is less clearly defined; but is in a general orientation normal to the plane of the others. Two orientations are parallel to the maximum shear stress, as observed by Rigsby (1960), two are in the direction of 15° and

$75^{\circ}$ with the others to give the maximum value of the CR ($\alpha$m) function. Fig. 2.37 shows the measured crystallographic orientations of recrystallized ice that was subjected to large deformations by torsion of a cylindrical core of glacier ice (Duval, 1976). A concentration of preferential oriented crystals at the top of a lozenge was found. The difference in position with the axes of the main shear stresses in these tests might possibly be explained by rotation of the sample, which produces a permanent but unsteady state at the level of the crystals.

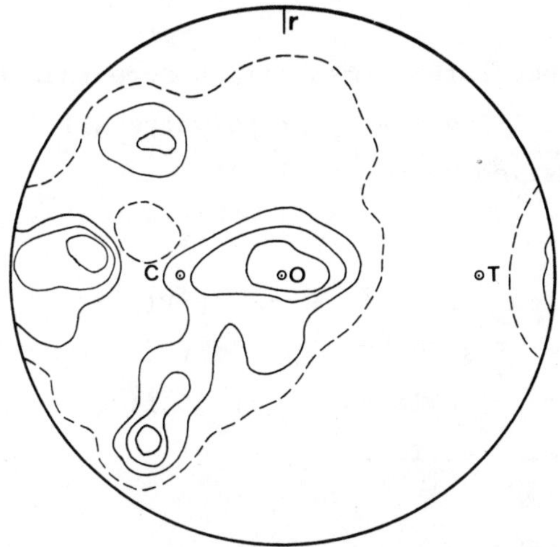


*Fig. 2.37. C-axis fabric of artificial glacier ice tested under combined torsion and compression until complete syntectonic recrystallization (Duval, 1976). The axes of principal stresses are C (compression), T (torsion) and O is the axis of torsion and compression. The contours correspond to densities of 2, 4 and 6% of 1% of the hemisphere area. The doted contours are at ½% area.*

With this model, the permanent creep rate of recrystallized ice can be computed from Eq. (2-104), with the average stress for each crystals, which make an angle of either $90^o$ or $15^o$ ($75^o$) with one main shear direction. This gives:

$$\dot{\gamma}_\nu = \frac{1}{2} C (\tau_o)^n = \frac{1}{2} C (\tau_{15})^n CR (15^o) \qquad (2\text{-}149)$$

The value of $\tau_o$ is then:

$$\tau_o = \frac{2 \tau_{90} + \tau_{15} + \tau_{75}}{4} \qquad (2\text{-}150)$$

For n = 3 this gives:

$$\dot{\gamma}_\nu = .257 C \tau_o^{\,n} \qquad (2\text{-}151)$$

The creep rate is many times more than that of unorganized polycrystalline ice. For polycrystalline ice, the steady creep is given by Eq. (2-127):

$$\dot{\gamma}_\nu = \frac{1}{2} C (\Omega')^n \tau_o^{\,n} \qquad (2\text{-}152)$$

The value of $\Omega'$ is shown in Fig. 2.23. For all angles $\alpha_m > 9^o$ the value of $\frac{1}{2} (\Omega')^n$ is smaller than .025 and the creep rate of unorganized ice is more than 10 times smaller than that of fully recrystallized ice.

## 2.2.8 - YIELD STRENGTH OF ICE

### 2.2.8.0 - INTRODUCTION

The yield strength of ice in compression is obtained, in the ductile range, by loading ice samples under a constant strain rate. This condition may be used in nature for many cases of interaction of ice with structures. A typical example of such a test is shown on Fig. 2.38 where the stress is given in function of strain for $S_2$ ice at $-10^oC$ (Paradis, 1978). At the beginning of loading the stress increases rapidly when the ice is deforming mainly elastically; then the stress attains a maximum at the yield point when the plastic deformations become dominant. From thereon the stress either goes to zero if the sample fails in a ductile manner or attains a constant lower level.

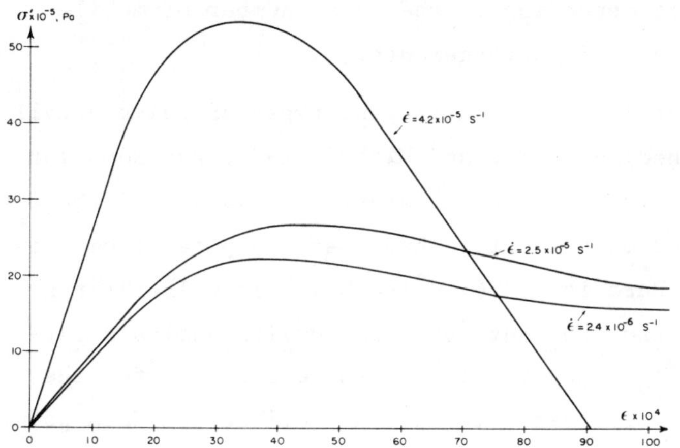


*Fig. 2.38. Creep curves of $S_2$ ice at constant strain rate, θ = -10°C.*

The yield strength obtained in this manner does not correspond to a state of steady creep, this state being obtained only at lower strain rates after the peak stress has occured and only when the number of mobile dislocations becomes a constant.

Under a constant load the steady creep correspond to the second stage of deformation of the ice.

### 2.2.8.1 - YIELD STRENGTH REPRESENTATION

The yield strength of ice in the ductile range can be represented in tension or compression at constant speed with Eqs (2-141) and (2-142):

$$\dot{\gamma}_y = \frac{1}{2}\, Co\, (1 + \alpha\, \gamma_y^m)\, (\Omega')^n\, \tau_y^{\,n} \qquad (2\text{-}153)$$

Changing to longitudinal deformation and taking into account the effect of temperature in Eq. (2-75) we obtained the well know Glen's equation for permanent creep of ice:

$$\dot{\varepsilon}_o = A\ \sigma_y^{\ n}\ \exp\ [-Qc/R\theta^*] \qquad (2\text{-}154)$$

when A is a constant depending on the total number of mobile dislocations and the grain arrangements.

Experimental data for different types of polycrystalline ice has been obtained by Carter and Michel (1971) and Ramseier (1976), for a wide range of strain rates. The values of the activation energy $Q_c$ for each ice type have been computed from this data by Michel and Paradis (1976). The best fit values for n have been computed from the data by a regression analysis between log $\dot{\varepsilon}$ and log $\sigma_y$. The results of these analysis for the different types of polycrystalline ice as well as the data points reduced to a temperature of $-10^{o}C$ are shown in Figs 2.39, 2.40, 2.41, and 2.42 with $\sigma$ in Pa.

<u>Snow ice - $T_1$</u> ($d^* = 1.5$ mm, $\rho' = 0.875 g\ cm^{-3}$)

The results have to be separated in two fits:

For $\dot{\varepsilon} < 5 \times 10^{-6}\ s^{-1}$

$A = 1.22 \times 10^{-15}$

$Q_c = 17{,}430$ cal/mole

$n = 3.713$

The coefficient of determination $r^2$ between log $\dot{\varepsilon}$ and log $\sigma_y$ is 0.962. The results are shown in Fig. 2.39.

For $\dot{\varepsilon} > 5 \times 10^{-6}\ s^{-1}$

$A = 1.02 \times 10^{-25}$

$Q_c = 17{,}430$

$n = 5.28$

The coefficient of determination $r^2$ is .956 and the results are also shown in Fig. 2.39. They compare fairly well with results obtained by Hawkes and Mellor (1972).

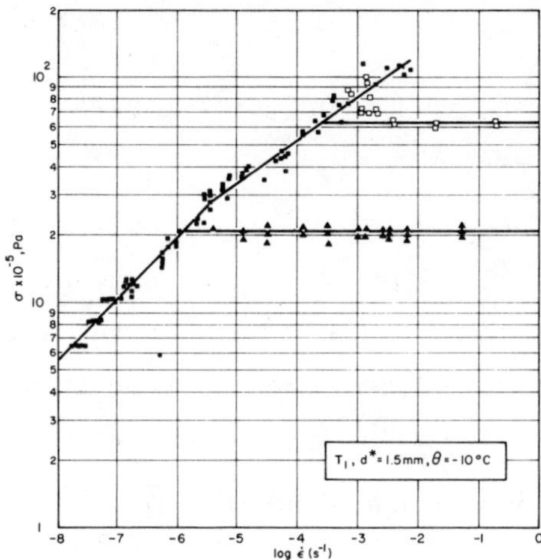


*Fig. 2.39. Strength diagram for snow ice ($T_1$) in compression and tension, in the ductile and brittle ranges ($\theta$ = -10°C).*

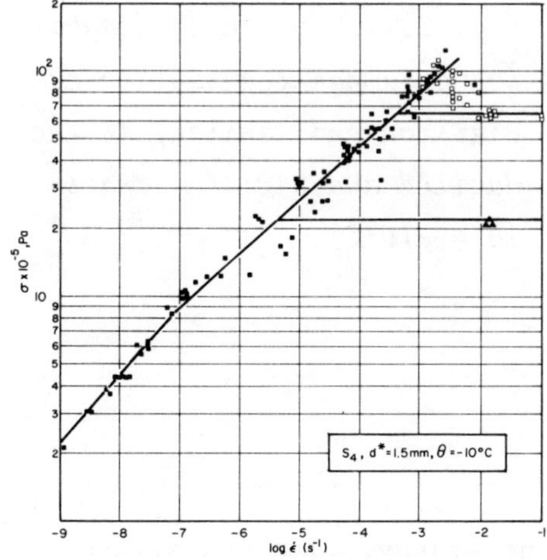


*Fig. 2.40. Strength diagram for frazil ice ($S_4$) in compression and tension, in the ductile and brittle ranges ($\theta$ = -10°C).*

Frazil ice - $S_4$ (d* = 1.5 mm)

Also two curves have been used. The best fit values are:

For $\dot{\varepsilon} < 1 \times 10^{-7}\ s^{-1}$

$A = 1.54 \times 10^{-10}$

$Q_c = 18770$

$n = 3.05$

$r^2 = .969$

For $\dot{\varepsilon} > 1 \times 10^{-7}\ s^{-1}$

$A = 8.06 \times 10^{-17}$

$Q = 18770$

$n = 4.156$

$r^2 = .899$

The results with these curves are shown in Fig. 2.40.

Columnar ice - $S_2$ (d* = 5 mm)

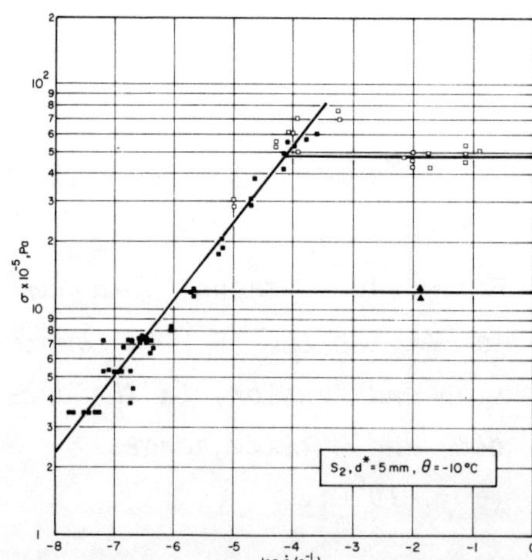


*Fig. 2.41. Strength diagram for columnar ice ($S_2$) in compression and tension, in the ductile and brittle ranges (θ = -10°C).*

Only one fit was found necessary. Its values are:

$A = 1.34 \times 10^{-6}$

$Q = 21,660$

$n = 2.96$

$r^2 = 0.9665$

The results are shown in Fig. 2.41. They compare fairly well with those obtained by Gold (1972).

Pseudo-monocrystalline - $S_1$ ice (H = 1.2 mm)

One fit only was used. The values are:

$A = 7.143 \times 10^{-7}$

$Q = 14090$

$n = 2.18$

$r^2 = 0.9917$

The results are shown in Fig. 2.42.

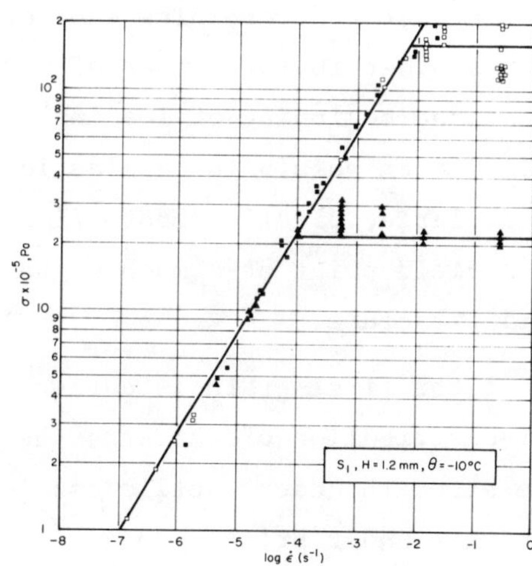


*Fig. 2.42. Strength diagram for pseudo-monocrystalline ice ($S_1$) with $\alpha = 45°$, in compression and tension, in the ductile and brittle ranges ($\theta = -10°C$).*

## 2.2.8.2 - TRANSITION FROM DUCTILE TO BRITTLE FAILURE

The strength of polycrystalline ice in the brittle range becomes independent of the strain rate at higher rates. Data points both for tension and compression at a temperature of $-10^{\circ}C$ are shown in Figs 2.39, 2.40, 2.41 and 2.42 so these figures represent complete diagrams of the strength of polycrystalline ice.

It can be seen from these diagrams that the maximum strength of ice in compression occurs at the limit of the ductile range and can be as high as 60% higher than the value of the corresponding brittle strength at higher rates. From thereon the strength decreases with increasing strain rates as has been observed before by many authors (Butkovitch, 1954; Korzhavin, 1955; Voytkovskii, 1960). The passage from ductile failure to brittle fracture in compression, is not a clear cut even. Tests indicate a general dispersion of strength values in the transition zone. For the same strain rate one sample might fail in a brittle manner at a much lower stress than another sample going up to the yield point. It seems to us that the main factor determining whether the ice will fail in one manner or another is the number of mobile dislocations in the ice crystals at the beginning of loading. If this number is small, the ice will deform mainly in an elastic manner and fail brittely. If it is large, plastic deformations will occur and the ice will show a yield point at a much higher load than that which causes brittle fracture.

In tension the failure of the first grain is catastrophic. If there is time for redistribution of the stresses by plastic deformations, the tensile strength under ductile conditions may somewhat be higher than that under brittle conditions.

## 2.2.8.3 - FAILURE CRITERION IN THE DUCTILE RANGE

Up to the value of the brittle strength in tension of polycrystalline ice, the yield strength is the same in tension as in compression. Because of the plastic flow of the material at yield point and the continuity of the failure surface under the main shear stresses, it may be expected that the Von Mises failure criterion might apply with minimum octahedral shear stress, so that the failure surface is cylindrical in space.

When the stresses are at a level situated between the elastic strength in tension and the yield strength in compression, there is unsymmetrical behaviour of the ice. It will flow in compression but fail in a brittle manner in tension, so that there is a mixed failure process depending on the type of stress which attains its limiting value first.

Finally for very quickly applied loads, there will be brittle fracture in tension and compression, as discussed in section 2.1.2.

## REFERENCES

Brill, R., Camp, P.R. (1961) - "Properties of ice" U.S. Army SIPRE, R.R. 68.

Brown, J.H. (1963) - "Elasticity and strength of sea ice" in Ice And Snow: processes, properties and applications, MIT Press, Cambridge, Mass.

Butkovitch, T.R. (1954) - "Ultimate strength of ice" U.S. Army Snow Ice and Permafrost Research Establishment. (USA SIPRE) Research Paper II.

Butkovitch, T.R. (1956) - "Strength studies of sea ice" USA SIPRE Research Report 20.

Carter, D., Michel, B. (1971) - "Lois et mécanismes de l'apparente rupture fragile de la glace de rivière et de lac" Rapport S-22, Dép. Génie Civil, Université Laval, Québec.

Dantl, G. (1968) - "Die elastichen Moduln von Eis-Einkristallen" Physik der kondersierten Materie. Bd. 7, Ht. 5, p. 390-397.

Drouin, M., Michel, B. (1972) - "La résistance en flexion de la glace du Saint-Laurent déterminée en nature" Rapport GCT-72-09-26, Dép. Génie Civil, Université Laval, Québec.

Duval, P. (1976) - "Lois de fluage transitoire ou permanent de la glace polycristalline pour divers états de contrainte" Ann. Geophys., T. 32, fasc, 4, p. 335-350.

Ewing, M., Grary, A.P., Thorne, A.M. (1934) - "Propagation of elastic waves in ice" Pt. 1, Phys. 5, p. 165-168.

Fletcher, N.H. (1970) - "The chemical physics of ice" Cambridge: Cambridge, University Press.

Friedel, J. (1956) - "Les dislocations" Gauthier-Villars Paris.

Fukuda, A., Higashi, A. (1969) - "X-ray diffraction topographical studies of the deformation behavior of ice single crystals" In Physics of ice, pp. 239-50.

Glen, J.W. (1953) - "The creep of polycrystalline ice" Proceedings of the Royal Society, Ser. A., Vol. 228, no. 1175, p. 519-538.

Glen, J.W. (1975) - "The mechanics of ice" USA CRREL Monograph II-C2b.

Gold, L.W. (1958) - "Some observations on the dependence of strain on stress for ice" Can. Journ. Phys., 36, p. 1263-1275.

Gold, L.W. (1960) - "The cracking activity of ice during creep" Can. Journal of Physics, Vol. 38, No. 9, p. 1137-1148.

Gold, L.W. (1970) - "Process of failure in ice" Canadian Geotechnical Journal, 7, p. 405-413.

Hawkes, I., Mellor, M. (1972) - "Deformation and fracture of ice under uniaxial stress" Journal of Glaciology, Vol. 11, No. 61, p. 103-131.

Hayes, C.E., Webb, W.W. (1965) - "Dislocations in ice" Science 147, 44-5.

Higashi, A. (1968) - "Mechanisms of plastic deformation in ice single crystals" In Physics of snow and ice, pp. 227-89.

Higashi, A., Koinuma, S., Mae, S. (1964) - "Plastic yielding in ice single crystals" Japanese Journal of Applied Physics, Vol. 3, No. 10, p. 610-616.

Johnston, W.G. (1962) - "Yield points and delay times in single crystals" J. Appl. Phys. Vol. 33, pp. 2716-30.

Jones, S.J., Glen, J.W. (1969) - "The mechanical properties of single crystals of pure ice" Journal of Glaciology, Vol. 8, No. 54, p. 463-473.

Ketcham, K.M., Hobbs, P.V. (1969) - Phil. Mag. 19, No. 162, p. 1161-1173.

Korzhavin, K.N. (1955) - "The effect of the speed of deformation on the ultimate strength of river ice subject to uniaxial compression" Trudy Novosibirskogo Institute Inzhenirov Zheleznodoro Transporta, Tome 11, p. 205-216.

Krausz, A.S. (1968) - "A rate theory of dislocation mobility" Acta Metallurgica. Vol. 16, No. 7, p. 897-902.

Kuroiwa, D. (1964) - "Internal friction in ice" Constr. Inst. Low. Temp. Sci., Hokkaido Univ. A18, 1-62.

Langleben, M.P., Pounder, E.R., (1963) - "Elastic parameters of sea ice" in 'Ice and Snow: processes, properties and applications' MIT Press, Cambridge, Mass.

Lliboutry, L. (1964) - "Traité de Glaciologie" Tome 1, Masson et Cie, Paris.

Mellor, M., Testa, R. (1969) - "Creep of ice under low stress" Journal of Glaciology, Vol. 8, No. 52, p. 147-153.

Michel, B. (1970) - "Ice pressure on engineering structures" USA CRREL Monograph III-Blb.

Michel, B. (1978) - "The strength of polycrystalline ice" Can. Journ. of Civil Eng., Vol. 5, No. 3, p. 285-300.

Michel, B. (1978) - "A mechanical model of creep of polycrystalline ice" Can. Geotechn. Journal., Vol. 15, No. 2, p. 155-171.

Michel, B., Ramseier, R. (1969) - "Structural and textural analysis of river ice" Rapport T-9, Dép. de Génie Civil, Université Laval, Québec.

Michel, B., Drouin, M. (1971) - "Les propriétés mécaniques à l'impact de la glace du Saint-Laurent" Rapport T-19, Dép. de Génie Civil, Université Laval, Québec.

Michel, B., Paradis, M. (1976) - "Analyse statistique du fluage secondaire de la glace de rivière et de lac" Rapport GCS-76-02, Dép. de Génie Civil, Université Laval, Québec.

Michel, B., Toussaint, N. (1978) - "Mechanisms and theory of indentation of ice plates" Journal of Glaciology, Vol. 19, No. 81, p. 285-301.

Nabarro, F.R.N. (1967) - "Steady state diffusional creep" Phil. Mag., 16, 231.

Nakaya, U. (1959) - "Visco-elastic properties of snow and ice in the Greenland ice" Cap. SIPRE Res. Rep., 46, 29 p.

Northwood, T.D. (1947) - "Some determination of the elastic properties of ice" Can. Journ. Res. A., 25, p. 88-95.

Nye, J.F. (1957) - "Physical properties of crystals: their representation by tensors and matrices" Oxford, Clarendon Press.

Orowan, E. (1934) - "Crystal Plasticity III; On the mechanism of the glide process" Physik 89, 614 and 635.

Paradis, M. (1978) - "Etude expérimentale et théorique du comportement mécanique de la glace sous différentes formes cristallines" Thèse M.Sc., Université Laval, Québec.

Poirier, J.P. (1976) - "Plasticité à haute température des solides cristallins" Eyrolles, Paris.

Ramseier, R. (1976) - "Growth and mechanical properties of river and lake ice" Thèse D.Sc., Université Laval, Québec,

Ready, D.W., Kingery, W.D. (1964) - "Plastic deformation of single crystal of ice" Acta Metallurgica, Vol. 12, No. 2, p. 171-178.

Rigsby, G.P. (1960) - "Crystal orientation in glacier and in experimentally deformed ice" Journal of Glaciology, Vol. 3, No. 27, p. 589-606.

Sadowsky, M.A., Sternberg, E. (1949) - "Stress concentration around a triaxial ellipsoidal cavity" J. Appl. Mech. p. 149-157.

Steineman, S. (1958) - "Experimentelle Unter sunchungen zur Plastizität vos Eis" Beiträge zur Geologie des Schweiz, Geotechnische Serie, Hydrologie, No. 10, 72 p.

Voytkovskii, K.F. (1960) - "Mechanical properties of ice" Moscow Izdatel'stvo Akademii Nauk. SSR. (U.S. Dept. of Commerce, Office of Technical Services, Translations ANS-T-R-391).

Wakahama, G. (1962) - "On the plastic deformation of ice" Low Temperature Science, ser. A, Vol. 20, p. 57-100.

Wakahama, G. (1964) - "Plastic deformation of polycrystalline ice" Low Temperature Science, ser. A, Vol. 22, p. 1-24.

Weeks, W.F., Anderson, D.L. (1958) - "An experimental study of the strength of young sea ice" Trans. Am. Geophys. Un. 39, p. 641-647.

Weeks, W.F., Assur, A. (1967) - "The mechanical properties of sea ice" USA CRREL Monograph II-C3.

Weertman, J. (1972) - "High temperature creep produces by dislocation motion" John. E. Dorn Memorial Symposium (Cleveland, Ohio, oct. 17).

# CHAPTER 3

# THE BEARING CAPACITY OF ICE

---

## 3.0 - INTRODUCTION

The ice covers of rivers and lakes have been utilized since very early times for transportation and water crossing purposes. In Canada, temporary ice roads were used across the St. Lawrence river ever since the beginning of the French colony. With the advent af the railroads at the end of the XIX century, tracks were set on the ice across the river in front of the City of Montreal. Similar use was made of the ice in the U.S.S.R. It is a well known fact, that during World War II, the ice covers of rivers and lakes were extensively used for transport of armies and military equipment. The successful defence of the City of Leningrad was greatly facilitated by the ice road on Lake Lagoda.

The recent explorations for natural resources in the Arctic regions have considerably increased the interest in the utilization of ice covers for winter roads, air landings and the support of material and equipment for construction and even for drilling purposes. It is presently an established practice to construct ice bridges for the transportation of heavy construction equipment and to use thickened ice platforms for setting offshore drilling oil rigs. The use of the natural ice cover is considered for the lay-out of pipelines and other construction pruposes.

The bearing capacity of ice covers is presently determined from field experience and simplified analysis. This has given a semi-empirical basis for the computation of the ice thickness which is required to support a given load for over-ice operation.

There is however a strong need to place design, construction and the use of load supporting ice structures on a more accurate engineering basis to ensure a better level of safety for men and equipment on the ice. Analytical techniques are well advanced to solve most of the problems dealing with the mechanics of plates on elastic foundations of the Winkler type; that is, those, like water, which exert only a normal reaction to the plate. The main drawback, up to now, has been the poor understanding of the properties and mechanical behavior of the various types of ice. We will thus treat this question with the help of the fundamental knowledge on the mechanical behavior of ice, that was discussed in Chapter 2, and with recent experimental results on the flexural strength of various types of ice.

## 3.1 - MECHANICAL BEHAVIOR OF FLOATING ICE

### 3.1.1 - FLEXURAL STRENGTH OF ICE BEAMS

The mechanical behavior of fresh water ice under bending stresses can be ascertain from the experimental works of Lafleur (1971) and Nadreau (1978) both working with beams loaded symmetrically at four points. This produces uniform pure bending in the region between the central loading points. Such a beam is shown in Fig. 3.1. The pure bending moment, in the central section of the beam, is given by:

$$Ma = Fd \tag{3-1}$$

where:

Ma - external applied moment

F - force applied at two points

d - distance between the points

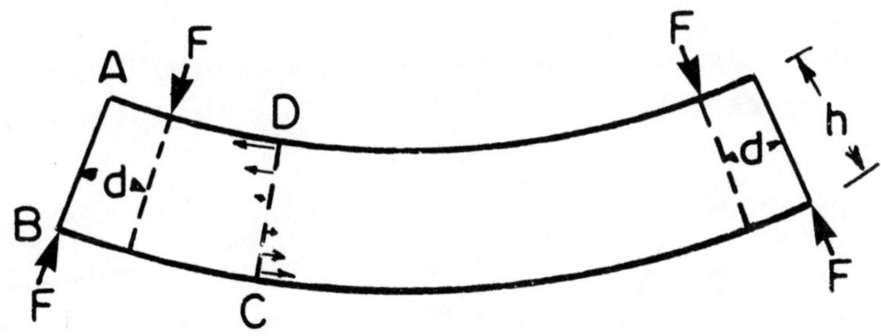


*Fig. 3.1. Symmetrical four-point loading.*

For small deflections of the beam compared to its length, the deflection is related to the radius of curvature ρ, at a given point, by:

$$\frac{d^2z}{dx^2} = \frac{1}{\rho} \tag{3-2}$$

One of the major findings of Nadreau is that the strains within a cross-section of this type of beam keep a linear distribution, at all times, during ductile deformation of the beam. This enables us to state the fundamental relationship that holds for ice beams and ice plates, which is an immediate consequence of the geometry, as can be seen in Fig. 3.2.

$$\varepsilon = \Delta\ell/\ell = \frac{z - z_o}{\rho} \tag{3-3}$$

where:

$\varepsilon$ - is the longitudinal strain in tension or compression

$z$ - is the vertical distance in the beam, counted downwards from the top of the beam

$z_o$ - is the position of the neutral axis, from the top.

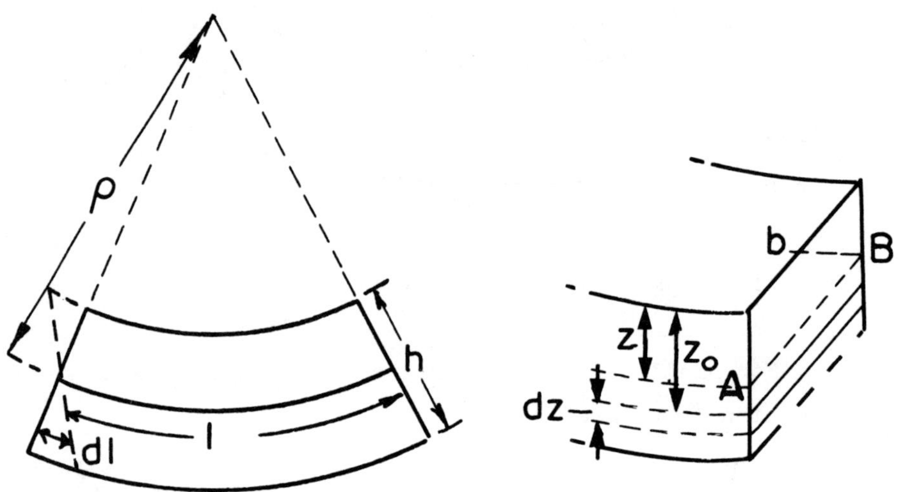


*Fig. 3.2. Bending of a beam.*

In particular, at the extreme layer, where z = h, we have, in tension:

$$\varepsilon_o = \frac{h - z_o}{\rho} \tag{3-4}$$

This longitudinal strain $\varepsilon_o$, which is a function of x and time t is then related to the deflection w of the beam by Eq. (3-2).

$$\frac{d^2w}{dx^2} = \frac{\varepsilon_o\ (x,\ t)}{(h - z_o)} \tag{3-5}$$

The resisting moment $M_r$ of a beam of uniform width b is given by:

$$M_r = b\int_o^h \sigma\ (z)\ zdz \tag{3-6}$$

And the position of the neutral axis, can be obtained from:

$$\int_o^{z_o} \sigma\ (z)\ dz = \int_{z_o}^h \sigma\ (z)\ dz \tag{3-7}$$

For a pure elastic stress distribution,

$$\varepsilon = \sigma/E \tag{3-8}$$

$$z_o = h/2 \tag{3-9}$$

$$M_r = \frac{\sigma_f\ h^2\ b}{6} \tag{3-10}$$

where $\sigma_f$ is the stress on the bottom layer, in tension, b is the width of the beam. In the case of a non-linear stress distribution, if we write:

$$\zeta = z/h \tag{3-11}$$

$$\overline{\sigma} = \sigma/\sigma_f \tag{3-12}$$

where $\sigma = \sigma_f$ at the lower layer for the corresponding elastic stress distribution, so that $\zeta = 1$ for $z = 1$, then we may write:

$$M_r = \frac{\lambda\ \sigma_f\ h^2\ b}{6} \tag{3-13}$$

where $\lambda$ depends only on the stress distribution and is given by:

$$\lambda = 6\int_o^1 \bar{\sigma}\ (\zeta)\ \zeta\ d\ \zeta \tag{3-14}$$

The resisting moment, in a section, must be equal at all times to the applied moment Ma, so that Eq. (3-13) gives:

$$\sigma_f = \frac{Ma\ (x)}{\lambda\ S} \tag{3-15}$$

with

$$S = bh^2/6 \tag{3-16}$$

The stress variation at a given section of a beam is similar to that of uniaxial loading where the stress will be progressively reduced because of an increase in the coefficient $\lambda$ caused by plastification of the section of the beam.

For uniaxial loading under tension, the relationship between stress and strain under a constant load is represented by the equivalent of Eq. (2-99):

$$\varepsilon_o = \varepsilon_\varepsilon + \varepsilon_{\varepsilon\alpha} + \varepsilon_\nu \tag{2-99 b}$$

The instantaneous elastic deformation is given, when $\lambda = 1$, for pure elastic stress distribution, by:

$$\varepsilon_\varepsilon = \sigma_f/E = Ma/ES \tag{3-17}$$

The permanent creep $\varepsilon\nu$ is given by the integration of Eq. (2-155):

$$\varepsilon_\nu = A\sigma_f{}^n \exp\left[-Qc/R\theta^*\right]\ t \tag{3-18}$$

with

$$\sigma_f = Ma/\lambda S \tag{3-19}$$

where $\lambda$ is given here by a state of full plastification of the section.

The transitory creep $\varepsilon_{\varepsilon\alpha}$ will not be explicitated here for this particular condition.

With these expressions for the deformation $\varepsilon_o$, Eq. (3-5) becomes:

$$(h - zo) \frac{d^2w}{dx^2} = \frac{Ma(x)}{ES} + \varepsilon_{\varepsilon\alpha} (x, t)$$

$$+ A [Ma(x)/\lambda S]^n \exp [-Qc/R\theta^*]t \quad (3-20)$$

If this equation is integrated for the deflection for a given loading condition, it is obvious that the deflection will follow a time dependent law that is similar to that of an uniaxial test under a constant load.

Fig. 3.3 shows the central deflection for two beams made of $S_2$ ice and loaded at four points at $-3^oC$ (Nadreau, 1978). For beam A, with a load of 66 kg, there is an instantaneous deflection, followed by transitory creep of short duration of the $\varepsilon\alpha$ type and permanent creep. This is the same as for an uniaxial test under a constant load, as can be observed in Fig. 2.18, and which is expressed by Eq. (3-20). If the load is removed at time t, there is a visco-elastic recovery as for the uniaxial test. For beam B, the load is higher at 99 kg so that the beam failed during transitory creep when a first crack was formed in the lower extreme layer. This indicates the importance of delayed elasticity when the elastic component of the stress is delayed up to that required for brittle fracture of a grain. This fracture is catastrophic for the whole ice beam, exactly as in the case of uniaxial tensile testing.

Fig. 3.4 shows the central deflection of three beams of $S_2$ ice at $-10^oC$, loaded in the same manner but where the loads were progressively increased. On beam A the initial load, 99 kg, was increased by 10, 11 and 12 kg until fracture. For beam B the initial load 99 kg was increased by 10 and 11 kg and on beam C the initial load was 109 kg, increased by 11 kg. In all cases, transitory creep was observed at the beginning of loading followed by pure permanent creep. With the increases in loading, there was a very short stage of transitory creep followed immediately by permanent creep. In all cases, fracture occured without accelerated creep. We believe that this is because none of the tertiary creep mechanisms intervene in bending, as for the case of uniaxial testing in tension. As κ creep does not exist on the tensile side of the beam and δ creep may exist only for very long periods of loading, the appearance of the first crack leads to catastrophic fracture of the ice beams.

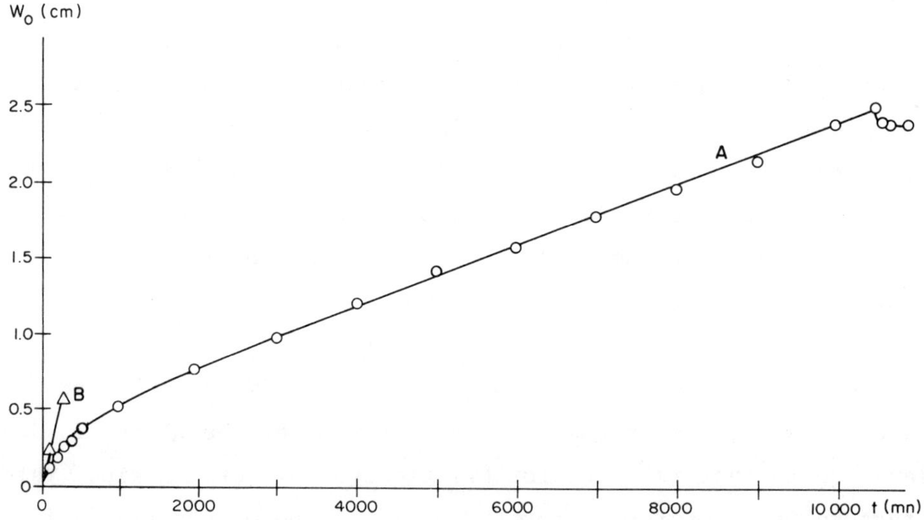


*Fig. 3.3. Results of tests giving central deflection of $S_2$ ice beams at -3°C (Nadreau, 1978).*

The value of the plastification coefficient λ has been computed in various cases. For full plastification of a homogeneous ice sheet at constant temperature, we have, because of the hypothesis of linear strain distribution:

$$\bar{\varepsilon} = \varepsilon/\varepsilon_o = \frac{\zeta - \zeta_o}{1 - \zeta_o} \tag{3-21}$$

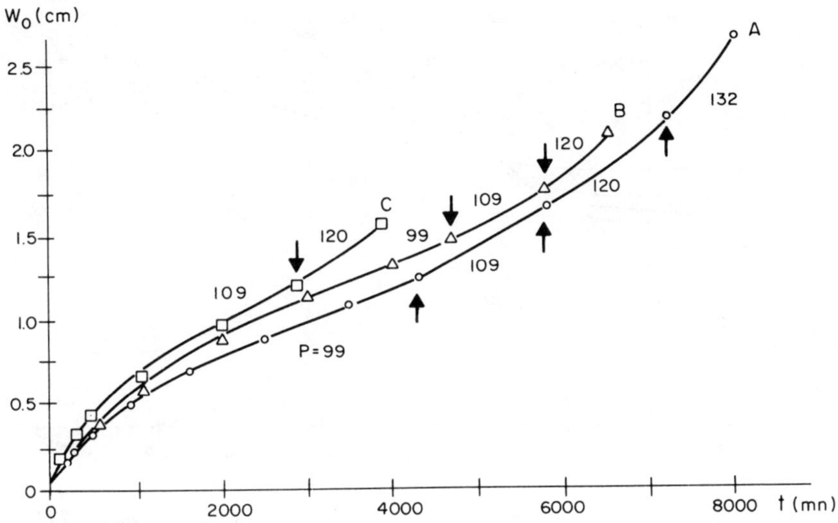


*Fig. 3.4. Results of tests giving central deflection of $S_2$ ice beams at -10°C (Nadreau, 1978).*

The value of $\bar{\sigma}$ can be obtained from Eq. (3-18):

$$\bar{\sigma} = \sigma/\sigma_f = \left(\frac{\zeta - \zeta_o}{1 - \zeta_o}\right)^{1/n} \tag{3-22}$$

With these values in integral (3-14), we obtain:

$$\zeta_o = z_o/h = 0.5$$

$$\lambda = 3n/(2n + 1) \tag{3-23}$$

The stress distribution for this case is shown in Fig. 3.5. For the usual value of n = 3, the resisting moment is increased by a value of λ=1.3, if use is made of the same value of

$\sigma_f$ in the bottom layer as in the case of an elastic stress distribution.

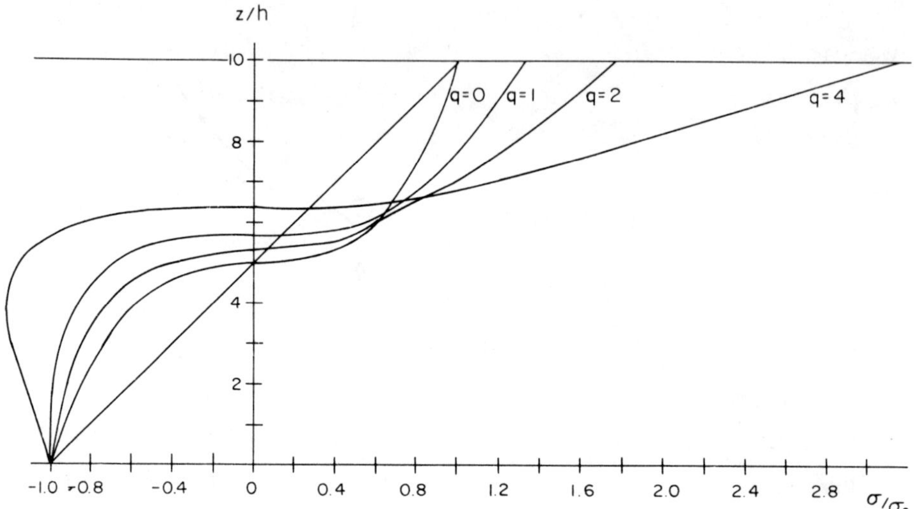


*Fig. 3.5. Stress distribution $\sigma/\sigma_o$ in an ice beam with ductile behavior and various characteristics q of the temperature distribution.*

If we now examine the more usual case of plastification in a natural ice cover, where there is a linear temperature distribution from $-\theta_s$ in $^oC$ at the surface to $0^oC$ at the bottom of the ice sheet, we then have:

$$\varepsilon = A\,\sigma^n\, t\, \exp\left[-Qc/R\,(273 + \theta\,)\right] \qquad (3\text{-}24)$$

with

$$\theta = \theta_s\,(1 - \zeta) \qquad (3\text{-}25)$$

Because $\theta_s$ is always small compare to 273 in Eq. (3-24) we can write:

$$-Qc/R\,(273 + \theta) \simeq -q\,(\theta/\theta_s) \qquad (3\text{-}26)$$

with $$q = Qc\,|\theta_s|/R\,(273)^2 \tag{3-27}$$

The parameter q will vary from 0 to a maximum of around 4 for very cold ice at the surface. The value of q = 1 might be considered, more or less, as a normal condition of temperature distribution in the ice cover. With this parameter, the stress function in Eqs (3-24) and (3-26) becomes with Eq. (3-21):

$$\sigma. = \left\{\left[\frac{\zeta - \zeta_o}{1 - \zeta_o}\right] \exp\,[q\,(1 - \zeta)]\right\}^{1/n} \tag{3-28}$$

With this value in integral (3-14), the value of λ has been computed with Simpson's rule for values of q equal to 1, 2 and 4 and a value of n = 3. The results are given in Table 3.1. The stress distribution which corresponds to these conditions is shown in Fig. 3.5. For the same tensile stress at the lower fiber $\sigma_f$ it can be seen that the effect of plastification is to increase considerably the resisting moment in the ice. This increase is more and more pronounced as the temperature at the surface of the ice gets colder.

| q | $z_o/h$ | λ |
|---|---|---|
| 0 | 0.500 | 1.29 |
| 1 | 0.464 | 1.57 |
| 2 | 0.430 | 1.92 |
| 4 | 0.365 | 2.90 |

Table 3.1 - Plastic resisting moment of an ice section with variable temperature of the surface, for n = 3.

The flexural strength of an ice beam, may be expected to be obtained from the uniaxial tensile strength at the bottom layer of this beam. For conditions of loading under a constant rate of deformation of the beam, corresponding to constant strain rates in the bottom layer, we should expect the strength to depend on the uniaxial tensile strength for this strain rate and the value of the coefficient $\lambda$ in Eq. (3-15). Thus Eq. (3-15) should give the flexural strength of an ice beam:

$$\sigma_f = \sigma_o(\dot{\varepsilon}) = Mmax/\lambda(\dot{\varepsilon})S \qquad (3\text{-}15)$$

where $\sigma_o$ is the uniaxial tensile strength of the ice obtained from (2-50) for fresh water ice:

$$\sigma_o = 7.94 \times 10^4 \sqrt{(1 - 0.9 \times 10^{-3}\theta)/d^*} \qquad (2\text{-}50)$$

for brittle crack formation and:

$$\sigma_o = (\dot{\varepsilon}/A \exp[-Qc/R\theta^*])^{1/n} \qquad (2\text{-}155)$$

in the ductile range.

Figs (3.6 a and b) show the maximum resisting moment $M_{max}$ on beams of dense snow ice, with four points loading, as a function of strain rates (Lafleur, 1970) at temperatures of $-5^\circ C$ and $-25^\circ C$. The strain rates were computed with the formula for elastic deformation at the bottom of such beams, as a function of the rate of deflection at the beams'center.

From this figure it can be seen that at lower rates of deformation, the resisting moment increases with the strain rate in a manner similar to that of uniaxial testing and given by Eqs (3-15) and (2-155). For complete ductile behavior of the beam it can be expected that the value of $\lambda$ in (3-15) will be given by a state of full plastification of the section. In this range the beams do not fail but deform continuously.

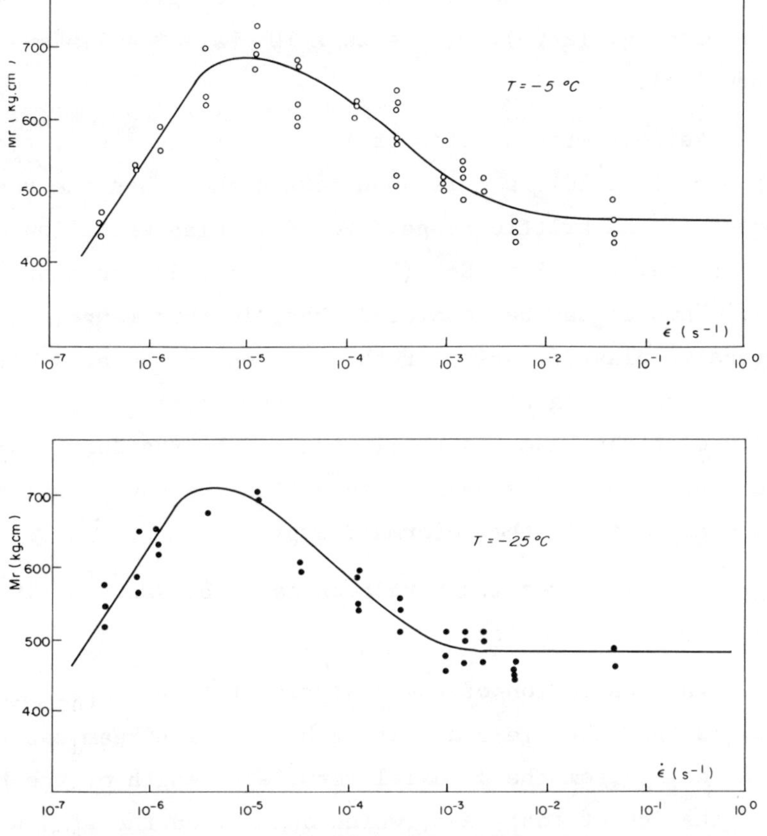


*Fig. 3.6 a, b. Resisting moment at failure of beams of snow ice at -5°C and -25°C, for different strain rates.*

The maximum resisting moment comes to a maximum value at a strain rate of around $10^{-5}$ $s^{-1}$, but which varies somewhat with the temperature; it then decreases progressively to a constant value for strain rates higher than $10^{-3}$ $s^{-1}$.

If the stress $\sigma_f$ is computed from the elastic theory with $\lambda = 1$ in Eq. (3-15); then it was found exactly that:

$$\sigma_f = \sigma_o$$

This corresponds to the value of $\sigma_o$ given for brittle fracture with Eq. (2-50). ($\sigma_o = 20 \times 10^5$ Pa, d* = 1.5 mm, for this snow ice).

Between strain rates of $10^{-5}$ $s^{-1}$ to $10^{-3}$ $s^{-1}$, the resisting moment at $10^{-5}$ $s^{-1}$ is about 40% higher than the resisting moment in the brittle range. For full plastification of the section the value of $\lambda$ in Eq. (3-15) is $\lambda = 1.33$ for snow ice ($n = 4$). Thus it can be concluded that, in this range, there is a progressive plastification of the section of the beam with the value of $\lambda$ increasing progressively as the strain rate decreases. The failure of the beam, is still obtained by the formation of a brittle crack that comes unexpectedly with the contribution of delayed elasticity in the deformation process.

Identical results have been found by Nadreau (1978) working on beams of $S_2$ ice.

The conclusion of the analysis of these experimental results, is that the flexural strength of an ice beam can be obtained directly from the uniaxial tensile strength of the bottom layer of the ice of that beam, which depends on the strain rate in the ductile range and on the value of the coefficient $\lambda$ of plastification of the ice section.

Comparing the values of the critical strain rates with those obtained for different types of ice, we can say that fresh water ice will behave approximately in the following manner, under flexural stresses:

$\dot{\varepsilon} > 10^{-3}$ $s^{-1}$, full elastic behavior, with brittle fracture;

$10^{-3}$ $s^{-1} > \dot{\varepsilon} > 10^{-5}$ $s^{-1}$, partial plastification of the section, with brittle fracture;

$\dot{\varepsilon} < 10^{-5}\ s^{-1}$, full plastification of the section, with permanent creep.

In general, under any condition of loading, the strength of ice beams is computed with the formula for elastic behavior. We may thus speak of an apparent flexural strength $\sigma_{fa}$ given by Eq. (3-15):

$$\sigma_{fa} = \frac{Ma}{S} = \sigma_o\ \lambda\ (\dot{\varepsilon}) \qquad (3\text{-}29)$$

Field tests conducted by Drouin and Michel (1972), on large beams, have shown conclusively that the variation of the apparent flexural strength, in the same ice cover, is due to the loading rates and thus to the variation of the coefficient $\lambda$ in function of the state of plastification, under load, of the ice cover.

### 3.1.2 - BEHAVIOR OF ICE COVERS UNDER LOAD

It can be expected that an ice cover under load will behave in a manner similar to that of an ice beam, and this is much so.

Fig. 3.7 shows the time dependent deflection of natural sea ice 1.33 m thick, loaded in a static manner, with a 7 m diameter pool filled with water (Kingery, 1962). The load was increased, by steps, at fixed intervals of time, over a period of 380 hours. The similarity between the creep curve with a natural cover and that of a beam, shown in Figs 3.3 and 3.4, is striking; except that the failure of the cover was not attained. After each increase in load, there is primary creep for a few hours and then, the rate of deflection becomes constant. This case of loading and deformation corresponds to very low strain rates where, as in a beam, there is full plastification of the ice sections and the establishment of a permanent state of creep.

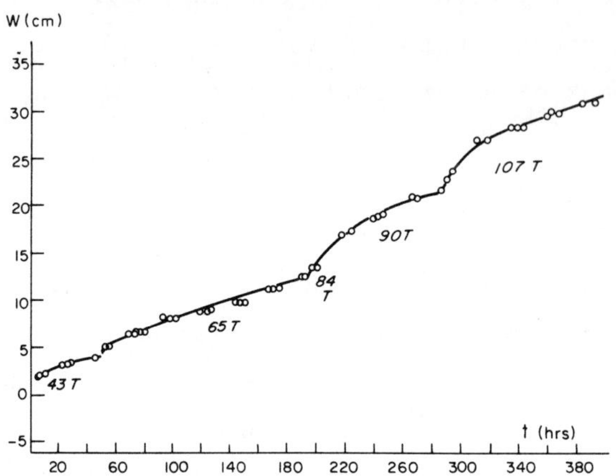


*Fig. 3.7. Deflection at the center of a 1.33 m thick natural sea ice sheet loaded with a 7 m diameter pool (Kingery, 1962).*

The mechanisms of failure of ice covers loaded dynamically, at a fixed point, have been well described in the literature (Lagutin, 1946; Assur, 1956; Gold, 1960). Under a load which is increasing at a fast rate, the first stage of failure is the formation of brittle radial cracks at the bottom of the sheet associated with the maximum transverse bending moment under the load. These cracks do not cause immediate failure if the load is kept constant at this time. A further increase in load causes the existing cracks to extend radially and vertically from under the ice sheet to the top. New radial cracks are also formed. It takes an increase in load more than twice that which produced the first cracks, to form a circumferential crack at a sizable distance from the load, followed by final collapse of the broken dish under the load.

In certain cases, and mainly during spring time, when the ice is rotten inside, the final stage might be preceded by the formation of a vertical circumferential shear plane immediately adjacent to the loaded surface.

Codbout (1967) tested circular ice plates welded at the periphery of a tank with a uniformly distributed pressure over the whole plate. He found that, in all cases, there was little and only elastic deformation of the plate up to the load that would form the radial cracks (total of 4 to 12) at the center of the plate. These cracks would appear only at the surface on the tensile side. Keeping the load pressure constant, from thereon, he observed that the plate would deform plastically, at a more or less uniform rate, and that the cracks would multiply and open progressively until they would cross entirely the ice plate. At that time the pressure was lost, and the test stopped.

Unfortunately, he did not study the structure of the ice crystals on the tensile side of the ice sheet, so it is difficult to discuss his results. He found for nine tests, an apparent flexural strength, computed from the elastic theory, of:

$$\sigma_{fa} = (15.1 \pm 5.7) \times 10^5 \text{ Pa}$$

If this value is equal to $\sigma_o$ for the quick loading rates that produced first cracks, then Eq. (2-50) gives en equivalent grain diameter d = 2.8 mm, which seems quite plausible for the condition of formation of a top layer of $S_2$ ice at -30°C which occured in these tests.

## 3.1.3 - CRITERIA FOR THE BEARING CAPACITY OF ICE COVERS

From the behavior of beams of ice under load it is evident that for quick loading rates the ice will behave like a perfect elastic material but for low rates of loading and static loading conditions, the ice sections will plasticize and the ice cover will slowly creep. Michel *et al.*, (1974) have estimated that when the speed of a vehicle is over 3 m/sec (10 km/h) the ice behaves elastically under the moving load. From 3 m/sec to 3 cm/sec there is progressive plastification of the ice and below the lower value, full plastic behavior.

For design purposes it is thus necessary to distinguish between the extremes of dynamic loading conditions when the ice behaves elastically and static loading conditions with plastification of the ice cover.

## Dynamic loading conditions

For vehicles moving over the ice, the speed is certainly over the limit which characterizes the elastic behavior of the material, and the bearing capacity should be computed accordingly. This view has been held by many early authors in the U.S.S.R. and strongly supported by Gold (1977).

The criterion for the bearing capacity of an elastic ice cover has been taken, in the literature, all the way from the load producing the first radial cracks (Michel *et al.*, 1974), to loads where there is some cracking activity but a low risk of breakthrough (Gold, 1971) and finally to the lower experimental limit of breakthrough loads (Kerr, 1975). As the breakthrough load, under elastic conditions, is at least twice that which produces the first crack, there is a sizable difference in the value of the allowable loads on the ice.

The choice of the criterion of first crack formation

is based on the following reasons:

a) If a moving load producing radial cracks has to stop over them, the cracks will have time to extend through the whole thickness of the ice cover.

b) If the load moves again on one of the thin wedges it has produced, it will breakthrough.

c) There is a factor of safety of around two, in normal operation, before breakthrough.

d) There is a factor of safety of barely one if the load crosses a wet crack that would have formed unexpectedly and be unseen.

e) The condition of first crack formation, can be easily verified in the field.

To determine the load that produces a first crack, the failure criterion which has been most widely used in the past is (Kerr, 1975):

$$\sigma_{fa} = \sigma_o \tag{3-30}$$

Which is the equivalent to Eq. (3-29) when $\lambda = 1$ for complete elastic behavior. There is ample evidence from the discussion in Sections 3.1.2 and 3.2.3 of this chapter, that this criterion is valid for the cracking of ice covers under a moving load, and for both tangential and radial stresses of the same sign, is equivalent to the Tresca failure criterion.

<u>Static loading conditions</u>

Under static loading conditions, should be considered both the cases of a vehicle which has moved on the ice and is allowed to park for some time and the gradual storage of stationary loads on the ice. In both cases, the ice sheet will have the time to plasticized completely and the full plastic resisting moment will be developped. Thus the loads that can be added on the ice under static conditions are much higher than those which could be safely transported individually, to the storage location.

Under static loads, the ice will deform in a ductile manner as shown on Fig. 3.7. After elastic deformation, we will have primary creep and then secondary creep.

One criterion for the determination of the bearing capacity for stationary loads, which has received large support (Panfilov, 1972; Meneley, 1974; Frederking and Gold, 1976) has been to limit the maximum deflection to that of the freeboard.

The main structural reason to use this criterion is the following. If the freeboard is exceeded, water can flood the surface through old thermal cracks or other kind of openings. This flood water must be considered as an added load, effectively neutralizing part of the buoyant force over the area it covers. This may increase the creep rate leading to collapse of the ice cover. Water on the surface will also increase the temperature of the top part of the ice cover and this may reduce considerably the plastic resisting moment as can be seen from Table 3.1.

There are also other practical reasons for limiting vertical deflection to this value. The flood water may interfere with operations, damage or freeze in the material and equipment and have a adverse psychological effect on the user.

This is why there is much to commend the use of this criterion to determine the safe bearing capacity of ice covers under stationary loads.

## 3.2 - DYNAMIC BEHAVIOR OF ICE COVERS

### 3.2.1 - EQUATIONS OF PLATES ON ELASTIC FOUNDATIONS

The basic equation describing the deflection, under vertical loading, of an infinite plate on an elastic foundation, when there is symmetrical bending from a center, is given by Timoshenko and Woinowsky-Krieger (1959). If we consider in Fig. 3.8 the deflection of the surface in a section along the radial direction, the curvature $\rho_r$ at any point along this line is given by:

$$\frac{1}{\rho_r} = - \frac{d^2w}{dr^2} = \frac{d\phi}{dr} \tag{3-31}$$

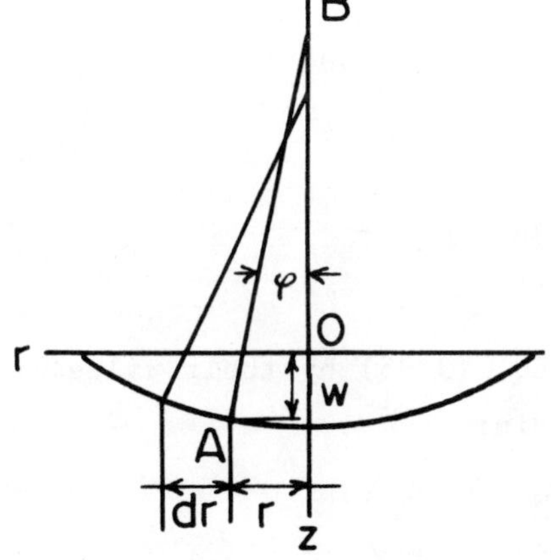


*Fig. 3.8. Geometric characteristics of flexure of a plate.*

The second principal curvature $\rho_t$ will pass in a section containing AB and the normal to the rz plane. From the figure:

$$\frac{1}{\rho_t} = - \frac{1}{r}\frac{dw}{dr} = \frac{\phi}{r} \tag{3-32}$$

A relation similar to Eq. (3-3) for the strains $\varepsilon$ in the cross-sections of the plate gives:

$$\varepsilon_r = z/\rho_r \qquad \varepsilon_t = z/\rho_t \tag{3-33}$$

With Hooke's law and taking into account the two-dimensionnal aspect of the deformation, the corresponding stresses are:

$$\sigma_r = \frac{E\ z}{(1 - \nu^2)}\left[\frac{1}{\rho_r} + \nu\,\frac{1}{\rho_t}\right] \qquad \sigma_t = \frac{E\ z}{(1 - \nu^2)}\left[\frac{1}{\rho_t} + \nu\,\frac{1}{\rho_r}\right] \tag{3-34}$$

The radial and tangential moments $M_r$ and $M_t$ are related to the stresses by:

$$\int_{-h/2}^{+h/2} \sigma_r\ z\ r\ d\theta dz = M_r\ rd\theta \qquad \int_{-h/2}^{+h/2} \sigma_t\ z\ dr\ dz = M_t\ dr \tag{3-35}$$

Replacing $\sigma_r$ and $\sigma_t$ in Eq. (3-35) by their value in (3-34), (3-31) and (3-32), we obtain:

$$Mr = -D\left(\frac{d^2w}{dr^2} + \frac{\nu}{r}\frac{dw}{dr}\right) \qquad Mt = -D\left(\frac{1}{r}\frac{dw}{dr} + \nu\,\frac{d^2w}{dr^2}\right) \tag{3-36}$$

where D is the flexural rigidity of the plate:

$$D = E\ h^3/12\ (1 - \nu^2) \tag{3-37}$$

Eq. (3-36) has one variable w that can be determined by considering an element of the plate abcd, as shown on Fig. 3.9.

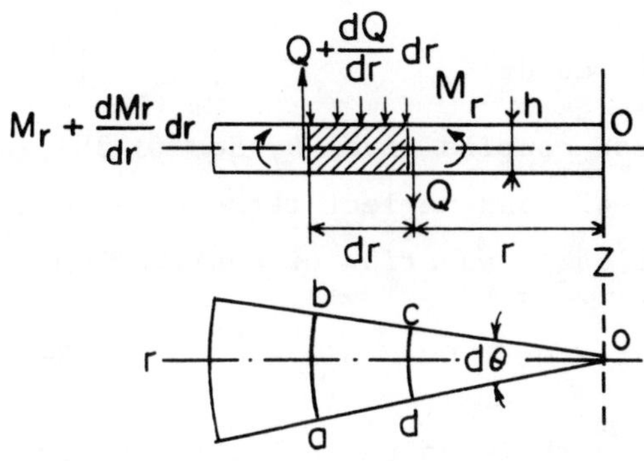


*Fig. 3.9. Forces and moments on an element of a plate.*

The couple acting on cd is:

$$M_r \, r d\theta \tag{3-38}$$

The corresponding couple on ab is:

$$(Mr + \frac{d\ Mr}{dr} dr)\ (r + dr)\ d\theta \tag{3-39}$$

The couples acting on sides ad and dc are both equal to $M_t dr$, but they give a resulting couple in the plane r o z:

$$M_t \ dr\ d\theta \tag{3-40}$$

Because of symmetry, the shear forces disappear in circumferential sections, but they persist in the cylindrical section cd and ab. If Q is the shear force per unit length in a cylindrical section of radius r, the total shear force on side cd is Q r dθ and on ab:

$$\left(Q + \frac{dQ}{dr}\, dr\right) (r + r\ dr)\ d\theta \qquad (3\text{-}41)$$

Neglecting the differences of smaller order of magnitude between the shear forces on both sides, the resulting couple is:

$$Qr\ d\theta\ dr \qquad (3\text{-}42)$$

If we add up all these couples given by Eqs (3-38), (3-39), (3-40), (3-41) and (3-42) and neglect terms of second order of smallness, we get the basic equation of equilibrium:

$$Mr + \frac{d\ Mr}{dr}\, r - Mt + Qr = 0 \qquad (3\text{-}43)$$

Replacing Mr and Mt by their value in Eq. (3-36) we get:

$$\frac{d^3w}{dr^3} + \frac{1}{r}\frac{d^2w}{dr^2} - \frac{1}{r^2}\frac{dw}{dr} = \frac{Q}{D} \qquad (3\text{-}44)$$

It is often preferable to represent the external shear force Q by a load intensity function q. To do so, both sides of Eq. (3-44) are multiplied by $2\pi r$. With,

$$2\pi r\ Q = \int_o^r 2\pi\ r\ q\ dr \qquad (3\text{-}45)$$

Eq. (3-44) becomes:

$$\left(\frac{d^2}{dr^2} + \frac{1}{r}\frac{d}{dr}\right)\left(\frac{d^2w}{dr^2} + \frac{1}{r}\frac{dw}{dr}\right) = \frac{q}{D} \qquad (3\text{-}46)$$

With the Laplacian operator:

$$\Delta = \left(\frac{d^2}{dr^2} + \frac{1}{r}\frac{d}{dr}\right)$$

The equation is usually written:

$$\Delta\Delta w = q/D \tag{3-47}$$

It is the biharmonic equation. With the Hamiltonian operator $\nabla$ it is also written:

$$\nabla^4 w = q/D \tag{3-48}$$

In rectangular coordinates, it simply becomes:

$$\frac{\partial^4 w}{\partial x^4} + 2\frac{\partial^4 w}{\partial x^2 \partial y^2} + \frac{\partial^4 w}{\partial y^4} = q/D \tag{3-49}$$

In the case of a plate on an elastic foundation, the reaction of the foundation is kw, k being the foundation's modulus which is, for water, $k = 1000\ kg/m^3$. The reaction of the foundation must then be substracted from the vertical load intensity q, so that the basic equation of a floating ice plate is:

$$\Delta\Delta w = \frac{q - kw}{D} \tag{3-50}$$

Using the notation:

$$\ell = \sqrt[4]{D/k} \qquad q' = q/k \tag{3-51}$$

and

$$\omega = w/\ell \qquad x = r/\ell \tag{3-52}$$

where $\ell$ is a characteristic length, often called the action radius of the floating plate. The basic Eq. (3-50) becomes with this non-dimensional notation:

$$\Delta\Delta\omega + \omega = q' \tag{3-53}$$

where $\Delta$ is here, the operator:

$$\Delta = \frac{d^2}{dx^2} + \frac{1}{x}\frac{d}{dx}$$

Biharmonic Eq. (3-53) is a linear differential equation of the fourth order, whose general solution may be written in the form:

$$\omega = A_1 F_1 (x) + A_2 F_2 (x) + A_3 F_3 (x) + A_4 F_4 (x) \quad (3\text{-}54)$$

where $A_1 \ldots A_4$ are four constants of integration and $F_1 \ldots F_4$, four independent solutions of Eq. (3-53).

In the case of a uniformly loaded plate, the value of q is a constant, within given boundary conditions, that can be eliminated by a change in variable $\omega$ in Eq. (3-53). The general solution can then be obtained from Bessel's functions. If we introduce the new variable $\xi = x\sqrt{i}$, we then get the equation:

$$\Delta'\Delta'\omega - \omega = 0 \quad (3\text{-}55)$$

where $\Delta'$ is the operator relative to the independent variable $\xi$. Eq. (3-55) is equivalent to:

$$\Delta'(\Delta'\omega + \omega) - (\Delta'\omega + \omega) = 0$$

and also to:

$$\Delta'(\Delta'\omega - \omega) - (\Delta'\omega - \omega) = 0$$

This is the same as Bessel's differential equation:

$$\Delta'\omega + \omega = \frac{d^2\omega}{d\xi^2} + \frac{1}{\xi}\frac{d\omega}{d\xi} + \omega = 0 \quad (3\text{-}56)$$

as well as the equation:

$$\Delta'\omega - \omega = 0 \quad (3\text{-}57)$$

which is identical to the preceding one when $\xi$ is replaced by $i\xi$. The solutions of both Bessel's equations are given in the real domain, by:

$$\omega = C_1 \text{ ber } x + C_2 \text{ bei } x + C_3 \text{ kei } x + C_4 \text{ ker } x \quad (3\text{-}58)$$

where ber, bei, ker, kei are the modified Bessel's functions of zero order.

### 3.2.1.1 - SOLUTION FOR CONCENTRATED LOAD

For a concentrated load P at point x = 0, the solution was obtained by Hertz (1884). Only the third function in Eq. (3-58) satisfies the boundary conditions, so that it is found that:

$$w = -\frac{P\ell^2}{2\pi D}\,\mathrm{kei}\,x \tag{3-59}$$

At the origin we have kei = $-\pi/4$, so that the deflection under the load is:

$$w_{max} = \frac{P\ell^2}{8D} \tag{3-60}$$

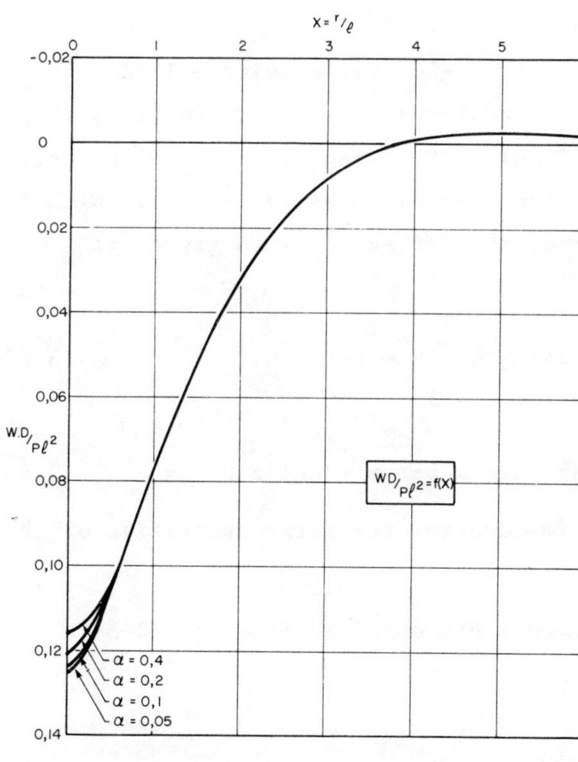


*Fig. 3.10. Deflections of a plate under a load uniformly distributed over a circular area.*

The bending moments can be computed from Eq. (3-36).

$$Mr = \frac{P}{4\pi} [(1 + \nu)\ (\ell n \frac{2\ell}{r} - \nu) - \tfrac{1}{2}\ (1 - \nu)] \qquad (3\text{-}61)$$

and

$$Mt = \frac{P}{4\pi} [(1 + \nu)\ (\ell n \frac{2\ell}{r} - \nu) + \tfrac{1}{2}\ (1 - \nu)] \qquad (3\text{-}62)$$

The moments are given in function of x on Fig. 3.11. The radial moments become negative at a distance equal to approximately 1.1 from the origin and their highest numerical value is close to -0.02 P. The positive moments are infinitely large at the origin, because of the punctual loading.

### 3.2.1.2 - SOLUTION FOR A UNIFORM LOAD OVER A CIRCULAR AREA

The solution of Eq. (3-50) for a uniform load of intensity q, applied over a circular area whose radius is a, has been obtained by Wyman (1950). It is made of a sum of two terms of the modified Bessel's function given by Eq. (3-58). With the proper boundary conditions, the deflection w is given by:

For $r > a$:

$$w = \alpha\ q'\ [\text{ber}'\alpha\ \text{ker}\ x - \text{bei}'\alpha\ \text{kei}\ x] \qquad (3\text{-}63)$$

and for $r < a$:

$$w = q' + \alpha\ q'\ [\text{ker}'\alpha\ \text{ber}\ x - \text{kei}'\alpha\ \text{bei}\ x] \qquad (3\text{-}64)$$

where $\alpha = a/\ell$ and the prime denotes the first derivative of the functions.

The bending moments are obtained from Eq. (3-36).

They are:

For r > a

$$\frac{Mr}{P} = \frac{\sqrt{2}}{2\alpha\pi}\left[\text{ker } x\ (\text{ber}_1\alpha - \text{bei}_1\alpha) - \text{kei } x\ (\text{ber}_1\alpha + \text{bei}_1\alpha) + \frac{\sqrt{2}\ (1 - \nu)}{x}\ (\text{ber}_1\alpha\ \text{kei}_1 x + \text{bei}_1\alpha\ \text{ker}_1 x)\right]$$

$$\frac{Mt}{P} = \frac{\sqrt{2}}{2\alpha\pi}\left[\nu\ \text{ker } x\ (\text{ber}_1\alpha - \text{bei}_1\alpha) - \nu\ \text{kei } x\ (\text{ber}_1\alpha + \text{bei}_1\alpha) + \frac{\sqrt{2}\ (1 - \nu)}{x}\ (\text{ber}_1\alpha\ \text{kei}_1 x + \text{bei}_1\alpha\ \text{ker}_1 x)\right] \quad (3\text{-}65)$$

and for r < a

$$\frac{Mr}{P} = \frac{\sqrt{2}}{2\pi\alpha}\left[\text{ber } x\ (\text{ker}_1\alpha - \text{kei}_1\alpha) - \text{bei } x\ (\text{kei}_1\alpha + \text{ker}_1\alpha) - \frac{\sqrt{2}\ (1 - \nu)}{x}\ (\text{kei}_1\alpha\ \text{ber}_1 x + \text{ker}_1\alpha\ \text{bei}_1 x)\right]$$

$$\frac{Mt}{P} = \frac{\sqrt{2}}{2\pi\alpha}\left[\nu\ \text{ber } x\ (\text{ker}_1\alpha - \text{kei}_1\alpha) - \nu\ \text{bei } x\ (\text{kei}_1\alpha + \text{ker}_1\alpha) + \frac{\sqrt{2}\ (1 - \nu)}{x}\ (\text{kei}_1\alpha\ \text{ber}_1 x + \text{ker}_1\alpha\ \text{bei}_1 x)\right] \quad (3\text{-}66)$$

where the $\text{ber}_1$, $\text{bei}_1$, $\text{ker}_1$ and $\text{kei}_1$ are the modified Bessel's functions of the first order, (Gagnon, 1978).

Fig. 3.11 shows the deflection and the moments in function of x for different values of $\alpha$ and $\nu = 0.33$.

The maximum deflection at the center of the load is given by:

$$w_{max} = \frac{P\ [1 + \alpha\ \text{ker}'\alpha]}{\pi\ k\ \alpha^2\ \ell^2} \quad (3\text{-}67)$$

with $P = \pi a^2 q$

$\alpha = a/\ell$

The maximum bending moment occurs under the load for x = 0. It is given by:

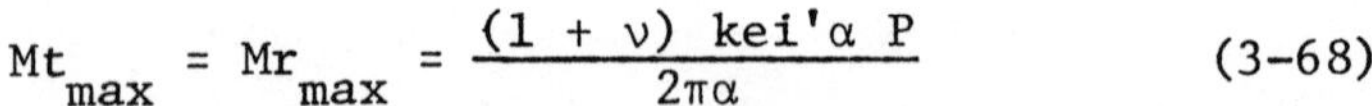

$$Mt_{max} = Mr_{max} = \frac{(1 + \nu)\ kei'\alpha\ P}{2\pi\alpha} \qquad (3\text{-}68)$$

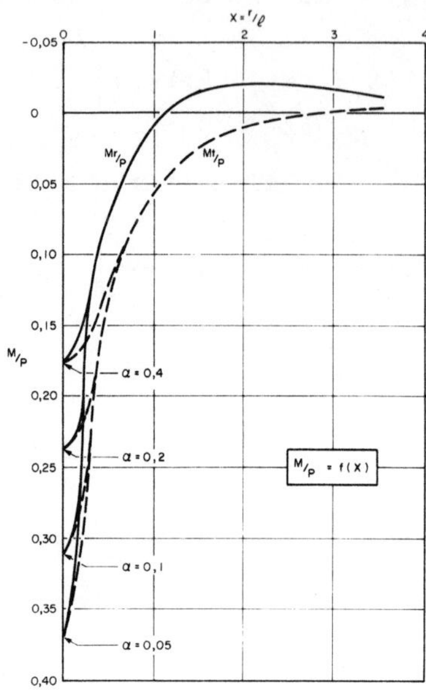


*Fig. 3.11. Tangential and radial moments in a plate under a load distributed uniformly over a circular area.*

For an elastic stress distribution, the load is given by:

$$P = \frac{\pi\ \alpha\ \sigma_f\ h^2}{3\ (1 + \nu)\ kei'\alpha} \qquad (3\text{-}69)$$

As can be seen from both Figs 3.10 and 3.11 the effect of the loading area α, does not extend appreciably outside the loading area, where the solution stays very close to that of a point load.

### 3.2.1.3 - SOLUTION FOR A THICK PLATE

The classical theory of plates is valid as long as the value of $\alpha > 0.05$, so that the radius of the loaded area is large in comparison to the thickness of the plate. For a small radius of loading, the thin plate theory predicts a stress which is too high. At the limit, Hertz's solution for a concentrated load gives a stress value of infinity under the load.

The above difficulty may be overcome by using the three-dimensionnal theory of elasticity, where the displacements w are not only vertical but also horizontal and designated by u. The equations for these displacements are then given by Nadai (1920):

$$(1 - 2\nu)\left(\Delta u - \frac{u}{r^2}\right) + \frac{de}{dr} = 0 \qquad (3\text{-}70)$$

and

$$(1 - 2\nu)\,\Delta w + \frac{de}{dz} = 0 \qquad (3\text{-}71)$$

where e is the volumetric strain. Operating on Eq. (3-70) with the operator $\Delta - \frac{1}{r^2}$ and on Eq. (3-71) with $\Delta$, we obtain:

$$\left(\Delta - \frac{1}{r^2}\right)\left(\Delta u - \frac{u}{r^2}\right) = 0 \qquad (3\text{-}72)$$

and

$$\Delta\Delta w = 0$$

The solutions of Eq. (3-72) are obtained in the same manner as that of the classical biharmonic Eq. (3-72) except that a Hankel's transform of the first order must be used for this operator. The results of the computations have been obtained by Nevel (1970), who gives:

$$u = \int_0^\infty [(A_1 + C_1 z)\exp(tz) + (B_1 + D_1 z)\exp(-tz)]\,tJ_1(tr)\,dt \qquad (3\text{-}74)$$

and

$$W = \int_0^\infty [(A_2 + C_2 z)\exp(tz) + (B_2 + D_2 z)\exp(-tz)]\,tJ_o(tr)\,dt \qquad (3\text{-}75)$$

$J_o$ and $J_1$ being the Bessel functions; $A_1 \ldots D_1$ and $A_2 \ldots D_2$ being constants of integration and t the independent variable of the integral.

The numerical values of the moments obtained from these functions have been obtained by Nevel and are shown in Fig. 3.12. A stress factor $\frac{\sigma\ h^2}{P\ (1+\nu)}$ is given in function of the load radius and the thickness of the plate. When $h/\ell \rightarrow 0$, the stresses are obtained from the thin plate theory. It can be seen that the effect of the plate thickness is to reduce considerably the maximum stress in the case of very concentrated loads.

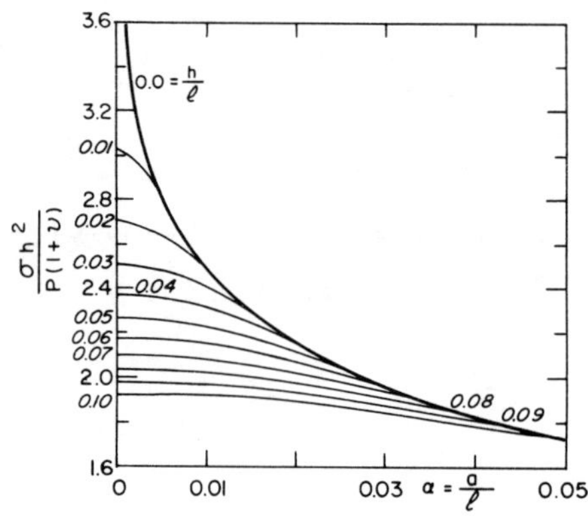


*Fig. 3.12. Stress intensity factor in a thick ice plate with various circular loading areas. (Nevel 1970).*

## 3.2.1.4 - EFFECT OF TRANSVERSE SHEAR STRAINS

Another effect which is not included in the classical theory is that of the deflections caused by the transverse shear strains arising in the plate. The corresponding basic equation of the so-called component theory is given by Panc (1975):

$$\Delta\Delta w - 2\chi\ \ \Delta w + w = q' - 2\chi\ \ \Delta q' \tag{3-76}$$

This equation reduces to the classical Eq. (3-53) for $\chi = 0$. This parameter is given by:

$$\chi = \frac{h^2}{10\ (1 - \nu^2)\ell^2} \tag{3-77}$$

Because $h/\ell$ rarely exceeds 0.1 in floating ice fields, the value of $\chi$ is at most $\chi = 1.5 \times 10^{-3}$. This is extremely small so that the component theory produces results very close to those given by the classical theory.

### 3.2.1.5 - ARBITRARY LOAD DISTRIBUTION ON AN INFINITE PLATE

The response of an ice cover to more complicated load distributions can be determined by superposition and some cases are discussed by Kerr (1975). The analysis may be greatly simplified by utilizing influence surfaces (Timoshenko, Woinowsky-Krieger, 1959). Influence surfaces for bending moments, suitable for ice plate problems were presented by Palmer (1971).

### 3.2.1.6 - SOLUTION FOR A SEMI-INFINITE PLATE

The solution for this condition with a rectangular load set on the edge of the plate has been obtained by Nevel (1965).

He used the differential equation of the plate in rectangular coordinates given by Eq. (3-49), but adding the effect of the elastic foundation and that of a point load:

$$\frac{\partial^4 w}{\partial x^4} + \frac{2\,\partial^4 w}{\partial x^2 \partial y^2} + \frac{\partial^4 w}{\partial y^4} = \frac{P}{D}\,\delta\,(y-c)\;\delta\,(x) - \frac{kw}{D} \qquad (3\text{-}78)$$

where the q function is replaced by the Dirac delta function which corresponds to a concentrated load P at position y = c, x = 0, as shown in Fig. 3.13.

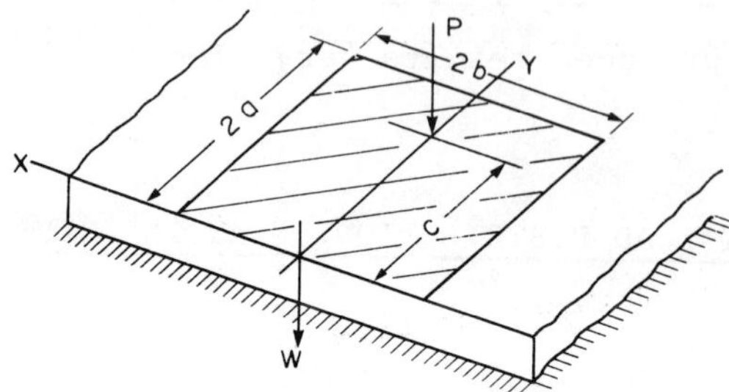


*Fig. 3.13. Geometry of loading at the edge of a semi-infinite plate.*

The boundary conditions for this problem, are:

$$w = 0 \text{ at } x = \pm \infty$$

$$w = 0 \text{ at } y = + \infty$$

$$Mr = D\left[\frac{\partial^2 w}{\partial x^2} + \nu \frac{\partial^2 w}{\partial y^2}\right] = 0 \text{ at } y = 0$$

and the shearing forces and twisting moments at the free edge:

$$D\left[\frac{\partial^3 w}{\partial x^3} + (2 - \nu)\frac{\partial^3 w}{\partial x \partial y^2}\right] = 0 \text{ at } y = 0$$

Nevel has solved the differential equation, by taking successively a Fourier transform relative to the y variable and a Laplace transform relative to the x variable. Solving algebraically for the transformed variable w and taking the inverse Laplace and Fourier transforms, he gets the solution for the deflection in the form of an integral.

The solution for a rectangular load, set at the edge, whose width is 2b and length 2a is obtained by integrating the previous integral for successive loads over the area. He computed numerically these values for the deflection at the edge of the load and computed the moments at the same place. These are given in Fig. 3.14 for various values of $\alpha = a/\ell$ and $\beta = b/\ell$.

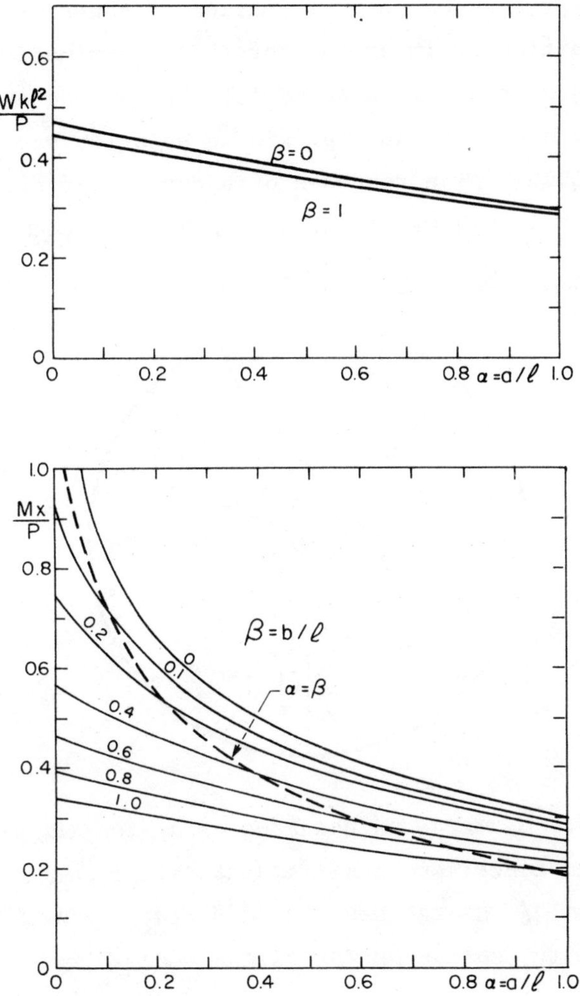


*Fig. 3.14. Deflections and maximum moments at the edge of a semi-infinite plate (Nevel, 1965).*

### 3.2.1.7 - SOLUTION FOR A NARROW WEDGE

The problem of the deformation of a narrow wedge on an elastic foundation is important for the determination of break-through loads on radially cracked ice sheets. No exact solution exist to this problem. An approximate solution was obtained by Kashtelian (1960). The response of a narrow-infinite wedge resting on a liquid was treated as a beam problem by Nevel (1961).

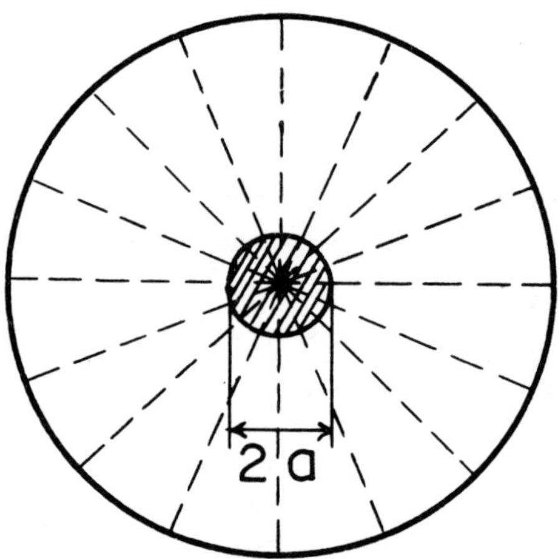


*Fig. 3.15. Geometry of cracking of a floating ice sheet. The dashed lines represents radial cracks due to tension on the bottom of the ice and the full line a circumferential crack due to tension on top of the ice.*

If we consider the equilibrium of moments on an element dr of the wedge shown on Fig. 3.16, the differential equation representing the problem can be obtained in a manner similar to that of a beam, Eq. (3-44). It gives:

$$\frac{d^4w}{dx^4} + \frac{2}{x}\frac{d^3w}{dx^3} + w = q' \tag{3-79}$$

Nevel has obtained the solution for a load of length a, uniformly distributed at the tip, by means of power series.

$$w = [A\ D_{no}\ (x) + B\ D_{n1}\ (x) + C\ D_{n2}\ (x)$$

$$+ D\ D_{n3}\ (x) + 2/\alpha^2]\ \frac{P}{2\ tg\ \phi/2\ k\ell^2} \tag{3-80}$$

The form and numerical values of the $D_{nm}$ functions can be found in Nevel (1961). The numerical computations have been made for the deflection at the tip of the wedge, which are shown on Fig. 3.17 for different values of the load geometry parameter α. On Fig. 3.18 are given both the values of the maximum radial moment and the distance from the tip where this moment occurs. It is a negative moment that will produce a circumferential crack at the upper surface of the ice.

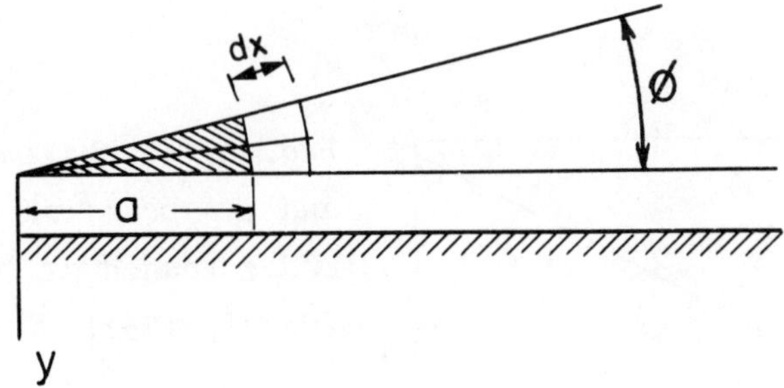


*Fig. 3.16. A narrow free infinite wedge on an elastic foundation.*

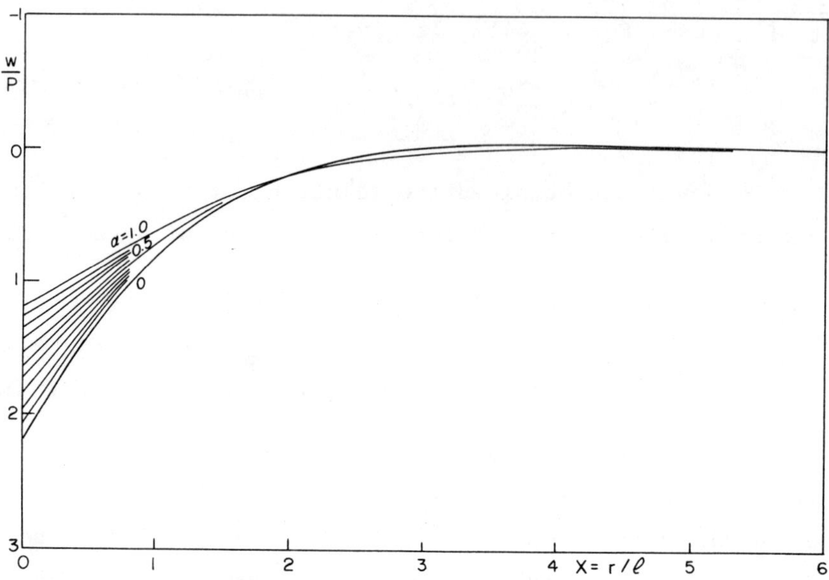


*Fig. 3.17. Deflection of a narrow wedge under a load at the tip. (Nevel, 1961).*

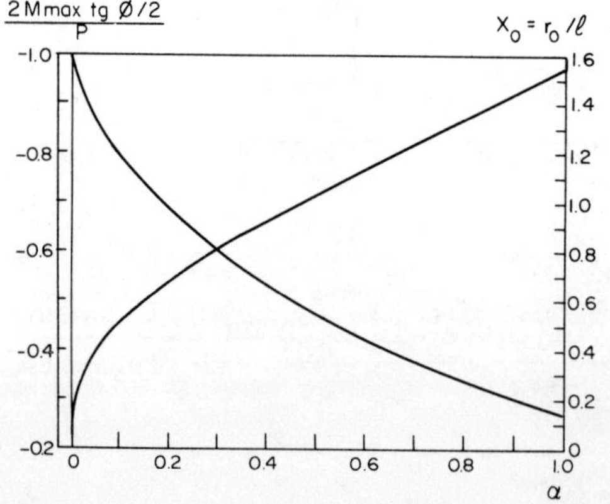


*Fig. 3.18. Maximum moment and its position for a narrow wedge loaded at the tip. (Nevel, 1961).*

## 3.2.2 - SPEED OF VEHICLES

One important factor affecting the bearing capacity of ice covers is the speed of travel of vehicles. On one hand, if the speed is very low, there is time for the ice plate to plasticize and the resisting moment increases. On the other hand if the speed of the vehicle becomes too high, the wave which it sets in the water underneath the ice, adds to the deflections of the cover and the ice may become seriously overstressed. Speeds should be kept below the critical value of wave celerity in the ice that would cause resonance in the cover.

The state of plastification of the ice cover may be determined if we consider that the ice passes quite abruptly from an elastic state to a full plastic state as discussed in Chapter 2. The rate of deformation of the lower extreme layers of the ice sheet can then be obtained simply from the elastic analysis (Michel *et al.*, 1974).

The strain is obtained in function of the deflection at the center of the loaded area by Eqs (3-31) and (3-33):

$$\varepsilon = h/2\rho_r = \frac{-h}{2}\frac{\partial^2 w}{\partial r^2} \qquad (3\text{-}81)$$

Neglecting Poisson's ratio effect, Eq. (3-36) gives:

$$\frac{\partial^2 w}{\partial r^2} = \frac{-Mr}{D} \qquad (3\text{-}82)$$

Using the Hertz's solution for the tangential moment in the ice sheet given by Eq. (3-61), and differentiating with respect to time, we obtain:

$$\frac{\partial\ Mr}{\partial t} = \frac{-P\ (1 + \nu)}{4\pi r}\frac{dr}{dt} \qquad (3\text{-}83)$$

The speed of the vehicle producing this moment is:

$$u = \frac{dr}{dt} \tag{3-84}$$

The relation between strain rate and vehicle speed is thus:

$$\dot{\varepsilon} = \frac{(1 + \nu)\ h\ P}{8\pi\ \ r\ D}\ u \tag{3-85}$$

With loads at the limit of those producing elastic cracks, we have from the results discussed in next Section 3.2.3.

$$P \simeq 0.5\ \sigma_o\ h^2 \tag{3-86}$$

With average values of $E \simeq 3.0 \times 10^9$ Pa, $\sigma_o = 12 \times 10^5$ Pa, $\nu = 0.33$, Eq. (3-85) becomes:

$$\dot{\varepsilon} \simeq 10^{-4}\ u/r \tag{3-87}$$

Thus the strain rate increases as the load approaches a given point. Under the wheel of a vehicle, at a distance $r = 0.3$ m, from the center of the wheel, the strain rate is:

$$\dot{\varepsilon} \simeq 3 \times 10^{-4}\ u \tag{3-88}$$

If we consider that the state of plastification in the ice cover is the same as that which occurs in a ice beam, as discussed in Section 3.1.1, we will have:

$u \geq 3$ m/s ($\dot{\varepsilon} \geq 10^{-3}\ s^{-1}$) full elastic behavior

3 m/s $\geq u \geq$ 3 cm/s partial plastification of the ice

$u \leq 3$ cm/s ($\dot{\varepsilon} \leq 10^{-5}\ s^{-1}$) full plastification of the ice cover.

Thus it can be concluded that vehicles travelling at speeds over 10 km/hr (3 m $s^{-1}$) engender a perfect elastic behavior in the ice cover. This is the usual case for traffic over ice. Full plastification of the ice would occur at such low speeds (0.1 km/hr) that this corresponds only to static loading conditions.

The critical speed of wave propagation in ice plates has been determined, in the general case, by Nevel (1970). He used the biharmonic equation but added to the hydrostatic water reaction kw a pressure caused by the acceleration of water particles, induced by the movement of the ice cover. This total pressure had the form:

$$p_o = k\ w + \rho\ u\ \left|\frac{d\phi}{dx}\right|_{z\ =\ w} \qquad (3\text{-}89)$$

where u is the velocity of the vehicle and $\phi$ is the velocity field set up in the water by the ice movement. Adding the equation of continuity of water flow, he looked for an instability in the solution of the biharmonic equation.

He obtained the critical velocity of resonance, giving an infinite theoretical deflection under the load, in the form:

$$\frac{u_c}{\sqrt{g\ell}} = \left(\frac{1 + \zeta^4}{\zeta}\right)\left(\frac{\text{tgh}\ \zeta\ y}{1 + u\ \zeta\ \text{tgh}\ \zeta\ y}\right) \qquad (3\text{-}90)$$

where $y = Y/\ell$, Y is the water depth and $\zeta$ is a parameter which is a solution of the transcendental equation.

$$\frac{1 - 3\ \zeta^4}{1 - \zeta^4} = 2\ \frac{\zeta\ y}{\sin\ (2\ \zeta\ y)} - \left(\frac{1 - \zeta^4}{1 + \zeta^4}\right) 2\ \mu\ \zeta\ \text{tgh}\ \zeta\ y \qquad (3\text{-}91)$$

where: $$\mu = \frac{\rho' h}{\rho \ell}$$

The value of the critical velocity $u_c$ is given in Fig. 3.19, in function of the water depth. One particular case is when y is small, which corresponds to shallow water. The hyperbolic functions can be simplified, and the critical velocity is then given by an expression which had been obtained by Kheishin (1964).

$$\frac{u_c^2}{g\ell} = \frac{2y}{1 + \sqrt{1 + (\mu y)^2}} \qquad (3\text{-}92)$$

For the case of an infinite water depth $y \to \infty$ and with an average value of $\mu = 0.05$, the critical speed is given by:

$$u_c = 1.25\sqrt{g\ell} \qquad (3\text{-}93)$$

It would be extremely dangerous for any vehicle to travel on ice at a constant speed equal to that of the critical velocity of wave resonance in the ice sheet.

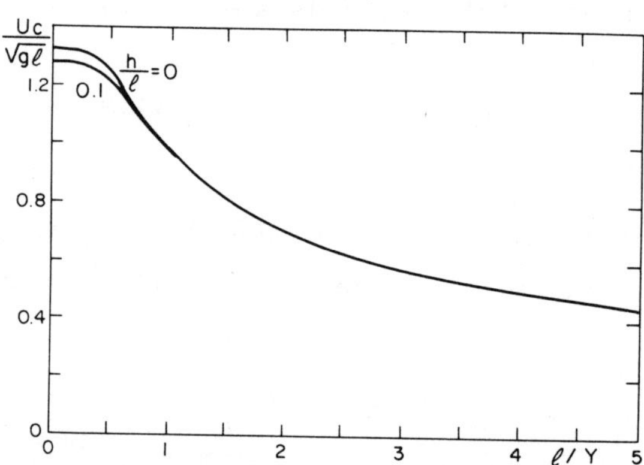


*Fig. 3.19. Critical speed for resonance on ice of a moving vehicle in function of the water depth (Nevel, 1970).*

## 3.2.3 - EXPERIMENTAL AND FIELD INVESTIGATIONS

There are very few experimental studies described in the literature where the bearing capacity of ice was measured with all the relevant parameters, including the flexural strength of the ice. Such tests were conducted by Panfilov (1960) in the laboratory as well as in the field.

The laboratory tests were conducted at -10°C with fresh and salt water ice. The ice plate thickness varied from 7 to

30 mm. The field tests were conducted on thicker ice sheets. The ice plates were loaded by placing metal water tanks on a structure which in turn rested on the ice plate and simulated the contours of wheel loads. The load for each test was placed statically at rates, which caused breakthrough within 5 to 20 sec. The ice strength was determined from floating cantilever tests with the load acting downwards.

The results for 56 laboratory tests for the infinite plate are shown on Fig. 3.20. The failures followed the usual pattern: first, the formation of radial cracks that emanated from the region under the load; then, the formation of circumferential cracks, at which time the load broke through the ice.

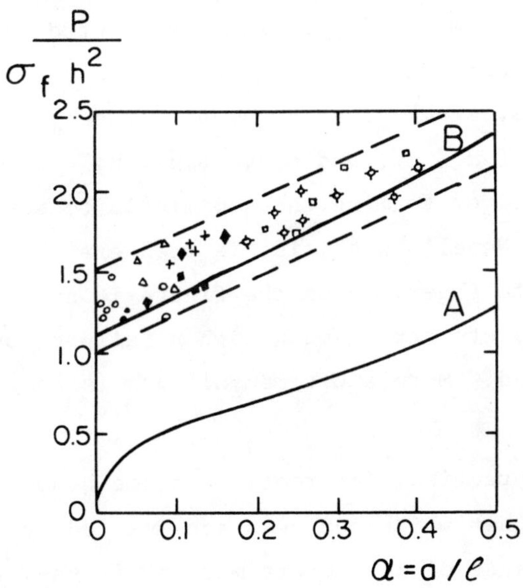


*Fig. 3.20. Bearing capacity obtained by laboratory tests for infinite ice plates (Panfilov, 1960).*

Panfilov gave the following formula to represent the average value of his observations:

$$\frac{P}{\sigma_f\ h^2} = 1.25\ (1 + 1.68\ \alpha) \tag{3-94}$$

With Panfilov results on Fig. 3.20 are also shown, in full line, two curves. Curve A corresponds to the load producing the first crack, which is given by Eq. (3-69). Curve B corresponds to the load producing breakthrough from Nevel analysis of wedges. It is computed from the values given in Fig. 3.18, when eight wedges are used to support the load ($\phi = \pi/4$). as observed in these tests. It can be seen that the theory of breakthrough by circumferential failure of narrow wedges predicts fairly well the ultimate bearing capacity of the ice plates.

In Fig. 3.21 are shown the results of similar tests made by Panfilov (1960) for semi-infinite plates subjected to an edge load. Also shown are curve A corresponding to the condition of formation of a first crack given by Nevel's theory and computed with Fig. 3.14 (for $\alpha = \beta$), and curve B giving the breakthrough load with 4 wedges forming the semi-infinite sheet. The bearing capacity, from Nevel's analysis, is then exactly half of that of the infinite sheet. From the figure it can be seen that the analysis predicts here a much higher failure load than that which was observed. More experimental work is needed to clarify this case.

The most useful practical information on the bearing capacity of ice covers is that which has been accumulated by Gold (1971) on the performance of ice covers made of fresh-water ice. This information pertains to recorded failures of isolated vehicles, where both the weight of the vehicles and the ice thicknesses were recorded. This data is shown on Fig. 3.22.

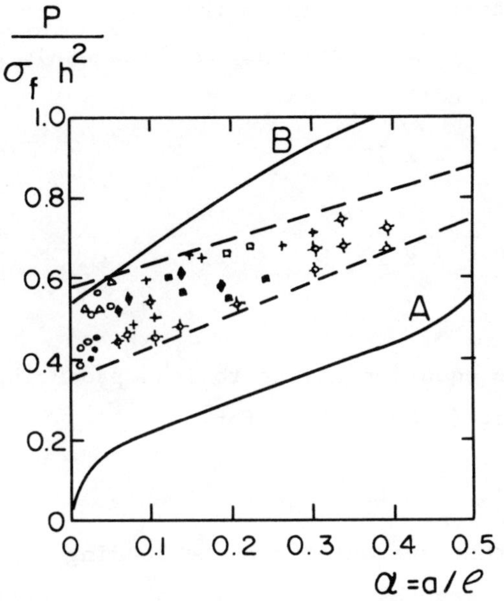


*Fig. 3.21. Results of tests of bearing capacity for semi-infinite ice plates (Panfilov, 1960).*

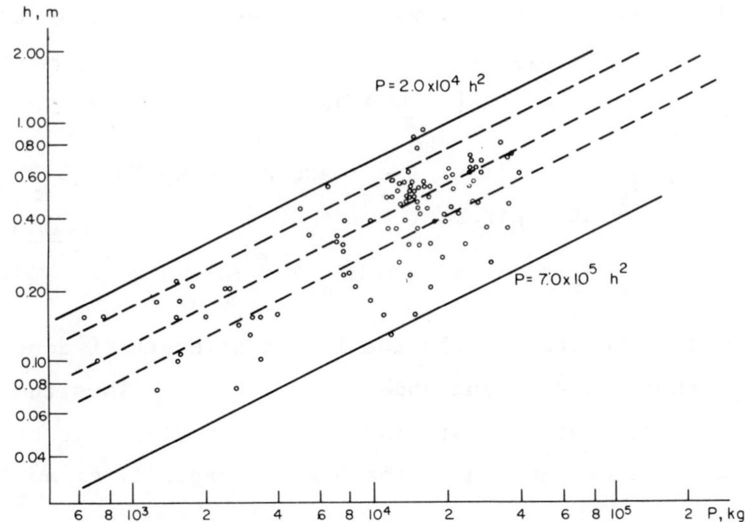


*Fig. 3.22. Field observations of ice failure by moving vehicles in function of the ice thickness (Gold, 1971, with the permission of the National Research Council of Canada).*

It can be seen that all points on the figure are limited by a lower curve (lowest load for a given thickness), under which no accident ever occured:

$$P = 2 \times 10^4 h^2 \qquad (3\text{-}95)$$

and by an upper curve:

$$P = 7 \times 10^5 h^2 \qquad (3\text{-}96)$$

where P is in kg and h in m.

We may write the equation giving to load producing the first crack in the ice, Eq. (3-69) in the form:

$$P = \beta \sigma_f h^2 \qquad (3\text{-}97)$$

where $\beta$ is a coefficient that characterizes the loading conditions for the vehicle. In general for travelling vehicles the value of $\alpha = a/\ell$ will be between 0.05 and 0.20. For these values of $\alpha$, the coefficient will be around $\beta \simeq 0.5$, so that we will have approximately:

$$P \simeq 0.5 \sigma_f h^2 \qquad (3\text{-}98)$$

The lower limit of the loads given by Eq. (3-95) then corresponds to a strength of the ice of:

$$\sigma_f \text{ min} = \sigma_o \text{ min} = 4 \times 10^5 \text{ Pa}$$

This is effectively the lowest strength of dense fresh water ice that can be found under an ice cover. Thus the condition of formation of a first crack is a very plausible criterion for the safe determination of the bearing capacity of an ice cover.

The highest bearing capacity corresponds to the upper limit in Fig. 3.22. For breakthrough loads the $\beta$ coefficient may be higher and attain a value close to 1.4. It may also be expected that the ice will have plasticized before collapse. The flexural strength is then given by Eq. (3-29):

$$\sigma_f = \lambda \sigma_o$$

For cold ice we may admit a value of $\lambda = 2$, as shown in Section 3.1.1. The value of $\beta\lambda$ in Eq. (3-97) would then be, for the most resisting ice covers: $\beta\lambda \simeq 2.8$. Eq. (3-96) then corresponds to a uniaxial tensile strength of:

$$\sigma_o \max = 25 \times 10^5 \text{ Pa}$$

This again is the maximum value for high strength fresh water ice made of fine crystals.

Thus, we may conclude that the bearing capacity of an ice cover under dynamic loading conditions is effectively given by:

$$P = \beta \sigma_f h^2 \qquad (3\text{-}99)$$

where $\beta$ can be computed from the loading conditions, $\sigma_f$ and h determined in the field.

## 3.3 - STATIC BEHAVIOR OF ICE COVERS

Ice covers deflect continuously under static loads. There has been many approaches to determine the time-dependent deformation and the breakthrough load on an ice plate by taking into account some viscous properties of the ice. The linear theory of visco-elastic plates has been utilized by many authors to study the ice plate problem (Panfilov, 1961; Kheishin, 1964; Nevel, 1966; Iakunin, 1970). The mechanical behavior of ice is essentially non-linear as discussed in Chapter 2 and Section 3.1; the linear analysis do not seem to apply well to the ice problem.

Many attemps have been made to obtain the breakthrough load with various theories on limit analysis. Persson (1948) and Assur (1961) used the yield line theory. Meyerhoff (1961) assumed that the ice plate is a thin, rigid and ideally plastic body that resist the full plastic flexural moment. The literature is extensive on plate analysis with rheological laws that do not correspond to the real behavior of polycrystalline ice. For $\alpha = 0.10$, these theories give breakthrough loads, within the following limits:

$$2.8 < \frac{P}{\sigma_f \; h^2} < 4.0 \qquad (3\text{-}100)$$

Even if we assume that the $\sigma_f$ value in this formula is 2/3 of that given for brittle fracture, because of plastification, it still remains that the breakthrough loads are between 33% to 90% higher than those obtained experimentally and given by Eq. (3-94). It seems dangerous, without close analysis taking in the real behavior of ice and without extensive experimental data, to rely on theoretical approaches that are over optimistic on the ultimate bearing capacity of ice.

A semi-empirical approach to the long term deflection of an ice sheet, based on simplified analysis and field measurements, has been made by Frederking and Gold (1976). From Eqs (3-5), (3-67) and (3-69), the elastic strain at the bottom of the ice cover is related to the deflection, in the elastic analysis, by:

$$\varepsilon = \frac{h^2}{4\ell^2} F(\alpha) \frac{w}{h} \qquad (3\text{-}101)$$

where:

$$F(\alpha) = \frac{\alpha \; \mathrm{kei}'\alpha}{1 + \alpha \; \mathrm{ker}\;\alpha} \qquad (3\text{-}102)$$

This function is plotted in Fig. 3.23.

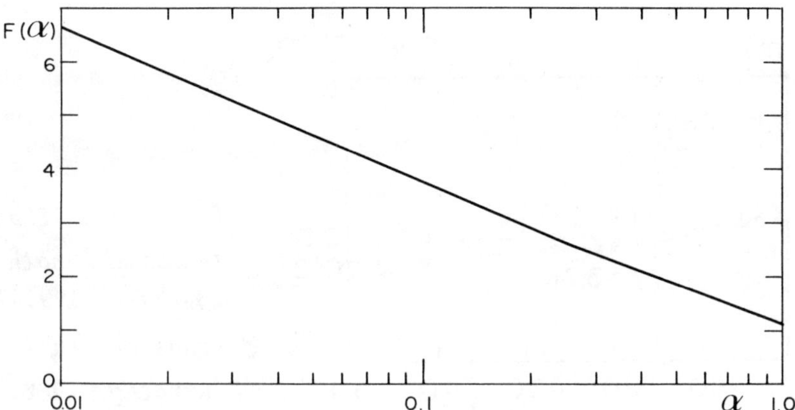


*Fig. 3.23. Load function F (α) in relation to load concentration ratio a.*

It can be seen that for the range encountered for load storage on ice, $0.1 < \alpha < 0.4$, this value decreases from 4.0 to 2.0.

It was said in Section 3.1.3 that a good criterion to obtain the long term bearing capacity of an ice cover was to limit the deflection to the freeboard. If the deflection is limited to the freeboard (i.e. $w/h = 0.08$), the maximum strain induced in ice as thick as 4 m, will still be less than 0.1%. Then the deformation of the ice will be well within the primary creep range.

The analysis of the extent of the deformation dish in function of time for tests done by Kingery (1962), indicates that the equation for elastic deformation can still be used, but with a characteristic length $\ell$ that decreases in a continuous manner with time. An example of the apparent time dependence of $\ell$ calculated from the results of Kingery by Frederking and Gold (1976) is shown in Fig. 3.24.

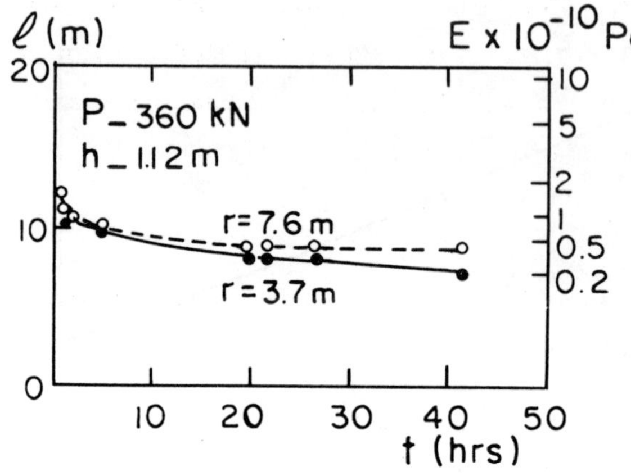


*Fig. 3.24. Characteristic length in function of time for loads on sea ice (Kingery, 1962).*

These results show that $\ell$ decreases very slightly with time, so that basic Eq. (3-101) can still used with a lower average value of $\ell_a$ during the time-dependent deformation. Differentiating Eq. (3-101) with respect to time.

$$\dot{\varepsilon} = \frac{h}{4\,\ell_a^2} F(\alpha)\,\dot{w} \qquad (3\text{-}103)$$

As the value of $\ell_a$ is related to the ice thickness by Eq. (3-51).

$$\ell_a \propto h^{3/4} \qquad (3\text{-}104)$$

Then we finally obtain:

$$\dot{\varepsilon} = \frac{A' F(\alpha)}{\sqrt{h}}\,\dot{w} \qquad (3\text{-}105)$$

Where A' is a constant. Frederking and Gold (1976) proposed to use this semi-empirical relationship to describe the static bearing capacity problem. They correlate the value of $F(\alpha)\,\dot{w}/\sqrt{h}$ obtained in field experiments with the maximum elastic stress $\sigma_f$ computed at the beginning of loading. For the bending of beams (Section 3.1.1) we have shown that the deflection rate was also proportionnal to the strain rate, but that it was related to the stress by Glen's formula:

$$\dot{\varepsilon} \propto \sigma_f^{\,n} \tag{3-106}$$

Which gives a non-linear relationship between the stress and the rate of deflection.

$$\sigma_f^{\,n} = \frac{A\ F\ (\alpha)\ \dot{w}}{\sqrt{h}} \tag{3-107}$$

In Fig. 3.25 the value of log. F ($\alpha$) $\dot{w}/\sqrt{h}$, determined from three published field cases (Beaudais *et al.*, 1974; Kingery, 1962), is plotted against the log of the initial maximum stress induced in the cover by the load, assuming elastic behavior. If this correlation can be extended to other cases, it may be used as a basic relationship to compute the time needed for the loads that are stored on ice to deflect the ice to the value of the freeboard.

The load P is directly proportional to $\sigma_f$ in the computation of Eq. (3-107). The time required to attain freeboard, at a constant deflection rate $\dot{w}$ may be given by:

$$0.08\ h = \dot{w}t + w_o \tag{3-108}$$

where $w_o$ is the total instantaneous and delayed elastic deformation, discussed in Chapter 2. With Eq. (3-107) and $w_o$ = BP, we get:

$$t = \frac{0.08\ h - BP}{C\ P^n} \tag{3-109}$$

where B and C are constants depending on loading conditions and the creep constants of Eq. (3-107). These constants may be obtained from field tests with a lower load $P_i$, from which the time of storage for larger loads can be computed. The time required for the deflection to attain freeboard increases considerably with the decrease in load, as shown by Eq. (3-109).

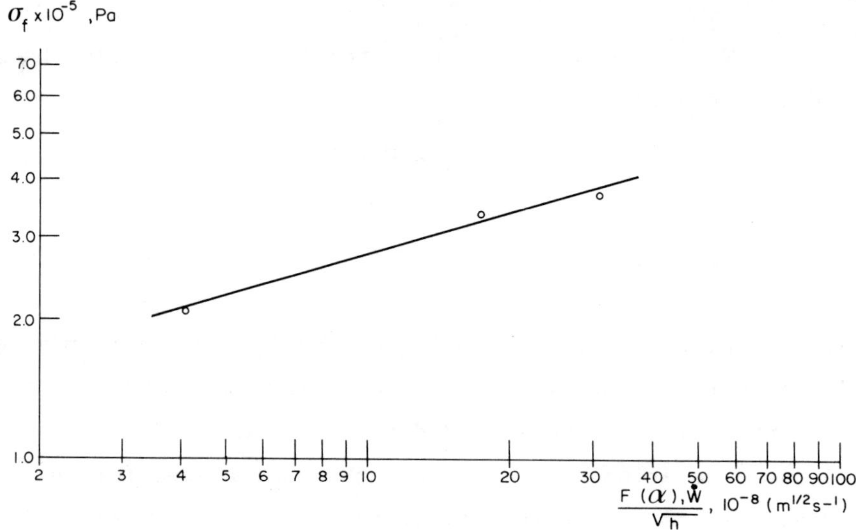


*Fig. 3.25. Stress dependence on rate of deflection of ice sheets computed from field data.*

## 3.4 - DESIGN CONSIDERATIONS

### 3.4.0 - INTRODUCTION

In view of the many factors that affect the bearing capacity of ice, and that cannot be taken into account readily in the computations, it has been usual to assume a very simple formula of the type:

$$P = Ah^2 \tag{3-110}$$

where A is a constant that is determined from experience and the conditions of the ice cover. The information collected by Gold (1971) and shown in Fig. 3.22 is very useful to discuss the values of A which are usually recommended.

The value of $A = 3.5 \times 10^4$ shown as curve A on the figure, has been largely used in Canada for fresh water ice and is recommended by Transport Canada (1976) for aircraft landing on fresh water ice in limited operations involving some risk of breakthrough. This can be confirmed from the figure as there has been very few accidents with this value; the lowest value for which there is no documented breakthrough is as low as $A = 2 \times 10^4$. This value would, indeed, be much lower for sea ice. The document by Transport Canada, recommends that a detailed ice survey should be made and that ice landings should be continuously monitored by ice experts.

Curves B and C in Fig. 3.22 represent the values given in the U.S.S.R. Instruction of the Engineering Committee of the Red Army (1946), which were used during the last war. The lowest curve, with $A = 7 \times 10^4$, is for wheeled vehicles and the highest, with $A = 1.23 \times 10^5$, for tracked vehicles. With these values the ice will often be subjected to considerable cracking, depending on its quality, so they can be used only in emergency conditions, during war for example, with strict control of the ice by qualified experts.

In general, for a civilian roadway or platform, absolutely no risk should be taken of loss of life or equipment and the ice thickness should be determined for the condition of no initial cracking under the design load. As discussed in Section 3.1.3, this criterion will give a factor of safety of over two for a breakthrough load, but just around one for the event where an unseen transversal wet crack would be present in the ice.

## 3.4.1 - USE OF NATURAL ICE COVERS

The bearing capacity of a natural ice cover depends on the ice thickness, the uniaxial tensile strength of the ice at the bottom layer and the load geometry as given by Eq. (3-99):

$$P = \beta \, \sigma_o \, h^2 \tag{3-99}$$

It also depends on macroscopic conditions in the floating ice sheet that we will now discuss. Without any serious ice survey, the bearing capacity should be limited to the minimum value for which no breakthrough has been observed.

### Ice characteristics

There are many types of fresh water and sea ice that can form the ice cover. The main factor affecting the mechanical properties of the ice is the crystallographic structure and, for sea ice, the brine volume. This has been extensively discussed in Chapter 2.

For design purposes, the tensile strength of the ice should be used. Its value varies within wide limits so that a survey of the ice is required if there is a need to attain safely the maximum bearing capacity of the ice cover. For fresh water ice, the tensile strength can be determined only from the crystallographic analysis and it depends very little on temperature. For sea ice, however, the tensile strength depends very much on brine volume, which in turn is highly dependent on temperature. The strength of sea ice is much smaller than that of fresh water ice, considering that the lower layer is weaker at the freezing temperature.

Effective ice thickness

The effective ice thickness is given by the thickness of good quality dense ice. Dense fresh water ice may be taken as ice with a specific gravity of at least 0.85. If the ice is layered and if one of those layers is of poor quality (e.g. light, drained snow ice, drained frazil ice, snow and frazil slush) then only the resisting portion of the continuous dense ice should be counted in the effective thickness. This is illustrated in Fig. 3.26, where the resisting moment of a non-homogeneous ice sheet formed by layers of various ice types can be computed from the elastic theory of composite beams. It is given by:

$$M\tau = \frac{\sigma_1 \sum E_n I_n}{E_1 c} \qquad (3\text{-}111)$$

where $\sigma_1$ and $E_1$ are the tensile strength and Young's modulus at the extreme unbroken layer at a distance c from the center of inertia; $E_n$ and $I_n$ are the Young's modulus and moment of inertia of all the ice layers in relation to the center of inertia of the section. Many cases may happen as shown in Fig. 3.26. In case A, all the ice is contributing to the resistance of the section; in case B, the light snow ice fails and only the upper stronger ice takes the moment; in case C, an intermediate layer of snow ice fails and the moment is taken by the remainding ice.

If a water layer is present within the ice cover, only the thickness of the upper layer of ice should be used to determine the effective thickness. An exception occurs when the lower layer has sufficient thickness to support the load at lower freezing temperatures.

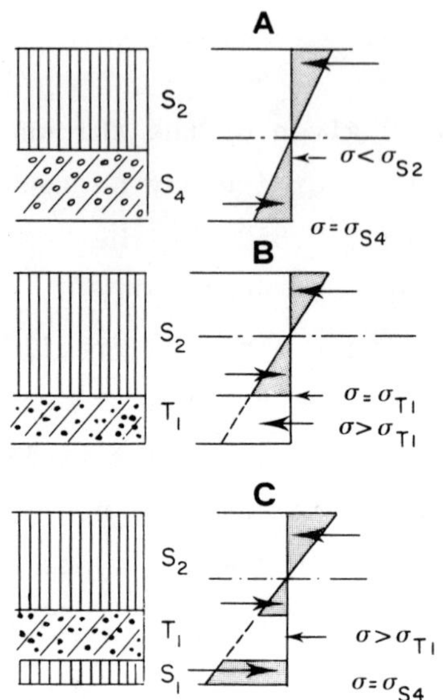


*Fig. 3.26. Modes of failure of a non homogeneous ice sheet made of columnar ice $S_1$ and $S_2$ and snow ice $T_1$.*

The effective thickness of an ice cover can vary within surprisingly wide limits. In particular, dangerously thin areas can occur in the covers of rivers, estuaries and on lakes near the inlet and outlet of rivers. In general, for a regularly used crossing, the thickness must be determined by holes spaced at a maximum of 15 m for a river, 30 m for a lake and 100 m for smooth sea ice. This thickness should be regularly checked.

<u>River ice</u>

It is very difficult to establish a safe crossing on river ice.

If the crossing is set at a place where the river flow is rapid, the ice cover will form later in the season by successive stages. The thickness might be important close to shores but decrease suddenly by steps as the cover gets closer to the faster river flow. During cold weather spells the ice might form over dangerous areas, but disappear suddenly and without prior notice with snow fall or warmer weather, because of the important thermal erosion underneath.

It is thus safer to establish crossings in lower velocity reaches (Michel *et al.*, 1974). But these will also present their own problems. One is the accumulation of frazil under the cover that will lift and break the solid part of the cover at many places. No effective method has yet been developed to take this condition into account. Another problem is the presence of restricted flow channels in the water mass, that seems to be favorized by the formation of ice covers. These channels may be narrow, change place and be unpredictable. Because they cause severe local thermal erosion, the ice thickness in a river crossing should be closely and frequently investigated.

Because of the variable water stages of a river, shore leads will usually form. Unless the ice is much thicker at the shores, it is advisable to look for crossing sites with long beaches so the broken ice will find a solid support underneath.

For deep rivers it is often found that the shorter crossings are the more dangerous ones.

## Cracks in the ice

There are various types of cracks that are present in an ice cover. Hair cracks are not more than 2 mm in width. Wider cracks are classified as "wet" or "dry" cracks. A wet crack is one in which water can be observed when it is formed. They then refreeze after their formation and eventually, the ice in the crack is as strong as the original ice. As it is difficult to ascertain the state of freezing of such a crack it is safer to consider it as a wet crack unless a core sample is taken to determine its state.

For a wet crack, the capacity of the ice is that of a semi-infinite plate and is thus reduced by half of that given for the infinite plate as discussed in Section 3.2.1. If another crack forms at an angle with this first one, the bearing capacity should be further reduced. It is thus a necessity to reduce by half the bearing capacity of a cracked ice sheet, in the case of a wet crack.

Hairline cracks are present at many places in the ice. If they are just made, they must be given a short time to heal. Otherwise they do not appear to affect the bearing capacity of the ice cover.

### Snow

Snow is both a load and an insultation on the ice.

If it is uniformly distributed it does not engender a bending moment in the ice cover. As soon as a track is made on the ice through the snow, snow banks on the sides will produce stresses that may even crack longitudinally the road; as has been sometimes observed. The crossing might then be more dangerous than the untouched ice cover. If snow is cleared on the ice, it must be cleared to a width at least 30 m wide. The wind rows should not be higher than two-thirds the thickness of the ice.

It is advisable to leave 5 to 10 cm of compacted snow on the ice. This will help against slippage for wheeled vehicles and is highly recommended for tracked vehicles; to protect the solid ice surface. This white layer is also useful to prevent inside deterioration of the ice by solar radiations.

If a wide track is cleared of snow, the ice in the crossing will thicken more than the snow covered remaining ice. The more buoyant way will then induce stresses at the side limits, that will eventually form lateral cracks.

Air temperatures

There is some evidence that the bearing capacity of an ice cover is reduced for a period following a marked sudden drop in air temperature. This effect has not been properly investigated, but may be due to induced thermal stress that will open up cracks on the upper part of the ice cover. Combined with cracks formed by vehicles on the lower part of the ice, this may increase the risks of failure.

The strength of fresh water ice does not depend much on temperature. However, at spring time, the sun radiations have a very fast devastating effect on the ice by dislocating the crystals at their boundaries and producing either corn snow ice or candle ice. This reduces practically to zero the shear strength of the ice in the upper part of the cover.

The strength of sea ice depends very much on the ice temperature. When it gets to melting point large volumes of brine are formed in the ice which then has very little resistance either to shear or to tension. This is a time when the loads will more easily break through the ice.

Recommendations by Transport Canada (1976) are to decrease aircraft weights by 10 percent per day, if the daily average air temperature is higher than $-1^{\circ}C$ for fresh water ice and higher than $-2^{\circ}C$ for sea ice. In any case, it is recommended to suspend operations after four days or if the maximum air temperature exceeds $4^{o}C$.

## Resonance

We have seen in Section 3.2.2 that if a vehicle travels at the critical speed of wave propagation in the ice and water underneath, the ice will be in a state of resonance and may be seriously overstressed. As an indication, for water 5 m deep and ice more than 0.5 m thick, the critical velocity is about 25 km/hr. Speeds should then be kept below this critical value. If higher speeds are used, they should be continuously varied in order not to induce resonance in the ice cover.

In order to minimize the effect of reflected waves it is also advisable not to travel at distances less than one characteristic length parallel to the shore and not to approach shores at right angle.

## First use of ice ways

The first time is the more dangerous time to travel on the ice. When the thicknesses have been determined, the first load should not be higher than the minimum given for no observed breakthrough. As an example, this would correspond to about 10 cm of ice for a single man, 15 cm for a snowmobile and 24 cm for a light 1000 kg vehicle.

The ice way may then be tested in the following manner. The loads on the way are progressively increased. Each time, the driver of the vehicle is tied (with door opened) to a man walking about 50 m from the vehicle. The vehicle advances at walking speed and the formation of radial cracks under the vehicle can be easily ascertained by the sharp audible noise they make. The safe capacity of the ice is obtained with the formation of these first cracks. When a crack is heard, the vehicle still continues to advance, as only one crack is not dangerous. If there are however more cracking activity in the near vicinity, the vehicle must retreat, as the safe bearing capacity is attained. If a few spaced cracks have been noticed, the time of an overnight is left for them to heal before the use of the ice way; which has attained its safe limit. Much care should be taken in crossing wet cracks because of the highly reduced ice bearing capacity.

## Operation of ice ways

Some safety rules for the operation of an ice bridge are given by Michel *et al.*, (1974):

a) Large and easily readable signs should be placed at both ends of the bridge, giving the speed limit (15 km/hr), the load limit and prohibiting the crossing of more than one vehicle at a time.

b) A line of poles should be set on one side of the roadway so that it could be easily identified.

c) It is important to measure regularly the thickness of the ice, observe cracks and make bearing capacity tests. Any new crack should be thoroughly investigated. If it is a wet crack the bearing capacity should be reduced accordingly.

d) During springtime, the top surface of the ice bridge should be protected from solar radiations. A good protection is a layer of about 5 cm of saw dust on the surface of the lanes.

## 3.4.2 - BUILT-UP ICE SURFACES

Because the growth of ice at a water surface is very much quicker than that at the bottom of an existing ice sheet, it is possible, in a given time, to increase considerably the total thickness of an ice cover by flooding the surface and producing successive ice layers on top of the existing ice sheet. A considerable increase in bearing capacity can be obtained in this manner.

A built-up ice apron for an aircraft was built as part of the "Project Ice Way" Kingery (1962) and is shown on Fig. 3.27. It had a circular shape, roughly 200 m in diameter. A submersible electric pumping unit was used to produce layered ice by flooding 10 cm lifts within 48 hours freezing cycles. Chipped-ice aggregates were first laid down before pumping in the sea-water for each lift. Artificial reinforcement was used consisting of three thicknesses of fiberglass matting, totaling 5 cm in thickness, which was set in with the first flood layer of water. The water was contained in the area with water-filled, 15 cm diameter, polyethylene plastic tubing, acting as a dike around the construction area. The total thickness of the built-up ice was 46 cm and the top was finished with one layer of fiberglass mat, averaging slightly over two cm in thickness.

This experimental reinforced apron did not prove very efficient in sea ice. The high salinity of the flood water and the delay in brine drainage caused by the lower reinforcement, increased the brine content of the ice and decreased its strength compare to natural sea ice. It would have been much more efficient with fresh water flooding.

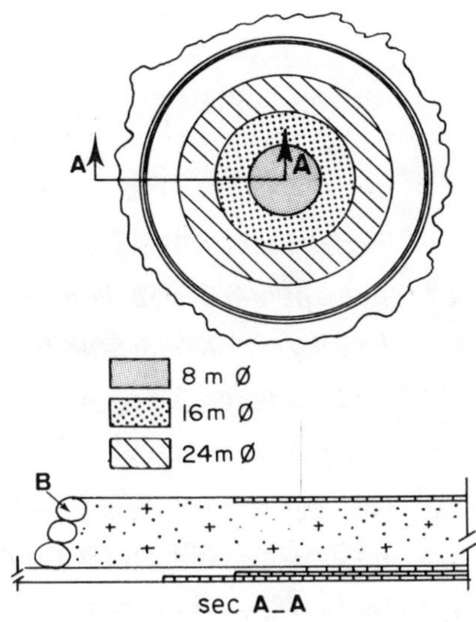


*Fig. 3.27. Configuration of a fiberglass reinforcement for a parking apron. At the bottom, in section A-A, we can see three layers of fiberglass. The built-up ice is held by 15 cm plastic tube dikes (B).*

Because of this difficulty in draining quickly enough the brine of sea water in built-up sea ice, it is recommended to design thickened ice platforms to take the full load by their natural buoyancy, as if the ice platform was a ship. This was done for the ice platforms used for supporting drilling rigs in the Canadian Arctic Archipelago.

Many built-up ice bridges have been used to accelerate the development of the James Bay Project in northern Quebec (Michel *et al.*, 1974). These bridges were designed with wooden logs reinforcement as shown on Fig. 3.28.

The first row of logs was placed when the ice cover (A) had a minimum thickness of 38 cm. The layered ice was built-up in layers of 1 to 2 cm at a time. Snow banks (C) were built, compacted and wetted to keep the water inside the 46 m wide ice surface. The ice was built-up to a total thickness of 160 cm to 254 cm, depending on the quality of the ice and reinforcement used.

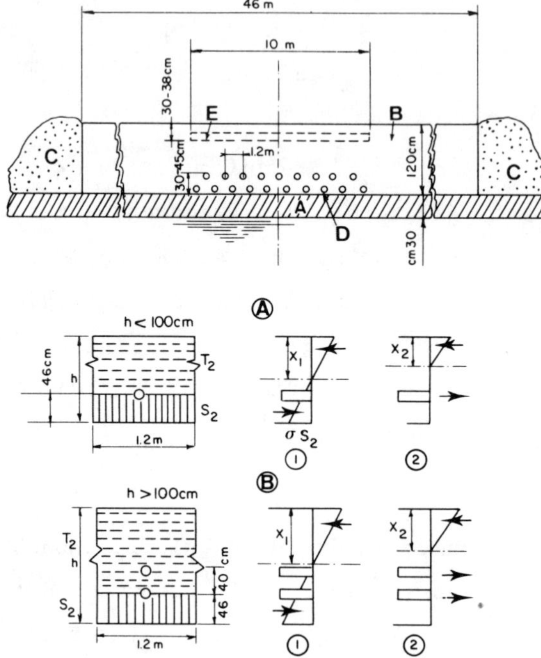


*Fig. 3.28. Diagram for design of James Bay ice bridges. A - natural and layered ice, B - man-made layered ice, C - snow bank, D - longitudinal 30-45 cm diam. wood logs, 1.2 m center to center, E - Transversal trunks, 12-24 m, center to center (Michel et al., 1974).*

The main reinforcement (D) was made of aspen and black spruce logs averaging 15 cm in diameter and set in the longitudinal direction at 1.2 m center to center. A transversal row (E) was also set at the top of the bridge at distances of 12 to 24 m, one from the other.

The resisting moment was computed according to the diagram shown in Fig. 3.28. Two cases were considered; case (1) when the ice would contribute to the moment on the tensile side and case (2) when it would fail completely on the tensile side. For thin ice covers (A), the failure occurs in compression in the upper layer in case (2). For thicker ice (B), the failure occurs either in tension in the ice (case 1) or in tension in the logs (case 2).

The design was made with a factor of safety of 1.2 against cracking in the ice in case (1) and 1.1 against failure either in the logs or by compression in the upper ice layer in case (2). Thus in case of a failure in tension at the bottom of the ice, there was always a second way to take the load, with the logs, so that these factors of safety are considered as good as the factor of two for the formation of the first crack in a non-reinforced ice bridge.

More than 3000 loads safely crossed the bridges during the two winters of 1972 and 1973, with the highest accepted load limited to 70 tons. The use of built-up ice permitted these loads to be crossed, which would not have been possible with the natural ice covers. The use of logs made possible a reduction in overall thickness of 25-35%. It was felt that this type of construction was the safest for ice bridges submitted to heavy service.

## REFERENCES

Assur, A. (1956) - "Airfields on floating ice sheets for regular and emergency operations" USA SIPRE, Technical Report 36.

Assur, A. (1961) - "Traffic over frozen or crusted surfaces" Proc. First. Int. Conf. on the Mechanics of Soil-Vehicle Systems, Torino, Italy, Edizioni Minerva Tecnica.

Beaudais, D., Masterson, D.M., Watts, J.S. (1974) - "A system for offshore drilling in the Arctic Islands" Proc. 25th Ann. Tech. Meet. Petrol. Soc. Can. Inst. Mining, Calgary, Alta.

Drouin, M., Michel, B. (1972) - "La résistance en flexion de la glace du Saint-Laurent déterminée en nature" Rapport GCT-72-09-26, Dép. de Génie Civil, Université Laval, Québec.

Frederking, R.F., Gold, L.W. (1976) - "The bearing capacity of ice covers under static loads. Can. J. Civ. Eng., 3, 288-293.

Gagnon, L. (1978) - "Le fluage de plaques circulaires de glace $S_2$" Thèse M.Sc., Université Laval, Québec.

Godbout, Y. (1967) - "Etude expérimentale de la ruptùre d'un champ de glace confiné de faible épaisseur sous l'action de forces verticales" Thèse M.Sc., Université Laval, Québec.

Gold, L.W. (1960) - "Field study on the load bearing capacity of ice covers" Woodlands Review, Pulp and Paper Magazine of Canada, Vol. 61, May.

Gold, L.W. (1971) - "Use of ice covers for transportation" Can. Geotechnical Journal, 8, p. 170-181.

Gold, L.W. (1977) - "Ice pressures and bearing capacity" Chapter 10 of 'Geotechnical Engineering for Cold Regions' Edited by D. Anderson.

Hertz, H. (1884) - "Uber das Gleichgewicth schwimmender elastis cher Platten" Wiedermann's Annalen der Phys. und Chem., Vol. 22.

Iakunin, A.E. (1970) - "The investigation of the effect of the loading time on the bearing capacity of an ice cover (In Russian)" Thesis Kandidat Tekhnicheskikh Nauk., Novosiberskii Institut Inzhenirov Ah/D Transporta, Novosibirsk.

Kashtelian, V.I. (1960) - "An approximate determination of forces which break up a floating ice plate (In Russian)" Problemy Arktikii Antarktiki, No. 5.

Kerr, A.D. (1975) - "The bearing capacity of floating ice plates subjected to static or quasi-static loads" CRREL RR 333, Hanover, USA.

Kheishin, D.E. (1964) - "On the problem of the elastic-plastic bending of an ice cover (In Russian)" Trudy, Leningrad, Arkticheskii i Antarkticheskii Navehno-Issledov, Inst., Vol. 267.

Kingery, W.D. (1962) - Editor of "Project Ice Way" Air Force Surveys in Geophysics, No. 145, Air Force Cambridge Research Laboratories, Office of Aerospace Research, U.S. Air Force.

Lafleur, P. (1971) - "Propriétés mécaniques de la glace de neige en flexion" Thèse M.Sc., Université Laval, Québec.

Lagutin, B.L., Shulman, A.P. (1946) - "Methods of calculating the load carrying capacity of ice crossings (In Russian)" Trudy Nauchno-Issledovatel'skikh Uchrezhdevici, Sverdlovsk, Moscow, seria 5, Vyp. 20, Translated by the Arctic Construction and Frost Effects Laboratory (ACFEL) 1954.

Meneley, W.A. (1974) - "Blackstrap Lake ice cover parking lot" Can. Geotech. J., II, p. 490-508.

Meyerhof, G.G. (1961) - "Bearing capacity of floating ice sheets" Trans. Amer. Soc. Civil Eng., 127, Pt. 1, p. 524-581.

Michel, B., Drouin, M., Lefebvre, L.M., Rosenberg, P., Murray, R. (1974) - "Ice bridges of the James Bay Project" Vol. II, No. 4, p. 599-619.

Nadai, A. (1920) - "Die Biegungsbeanspruchung von Platten durch Einzelkrafte" Schweizerische Bauzeitung, Band 76.

Nadreau, J.P. (1978) - "Etude du fluage de poutres de glace columnaire" Thèse M.Sc. Université Laval, Québec.

Nevel, D.E. (1961) - "The narrow free infinite wedge on an elastic foundation" USA SIPRE R.R. 79.

Nevel, D.E. (1965) - "A semi-infinite plate on an elastic foundation" USA CRREL R.R. 136.

Nevel, D.E. (1966) - "Time dependent deflection of floating ice sheet" USA CRREL R.R. 196.

Nevel, D.E. (1970) - "Concentrated loads on plates" USA CRREL R.R. 265.

Nevel, D.E. (1970) - "Moving loads on a floating ice sheet" USA CRREL R.R. 264.

Palmer, W.T. (1971) - "On the analysis of floating ice plates" Ph.D. Dissertation, New York University.

Panc, V. (1975) - "Theories of elastic plates" Noordhoff International Publishing.

Panfilov, D.F. (1960) - "Experimental investigation of the carrying capacity of a floating ice plate (In Russian)" Izvestia Vsesojuznogo Nauchno-Issledovatelskogo Instituta Iidrotekhniki, Vol. 64.

Panfilov, D.R. (1961) - "To the analysis of a floating ice cover subjected to loads of long duration (In Russian)" Izvestia Vuzov, Stroitel's tvo i Arkhitektura, No. 6.

Panfilov, D.F. (1972) - "On the determination of the carrying capacity of an ice cover for loads of long duration (In Russian)" USA CRREL Draft. Translation 67.

Perrson, B. (1948) - "Durability and bearing capacity of an ice layer (In Swedish)" Svenska Vögfäreningens Tidskrift. USA CRREL Translation 22.

Timoshenko, S., Woinowsky-Krieger, S. (1959) - "Theory of plates and shells" New York, Mc Graw Hill.

Transport Canada (1976) - "Recommanded minimum ice thickness for limited operations of aircraft" Document AK-68-14-001.

U.S.S.R. (1946) - "Instructions of the engineering committee of Red Army (In Russian)"

Wyman, M. (1950) - "Deflections of an infinite plate" Can. J. Res., A, 28, p. 293-302.

# CHAPTER 4

# FORCES EXERTED BY ICE ON STRUCTURES

---

## 4.0 - INTRODUCTION

One of the most important force in the design of an hydraulic or marine structure in northern regions is the thrust applied to it by a solid ice sheet. In fact, for marine works designed for the Arctic, the forces caused by an iceberg or an ice island are so high that it would be practically impossible to design a stable structure in a location exposed to them.

Until very recently, the analysis and design recommendations dealt with river and lake ice problems where the maximum thrust is, most often, that caused by the thermal expansion of a continuous ice sheet. The considerable magnification of the forces that occur in estuaries and in Arctic waters can readily be seen by considering in Fig. 4.1 the strength diagram of $S_2$ ice at $-10^{\circ}C$, as an example for Arctic fresh water ice (either old Arctic ice, icebergs or drained and refrozen pressure ridges). The yield strength of this ice is shown in the domain of thermal expansion where it does not exceed $8 \times 10^5$ Pa. For a large ice mass moving against a structure having an equivalent of 10 m in diameter, the velocities are given which correspond to different ice resistances. For a very slow speed of $2 \times 10^{-2}$ m $s^{-1}$, the ice will attain its maximum strength, which is close to tenfold of what could be expected with thermal expansion for the design of river works, and forces exerted by ice are the dominant forces for the design of Arctic marine works.

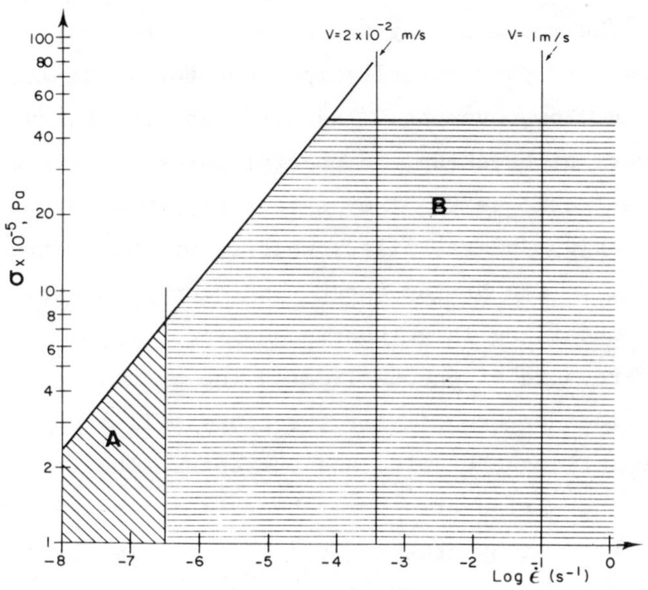


*Fig. 4.1. Domains of thermal expansion (A) and ice impact (B) in the crushing strength diagram σ of $S_2$ ice at -10°C. The velocities of ice fields are given for a structure 10 m in diameter.*

There are essentially four modes of ice action against marine and hydraulic structures that will be treated in this chapter. They are:

1°) Static pressure from expanding or contracting ice sheets. This type of ice movement caused by air temperatures changes induces important stresses in the ice and reactions to the structures. This pressure is particularly important in the case of constrained ice sheets and acts principally in lakes, reservoirs and sheltered coastal areas on dams, dikes and retaining structures.

$2^{o}$) Impact of moving ice sheets, pressure ridges and ice islands. The movement may be caused by wind and, or, water curents. They are particularly important for large ice masses of cold ice in slow movement. These forces are observed in very large rivers at break-up, in coastal waters, large estuaries and in the Arctic seas. They affect all types of marine structures and ships; piers, piles, wharfs, offshore structures, ship hulls, etc... One of the worst condition is slowly moving pressure ice containing large continuous ice masses.

$3^{o}$) Pressure of unconsolidated ice accumulations. When the sizes of floes are too small, the floes will accumulate together under the influence of wind and currents. The forces are transmitted to the marine structures by friction between the ice pieces instead of by crushing of the ice itself. The ice accumulations may be in the form of jams or pack ice acting on many types of hydraulic or marine structures.

$4^{o}$) Vertical forces exerted by ice. The main cause is the movement of an ice sheet adhering to a structure, due to a variation in water level. Vertical forces are also caused by the weight of ice caps adhering to the structures.

## 4.1 - FORCE OF AN EXPANDING ICE SHEET

### 4.1.0 - INTRODUCTION

The most important force for the design of a dam in northern regions is the horizontal thrust applied to the structure by a solid sheet of ice. Much progress has been made in estimating this force, but there is still a long way to go before we get an accurate design value for it.

With time, the analysis of this problem has been more and more refined and design values of this pressure have been successively lowered. At the beginning of the century, for instance, many dams were designed for the condition of crushing ice (Barnes, 1928; Shenehan *et al.*, 1931) which led to a value as high as 75 tons per meter. At present, the plastic properties of ice are being taken into account, and values used by Canadian engineers vary between 15 and 22 tons/m for rigid structures. For more flexible structures, like sluice gates, values in the vicinity of 7 tons/m are more commonly used (Drouin, 1968).

The importance of a proper allowance for ice thrust is apparent when it is appreciated that for a dam 12 m high, an ice thrust of 22 tons/m has the same tendency to tip the dam as the water pressure it is designed to resist.

There have been few reported dam failures caused by ice pressure (Eng. Record, 1899, 1912) and some engineers have doubted the advisability of making substantial allowances for this pressure in the design of hydraulic structures. But measurements made during the last few years leave no doubt as to its importance. A maximum recorded pressure of 5 tons/m was measured on a tainter gate of the Hasting Dam (Hill, 1935) on the Mississipi River during the winter of 1932-33. A panel instrumented (Willmot, 1952) and set flush with the face of Des Joachims Dam during the winter of 1951-52 showed a maximum pressure of 10.4 tons/m. Extensive measurements were made from 1946 to 1951 (Monfore, 1952) at several reservoirs in the mountains of Colorado. Indenter and strain gauges were set in thin mortar panels that could be installed on the face of the walls and at sites exposed to ice thrust. The maximum measured ice pressure at Eleven Mile Canyon was 24 tons/m in 1947-48, 21 tons/m in 1948-49 and 30 tons/m in 1949-50. In the winter 1950-51 tests were made at a number of reservoirs presenting a variety of shore conditions. At Antero and Shadow Mountains, where reservoirs have flat shores, maximum thrusts of 5.4 and 8.7 tons/m, respectively, were measured. At the Evergreen reservoir, with moderately steep shores, the greatest thrust attained 14.1 tons/m. The highest thrust of all, 26 tons/m, was recorded at Tarryall Reservoir which has steep rocky shores in the vicinity of the dam. Similar tests were carried out by Vniig (Korzhavin, 1962) from 1954-56 in the Ges Reservoir of the Dnieper River. Maximum recorded ice pressure was 12.5 tons/m.

The importance of ice thrust in the economic design of hydraulic structures requires it to be evaluated as accurately as possible at each site. Its value depends greatly on meteorological conditions and on the restraint caused by topography of the reservoir shores. It is a rule of thumb (Flinn, 1928) that this force can be ignored in regions where the mean January air temperature is above 4°C.

### 4.1.1 - COEFFICIENT OF THERMAL EXPANSION

Rising air temperature, with or without absorption of solar radiation, causes an ice sheet to expand.

The strains induced by a temperature change are relatively small. They can be calculated from the equation:

$$\dot{\varepsilon} = \alpha \, \dot{\theta} \tag{4-1}$$

where $\dot{\varepsilon}$ - time rate of change of thermal strain, $s^{-1}$

$\dot{\theta}$ - rate of change of the temperature, $^{o}C/s$

$\alpha$ - coefficient of linear thermal expansion.

For ice, the coefficient of expansion decreases with decreasing temperature and according to Drouin and Michel (1971) can be described by:

$$\alpha = (54 + 0.18\ \theta)\ 10^{-6} \text{ in } {}^{\circ}C^{-1} \quad (4\text{-}2)$$

Experimental measurements by Butkovitch (1957) are shown in Fig. 4.2. With an average value of $\alpha = 52 \times 10^{-6}$, $^{\circ}C^{-1}$ an ice sheet 1 km long would expend 104 cm for 20°C temperature rise, if it were free to expand. Considerable forces have to be exerted by the shores of a water body or by the structure to prevent, and compensate for, this type of expansion.

As shown by Eq. (4-1), the strain rate imposed on the ice cover is determined by the rate of temperature rise. A reasonable, practical range of rate of change in temperature to be considered for engineering calculations would be 1 to 10°C/h. This would correspond to a strain rate range of about $10^{-8}$ $s^{-1}$ to $1.5 \times 10^{-7}$ $s^{-1}$, which is way down in the ductile range as can be seen in Fig. 4.1.

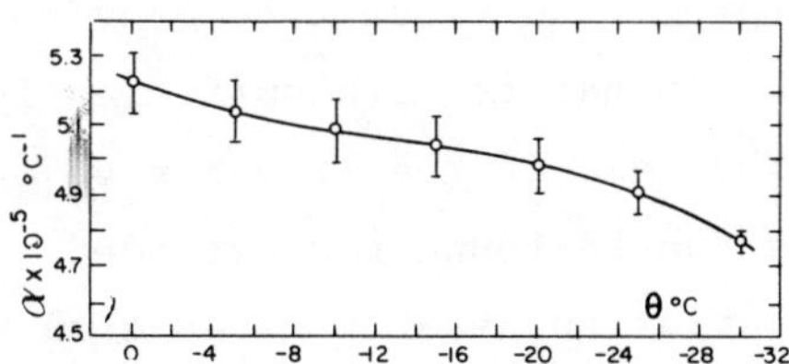


*Fig. 4.2. Mean coefficient α of linear thermal expansion of ice, with standard deviations of measurements (Butkovitch 1957).*

## 4.1.2 - THERMAL EXCHANGE

It is not easy to analyze the air temperature variations occuring in nature at a given site and to choose the values that are liable to produce the maximum thrust for design purpose. In most cold regions the mean rise of air temperature rarely exceeds $3^{\circ}C/hr$ for a long enough interval of time but cases have been known, mainly in the chinook belt, where the temperature rose $22^{\circ}C$ in 40 minutes, the rise being followed by an equally rapid fall with no perceptible interval in between. Many such variations are reported in the literature but because of their very short duration they are not representative of the temperature fluctuation inside the ice sheet itself which has a great thermal inertia.

The best way to approach the meteorological data of air temperatures in a region, for the purpose of obtaining the maximum ice thrust, is to separate the temperature rises in classes with simple laws of rise as a function of time. Three types of simplified analytical laws are representative of temperature fluctuations: the step function where the temperature rises suddenly and stays for a while at the new value, the linear increase for a long enough interval of time, and finally the diurnal sinusoidal variation of the air temperature. The first type of variation cannot be found at many places in nature with sufficiently high values of the step function. The case of the linear rise of the air temperature is universal and can be found for extended intervals of time. The sinusoidal variation usually gives a much larger thrust for the same total temperature rise, because it penetrates deeper into the ice.

Many factors influence the temperature distribution inside an ice cover for a given air temperature fluctuation. There is first a difference between the temperature at the ice surface and the temperature of the air because of the thermal boundary layer in the air. But the difference is small and measurements made for a whole winter (Drouin and Michel, 1971) on clear ice have shown that it rarely exceeds 1.7°C. It is thus safe and acceptable to neglect this effect.

The thermal conductivity of freshly fallen snow is from 10 to 30 times less than that of ice. A few cms of snow on an ice cover acts as a practically perfect insulator and no sensible temperature fluctuation is transmitted to the ice sheet. Fig. 4.3 shows the dramatic effect of 22 cm of snow on the temperature at a depth of 15 cm in a 30 cm ice cover at Eleven Mile Reservoir, Colorado (Monfore and Taylor, 1949). This is one of the reasons why young and relatively thin ice exerts the highest pressure on a dam at the beginning of winter. Little allowance for ice pressure need be made during the period that a snow cover is expected to stay on the ice. It must also be borne in mind that a uniform fresh snow cover will also, by its weight, drown an ice cover of approximately the same thickness, so that the mechanism of heat transfer is completely modified. At least, no thrust is developed while the ice sheet is forming again at the top.

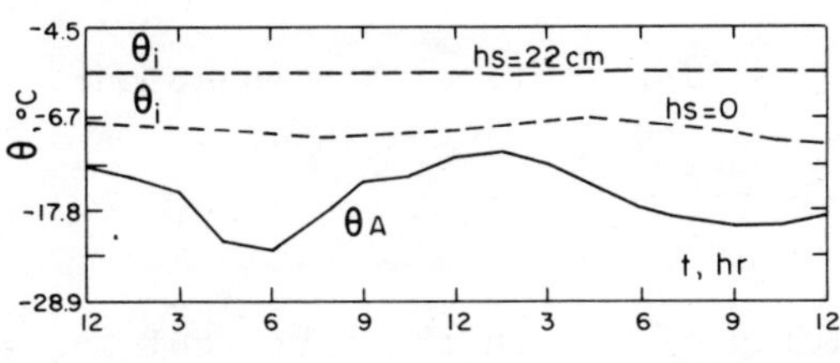


*Fig. 4.3. Effect of a snow cover on the ice temperature at a depth of 15 cm under the ice cover. The ice temperature θi is given for no snow cover (Hs = 0) or a 22 cm snow cover (Hs = 22 cm). (Montfore and Taylor, 1949).*

The computation of the probability of having or not having a snow cover on the ice is one of the main element in determining the thrust that can be exerted by thermal expansion.

The temperature distribution inside an ice cover can be obtained from the Fourier equation:

$$\frac{\delta\theta}{\delta t} = H_i^2 \frac{\delta^2\theta}{\delta z^2} \tag{4-3}$$

where: $\theta$ - temperature of the ice, $^{o}C$

$t$ - time, hours

$z$ - distance of a point from the lower ice surface, m

$H_i^2$ - thermal diffusivity of ice, which can be taken as $1.16 \times 10^{-6}$ $m^2/s$ (Dorsey, 1940).

The equation can be solved for the simpler boundary conditions discussed previously (Taylor, 1945). The simplest case is that of determining the temperature changes ($\Delta\theta$) in an ice sheet originally at a steady state when it is suddently exposed to surroundings below freezing temperature. The computation then gives:

$$\Delta\theta = \frac{K_c\ (h/K)}{1 + K_c}\ \theta_1 \left\{ \left[ \sum_{n=1}^{\infty} A_n \exp\left( - \frac{H_i^2 a_n^2}{h^2} \right) \sin \frac{a_n z}{h} \right] + \frac{z}{h} \right\} \tag{4-4}$$

where

$$A_n = \frac{-2\ (\sin a_n - a_n \cos a_n)}{a_n\ (a_n - \sin a_n \cos a_n)}$$

and the $a_n$ are the successive positive roots of the equation:

$$\tan a = \frac{Ka}{K_c\ h}$$

Here K is the thermal conductivity of ice, h the total thickness of the ice sheet, $\theta_1$ the value of the air temperature rise and

$K_c$ the film conductance of convection in cal/m$^2$ $^o$C hr.

The second case is that of a linear temperature rise at a rate $\dot{\theta}o$. It then gives:

$$\Delta\theta = \frac{K_c}{1 + K_c} \dot{\theta}o \left\{ \sum_{n=1}^{\infty} \left[ A_n \frac{h^2}{H_i^2 a_n^2} \left(1 - \exp\left(- \frac{H_i^2 a_n^2}{h^2} t\right)\right) \sin \frac{a_n z}{h} \right] + \frac{z}{h} t \right\} \quad (4\text{-}5)$$

The third and simplest case is that of a sinusoidal temperature variation at the surface. If we than assume that the ice is a semi-infinite solid, the temperature at any depth is given by:

$$\theta\ (z,t) = \frac{\theta m}{2} \left\{ 1 + \exp\left(- \frac{z}{H_i} \sqrt{\frac{\omega}{2}}\right) \cos\ \omega t - \frac{z}{H_i} \sqrt{\frac{\omega}{2}} \right\} \quad (4\text{-}6)$$

where:

$\theta\ (z,t)$ - temperature at depth z at time t

$\theta m$ - assumed minimum surface temperature

$\omega$ - angular frequency of the temperature variation $\frac{2\pi}{t_o}$

$t_o$ - time for one complete cycle of temperature change.

Eq. (4-6) shows that for the semi-infinite solid, the amplitude of the temperature disturbance decreases exponentially and its phase is retarded linearly with depth beneath the surface.

Differentiating Eq. (4-6) with respect to time gives:

$$\frac{\delta\theta}{\delta t} = \dot{\theta}\ (z,t) = \frac{\theta m}{to} \left[ \exp\left(- \frac{z}{H_i} \sqrt{\frac{\omega}{2}}\right) \sin \left(\omega t - \frac{z}{H_i} \sqrt{\frac{\omega}{2}}\right)\right] \quad (4\text{-}7)$$

If $\theta m$ is $-40^{\circ}C$, and the frequency $t_o$ is one day, the amplitude would be less than 10% of its surface value at a depth of 0.5 m, and the maximum rate would occur 10 h later than at the surface. If the ice cover were about 0.5 m thick, the depth and time dependence of the temperature change induced in it by a daily cyclical variation at the surface would only begin to differ significantly from the semi-infinite case at depths greater than about 0.3 m. An ice cover of thickness greater than 0.5 m is often considered as semi-infinite for temperature computations.

Another important factor affecting the rise of temperature inside an ice cover is the absorption of solar energy. This effect can be introduced in the computation. If we consider only the absorbed heat of solar radiation at the surface of the ice sheet $\psi_r$ in cal/m$^2$ hr, we then obtain:

$$\Delta\theta = \frac{\frac{\psi_r(t)h}{K}}{1 + (K_c + K_r)\frac{h}{K}} \left\{ \left[ \sum_{n=1}^{\infty} A_n \exp\left( - \frac{H_i^2 a_n^2}{h^2} t \right) \sin \frac{a_n z}{h} \right] + \frac{z}{h} \right\} \quad (4\text{-}8)$$

where $K_r$ is the film conductance of radiation in cal/m$^2$ $^{\circ}$C hr.

For the case of temperature change due to absorption within the ice sheet of part of that portion of the solar energy transmitted through the ice, and additional term is required in the heat equation which becomes:

$$\frac{\delta\theta}{\delta t} = H_i^2 \frac{\delta^2\theta}{\delta z^2} + \dot{\theta}_r (t) \exp \left[- \psi_r' h \left(1 - \frac{z}{h}\right)\right] \quad (4\text{-}9)$$

where $\psi_r'$ is the absorptivity of ice and $\dot{\theta}_r$ is the solar volume absorption coefficient of ice at the surface (i.e. temperature change per unit time).

The solution of this equation for the proper boundary conditions is:

$$\Delta\theta = \left\{ \left[ \sum_{n=1}^{\infty} B_n \exp\left( - \frac{H_i^2 \alpha_n^2}{h^2} t \right) \sin \frac{a_n z}{h} \right] + C\ (1 - \exp \psi_r' z) + D \frac{z}{h} \right\} \tag{4-10}$$

where:

$$B_n = \frac{2 \exp[-\psi_r' h]}{(\psi_r' h)^2 (a_n - \sin a_n \cos a_n)} \left\{ a_n (1 - \cos a_n) - \frac{a_n^2}{a_n^2 + (\psi_r' h)^2} \left[ \exp(\psi_r' h)(\psi_r' h \sin a_n - a_n \cos a_n) + a_n \right] \right\}$$

$$+ \frac{\left( \frac{2}{\psi_r' h} + \frac{2}{(\psi_r' h)^2} (K_r + K_c) \frac{h}{\theta_0} \right) (1 - \exp -\psi_r' h)(\sin a_n - a_n \cos a_n)}{1 + (K_r + K_c) \frac{h}{K} (a_n - \sin a_n \cos a_n)}$$

$$C = \frac{\theta_r(t) \exp(-\psi_r' h)}{H_i^2 (\psi_r')^2}$$

$$D = \frac{\frac{\dot{\theta}_r(t) h}{H_i^2 \psi_r'} + \frac{\dot{\theta}_r(t)}{H_i^2 (\psi_r')^2} \frac{(K_c + K_r) h}{K} (1 - \exp - \psi_r' h)}{1 + (K_c + K_r) \frac{h}{K}}$$

By means of the above equations it is possible to compute the change in temperature of an ice sheet for a step or linear function of temperature rise and for absorption of solar radiation. The heat equation being linear, the temperature rise at any point can be obtained by adding up the effects of separate components. This was done (Monfore and Taylor, 1949) for 22°C temperature rise that occurred in 1 hour and 15 minutes in January 1947 at Eleven Mile Reservoir. The results are shown in

Fig. 4.4. The ice temperature is that of ice free of snow at a depth of 15 cm, in an ice sheet 38 cm thick. The computed values are shown by the individual points near the smooth observed ice-temperature curve. The computed and measured values agree well.

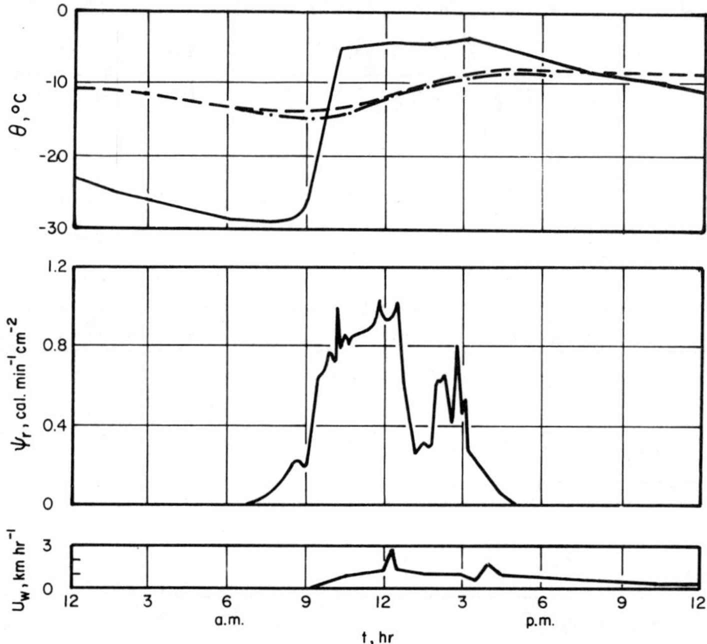


*Fig. 4.4. Computed and measured ice temperatures at a depth of 15 cm in the ice cover. The computed values are in dash line and the measured values in point and dash line. The full line is the air temperature. (Montfore and Taylor, 1949)*

## 4.1.3 - DUCTILE BEHAVIOR OF ICE SUBMITTED TO INCREASING TEMPERATURES

The most direct way of studying the behavior of ice under increasing temperature is to measure the force needed to keep a test sample at constant length between the loading blocks of a testing machine.

Early measurements are those of Monfore (1952) which were carried out on cylinders 10 cm in diameter and 10 cm long taken from lake ice, with the axes of the cylinders parallel to the surface of the ice sheet. Loads applied to the cylinders were thus in the same direction, relative to the crystal structure, as thrusts which develop in an ice sheet in the field.

Unfortunately, Monfore made little study of the crystallographic structure of this ice. Many different types of ice can coexist in the same lake and their creep properties vary over a wide range of values for the same loading conditions.

As far as can be seen from the crystallographic analysis. Monfore was working with pseudo-monocrystalline $S_1$ ice.

In these tests, ice temperatures were controlled by passing air through the test chamber and were measured inside the ice with copper-constantan thermocouples.

The ice in the testing machine was first brought to the initial temperature and was then held at this value until equilibrium was established. In order to obtain a given linear rate of temperature rise in the ice, it was necessary for the air temperature to rise at a much greater rate for about the first 15 minutes. As the temperature of the ice increased, the load on the sample was adjusted to maintain the ice at its initial length. Temperatures were read and the load on the sample was adjusted every 5 minutes for the first 30 minutes of test. Temperatures were read every 15 minutes and the load adjusted twice every 15 minutes thereafter.

The specimen length was kept constant with the help of

a gauge set between each of the loading blocks. The loads to equilibrate the zero deformation were applied with a small hydraulic ram. The ram was operated by a hand pump equipped with a large pressure gauge calibrated directly in total load.

More than 100 tests were carried out, and for each one of them the pressure rose to a maximum and then decreased. Ice temperatures and pressures for a typical test are shown in Fig. 4.5. In this test the ice temperature rose from -23.6°C at a uniform rate of 2.8°C/hr.

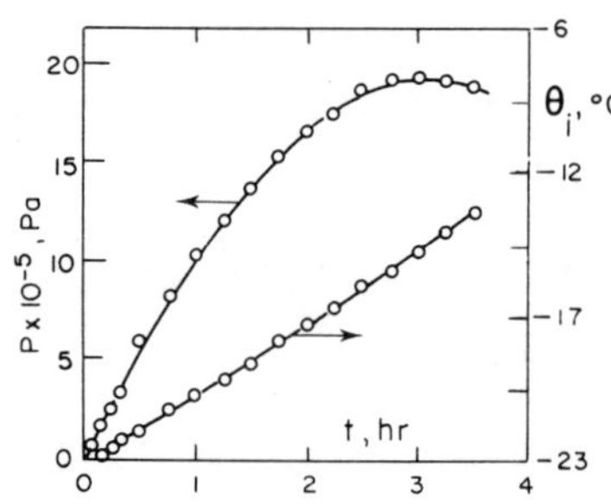


*Fig. 4.5. Ice temperatures and pressures with time.*

Check tests were run on several samples to determine the degree of reproduction of apparatus and techniques. The average deviation in maximum pressures for duplicate tests on a given sample was 6.0%, which was considered satisfactory. Much larger variations, averaging 25%, were found in the maximum pressures reached by different ice samples tested under the same conditions. This seems to depend on the arrangement of crystals in apparently similar samples and it was eventually found that the samples that reached low maximum pressures contained crystal surfaces or other sharp divisions along which slippage might have occurred.

Pressure tests were made with initial temperatures of -34.4°, -23.3° and -12.2°C and with rates of temperature rise in the ice of 1.1°, 2.8°, 5.6° and 8.3°C/hr. Curves similar to those of Fig. 4.5 were obtained for each test. For a given rate of temperature rise, the maximum pressure reached by the sample increased as the initial temperature decreased; for a given initial temperature the maximum pressure increased as the rate of temperature rise increased. The relationships between the pressures and time for various initial temperatures and rate of temperature rise are shown in Fig. 4.6.

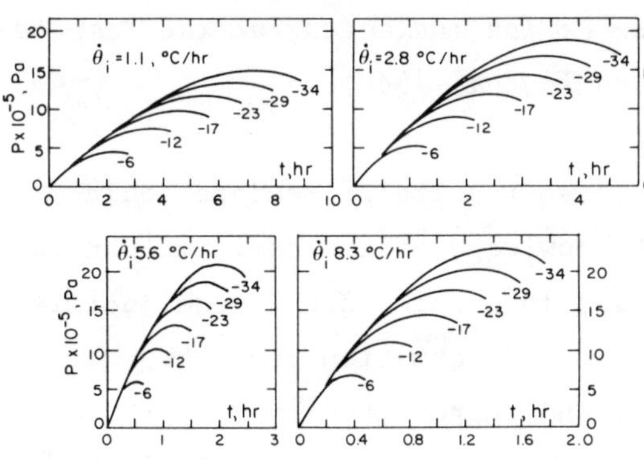

*Fig. 4.6. Average pressure-time of 1.1, 2.8, 5.6 and 8.2°C/hr temperature rises from indicated initial temperature in °C. (Montfore and Taylor, 1947).*

All these results have basically the same form if they are compared on a non-dimensional basis as shown in Fig. 4.7. The only two characteristic parameters are the maximum pressure and the time required to obtain this pressure (Fig. 4.8).

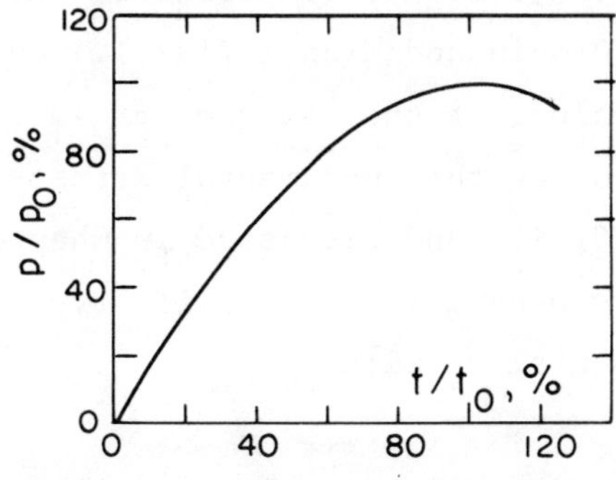


*Fig. 4.7. Curve showing percent of maximum pressure and percent of time to reach it.*

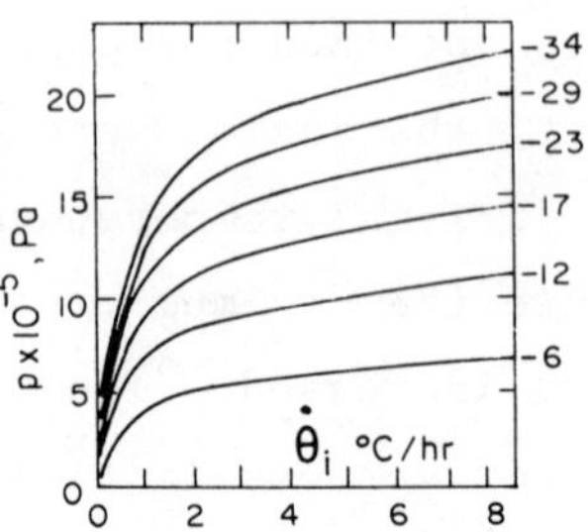


*Fig. 4.8. Curves showing maximum pressure* p *in function of rate of temperature rise* $\dot{\theta}i$ *for indicated initial temperatures in* °C *(Monfore and Taylor, 1949).*

In a broad sense, these results of Monfore could be interpreted with the present knowledge of the basic mechanical behavior of $S_1$ ice as discussed in Chapter 2. In the ductile range, the ice follows the law of deformation given by $\varepsilon\alpha$ type creep under a constant stress and represented by Eq. 2-95. At a given temperature, the stress-strain curve is entirely known. These curves for the range of temperature variation in a sample are shown in dashed lines in Fig. 4.10. For a linear variation of temperature, the corresponding average strain rate can then be computed from Eq. (4-11) and the stress-strain curve is obtained for the variable temperature or shown in Fig. 4.9. This has been done for a large range of initial ice temperatures and rates of temperature variation by Drouin and Michel (1971). On Fig. 4.10 are shown the computed values of the maximum stress obtained from basic computations, using the fundamental stress-strain relationships observed for $S_1$ ice and discussed in Chapter 2. The basic theory of the ductile behavior for this ice type applies very well to these experimental results.

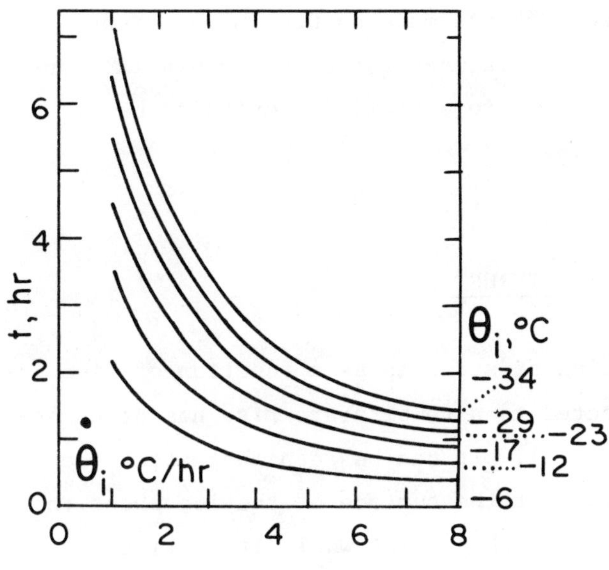


Fig. 4.9. Curves showing time to reach maximum pressure in function of rate of temperature rise $\dot{\theta}i$ for indicated initial temperatures in °C. (Monfore and Taylor, 1949).

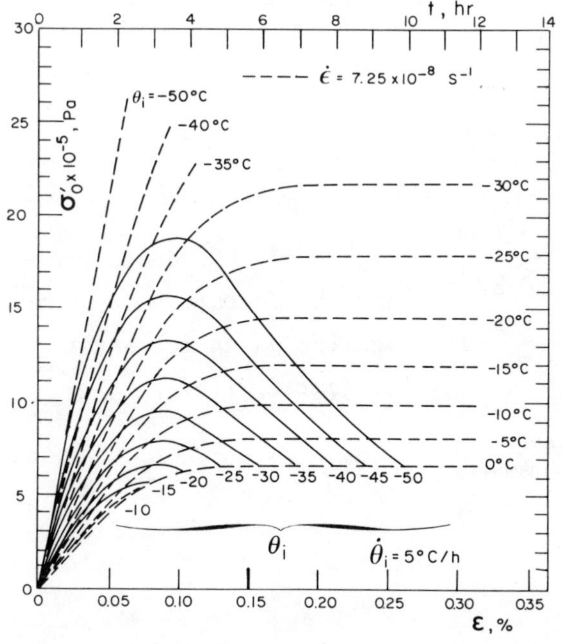


Fig. 4.10. Typical computed stress-strain curve at a constant temperature rise of $\dot{\theta}i$ = 5°C/h in full line from the creep behavior of $S_1$ ice at various temperatures (Drouin and Michel, 1971).

Laboratory measurements made by Drouin and Michel (1971) of the thermal stress induced for complete biaxial constraint indicates that the ratio σ biaxial/σ uniaxial has a maximum value of about 1.85.

## 4.1.4 - COMPUTATION OF ICE THRUST

The distribution of stresses as a function of time inside an ice cover subjected to a temperature rise has been meaured as shown on Fig. 4.11. It can be seen that after the maximum stress is attained at the surface of the ice sheet the stress starts to decrease at this point while it continues to increase underneath. Thus, a few measurements show generally that the total maximum thrust occurs slightly after the maximum stress is attained at the upper surface and its value may be from 10 to 20% higher.

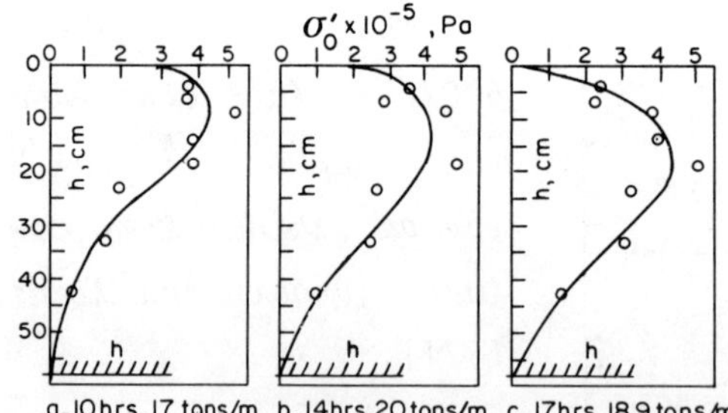


*Fig. 4.11. Measured pressure distribution in a cross section of an ice sheet (Lofquist, 1954).*

Two types of computations can be made to obtain the magnitude of the ice pressure. One analysis was made by Michel (1970) for the case of a linear increase in temperature, using Monfore experimental results, and limited to the thrust developed when the maximum stress was attained at the upper surface of the ice.

The total thrust was computed with the usual assumptions for this problem (Rose, 1947):

$1^{o}$) The initial air temperature is $-34.4^{o}$, $-23.3^{o}$ and $-12.2^{o}C$.

$2^{o}$) The initial ice temperature varies linearly from $-34.4^{o}C$ at the upper surface to $0^{o}C$ at the lower surface.

$3^{o}$) The air temperature changes at rates of $1.1^{o}$ to $8.3^{o}C/hr$.

$4^{o}$) The temperature of the ice at the surface is assumed to be the same as the air temperature.

$5^{o}$) The thickness of the ice sheet remains constant.

The temperature distribution and the rate of temperature distribution and the rate of temperature rise inside the ice sheet were computed with Eq. 4.5 for variable ice sheet thicknesses, air temperature rises and initial ice temperatures.

Fig. 4.12 shows that for an ice sheet of given thickness the maximum thrust is attained for an air temperature rise of only $1.7^{o}C/hr$. This can be explained by the fact that at this rate there is enough time for the ice to develop an important overall thrust before the maximum pressure is attained at the surface when the computation is stopped. The curves also show an inflection point at a rate of about $4.2^{o}C/hr$. This can be explained from the rheological behavior of Monfore's ice in this area (Fig. 4.8) where there is a sharp change in the variation of the maximum pressure that this ice can exert.

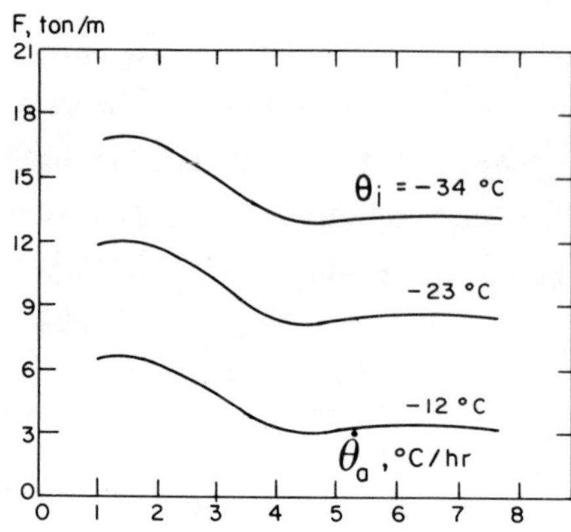


*Fig. 4.12. Thrust of an ice sheet in function of the rate of linear rise of air temperature (Michel, 1970).*

The effect of ice thickness is shown in Fig. 4.13. The thrust does not increase appreciably for ice sheets thicker than 30 cm, mainly because there is no sensible variation of temperature below this point before the stresses start to decrease at the upper surface. Furthermore, if we also consider that snow covers are related to thicker ice sheets, there is little doubt that the maximum thrust does not increase appreciably for ice sheets thicker than 30 cm.

When the ice sheet is free of snow and exposed to sunlight, its temperature may have an added increment by absorption of solar radiation, quite apart from the change caused by an increase in air temperature.

The magnitudes of ice thrusts were computed to include solar energy effects with the following additional assumptions:

$7^{o}$) The latitudes are $40^{o}$, $50^{o}$ and $60^{o}$.

$8^{o}$) The vernal equinox is the time limit for the formation of clear solid ice of adequate thickness.

$9^{o}$) The atmospheric transmission constant is 0.9, based on a clear atmosphere at a fairly high elevation.

$10^{o}$) The coefficient of heat transfer is 11.36 J/m s$^{o}$C, to allow for surface losses of the absorbed solar energy,

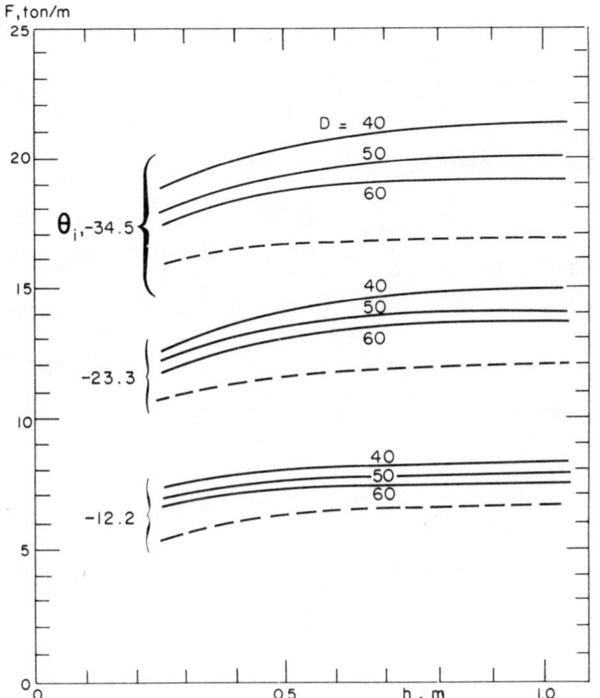


*Fig. 4.13. Maximum thrust exerted for a linear rise of temperature for ice of various thickness at three latitudes and three initial air temperatures (Michel, (1970).*

For these assumed conditions the solar energy that reaches the surface of the ice sheet during midday is about 789 Joules/$m^2$ s at $40^{\circ}$ latitude, 631 J/$m^2$ s at $50^{\circ}$ latitude and 473 J/$m^2$ s at $60^{\circ}$ latitude. The absorptivity of ice is small in the visible spectrum (Dorsey, 1940) but beyond a wave length of 1 micron it is great and is analogus to that of water. Consequently, about 40% of the solar energy reaching the surface, essentially all that of wavelength greater than 1 micron, is absorbed within 10 cm of the surface. For a clear ice sheet 30 cm thick an additional 15% of the energy is absorbed, whereas for a thickness of 120cm this additional absorption amounts to about 25%. Most of the remainder of the energy is transmitted although a small percentage is reflected.

The effect of solar energy was computed using Eq. (4-9) and the results (Fig. 4.13) indicate that the absorption of this energy may produce a considerable increase in thrust above that caused by the rise of the air temperature.

Fig. 4.13 is valid only for uniaxial conditions. If there is a complete biaxial restraint, the value of the thrust should be increased by up to 85% to take this effect into account.

Another computation of the ice thrust has been made by Drouin and Michel (1971) in the case of a sinusoidal air temperature variation. It was found that the condition giving the maximum thrust is for an air temperature variation of the form:

$$\theta a = \theta o + A \sin \frac{2 \pi t}{To} \qquad (4\text{-}11)$$

for

$$0 < t < T_o/2$$

where

$\theta o$ - is the initial air temperature

To - is the period of the cycle

As for the case of a linear temperature rise, it was assumed that the initial temperature varied linearly from $\theta o$ at the upper surface to $0^{o}C$ at the lower surface.

All temperature variations in function of time can then be easily computed inside the ice cover with Eq. (4-6).

The thrust can be then computed numerically at discrete levels in the ice by computing the temperature variation, the corresponding strain rate, and then by computing the stress. This was done by Drouin and Michel (1971), both for snow ice $T_1$ and pseudo-monocrystalline ice $S_1$, with a rheological model similar to that of α creep described in Chapter 2 giving the stress at constant strain rates with Eq. (2-146).

The results of the numerical, step by step, computations are given on Figs 4.14 and 4.15 for $T_1$ and $S_1$ ice.

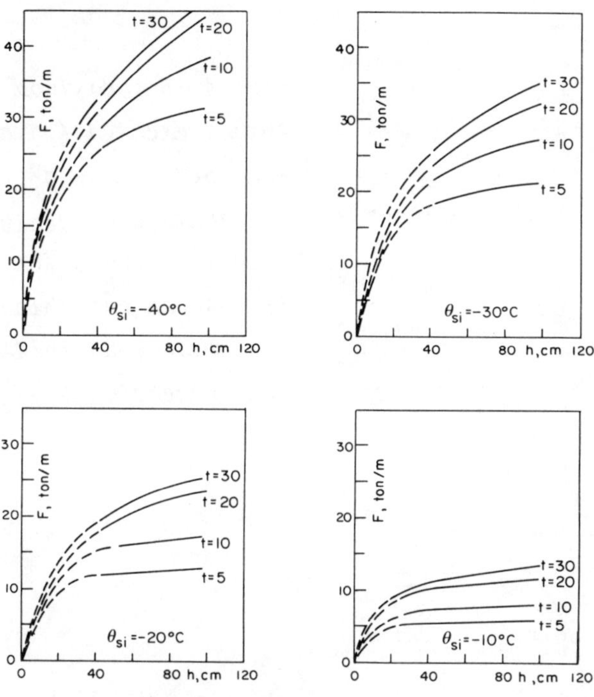


*Fig. 4.14. Maximum thrust exerted for a sinusoidal rise of temperature of duration t in hours for $T_1$ ice of various thickness with four initial air temperatures θsi.*

It can be seen, from these curves, that the highest thrusts are not the result of fast increases in temperature, but on the contrary, they are the results of the lowest rates of increase. This can be explained by the fact that the thermal wave gets deeper into the ice before there is a reduction of the stress over yield point at the surface. The main fundamental limitation of these computations is the fact that they are established from measurements on samples of new ice, where there are few initial mobile dislocations. In nature, once the ice cover has been stressed a few times by thermal movements, the number of mobile dislocations will increase considerably and the yield strength will drop accordingly. This will tend to reduce appreciably the thrust caused by thermal expansion.

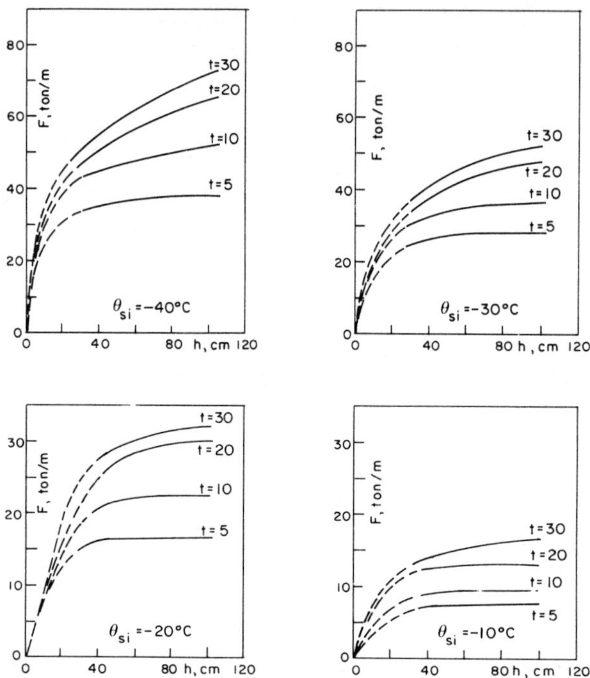


*Fig. 4.15. Maximum thrust exerted for a sinusoidal rise of temperature of derivation t in hours for $T_1$ ice of various thickness with four initial air temperatures $\theta_{si}$.*

It can also be seen, from these computations that the thrust does not increase very much with an increase in ice thickness, over thicknesses of about 30 cm. This is due to the very strong damping of temperature waves inside the ice cover, as discussed previously.

It is believed, by the author, that the maximum thrust exerted by thermal expansion will occur very early in the season after the formation of the first ice under a prolonged cold weather period, when the ice will be newly formed, uncracked, and still bare of snow.

The pressure of an expanding ice sheet on the tip of a protruding spillway pier set in the ice is somewhat higher than the pressure per unit length multiplied by the width of the pier. A first approximation of this value can be obtained with an empirical formula (Petrunichev, 1954):

$$F = p_{max} \, (Bo + B'/3) \qquad (4\text{-}12)$$

where:

F - maximum horizontal force, kg

$p_{max}$ - static pressure per unit length obtained for this case, kg/m

Bo - width of the pier, m

B' - span between piers, m.

For the case of a bridge pier this force has to be taken into account only if it is not balanced by an equal and opposite force on the downstream side of the pier. Because, in a general way, the turbulence of the flow in the wake of a pier might prevent ice from forming close to it, it might be safer in most applications to consider this force on the upstream edge.

Furthermore, the lateral ice pressures on a bridge pier usually balance each other on both sides. There are many cases, however, in high-velocity flow, where a channel may stay open in the winter between two bridge piers. If an ice sheet then forms on the other side of these particular piers, full allowance has to be made for lateral pressure. Such a state of loading may greatly affect the stability of piers.

## 4.1.5 - PROTECTION AGAINST STATIC ICE PRESSURE

At an existing dam, designed with no allowance for ice pressure, the following procedure may be used to protect against expanding ice. Generally, all dams are designed for the full reservoir condition with the water surface at or very near the top of the dam. It is possible to compensate for the effect of ice thrust by lowering the reservoir level during the season when ice pressure may exist, so that the total load at the critical section is not larger than for the full reservoir condition.

Fig. 4.16 (Rose, 1947) shows the required lowering of a reservoir to balance the effect of the indicated ice thrusts. The criterion used for evaluation in Fig. (4.16 a) is the moment at the base, whereas Fig. (4.16 b) shows results with horizontal shear as the governing factor. The moment effect governs (that is, yields larger reductions in the water level) for all heights greater than those at the maximum points of the curves in Fig. 4.16 a. The height above which the moment governs and below which shear governs is shown by the lines of "moment-shear". The base of a dam is usually the critical section, but the curves can be applied to any other elevation by considering the height above the elevation in question. The curves are applicable to a gravity-type dam.

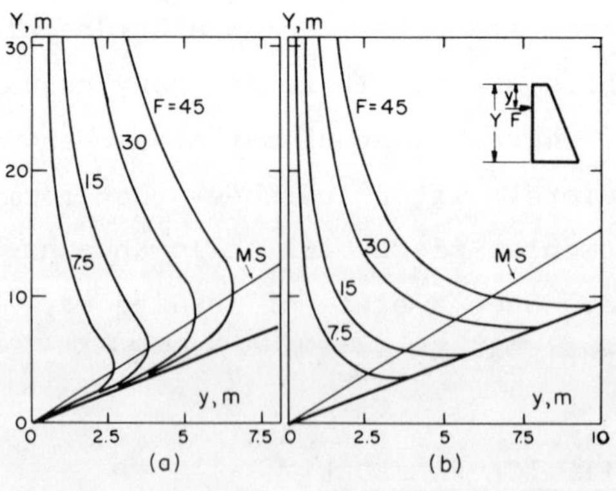


*Fig. 4.16. Reduction in water level behind a dam (y) for various ice thrust F (tons/m) and height of dam (Y). The M or S line is the limiting moment shear line.*

The curves indicate that in the case of high dams the ice thrust is of relatively minor importance in its effects on forces at the base of the section, whereas for low dams it is of considerable importance. For high dams, it should be borne in mind that an elevation at, say, mid-height might give a condition more severe than at the base elevation, even though the latter is the location of maximum stress for full reservoir. Complete consideration of this factor requires actual stress and stability analysis.

The best-known method of preventing an ice sheet from pushing on a dam is to maintain an open water gap in front of the structure. This can be done in various ways. A permanent solution is to install a system of piping along the edge of the wall at the bottom of the reservoir so that compressed air can be used to create a vertical current bringing the warm water at the reservoir bottom to the top, thus preventing ice formation. This system works well as long as there is an adequate vertical temperature gradient in the reservoir. In turbulent flows this gradient is very small and there is normally not enough heat stored in the bottom layer of water to prevent ice formation at the top.

A wooden boom laid along the perimeter of the wall of a reservoir has been found effective in protecting it against ice pressure (Cook, 1941). Although no complete explanation was ever given for this beneficial effect it seems that the boom acts as a hinge when the ice rises and falls and as a roller when it expands and contracts. Another solution, which has never been tried, would be the setting of plastic foam, or rubber panels at the water line on the walls of the structure. If thick enough, these materials could take a good amount of deformation without transmitting any sensible thrust to the wall.

In an emergency, a gap a few inches wide can be made by hand with ice chisels, saws or steam heating. Chemicals might also be used to melt and considerably weaken the ice in front of the endangered structure. No unassisted relief to the situation can be expected before an adequate snow cover has formed on the ice sheet.

## 4.1.6 - OTHER EFFECTS

Observations on lakes free of snow have shown that reefs and pressure ridges formed by the expansion and buckling of an ice sheet seldom occur when the ice is thicker than 30 cm, although they have been known to form in ice 50 cm thick (Rose, 1947). This phenomenon might be interpreted to indicate that for ice sheets less than 30 cm thick the critical buckling load is the limiting factor governing the ice pressure on a dam, whereas for thick ice sheets the pressure may develop to its maximum without buckling of the sheet. These observations have been somewhat correlated by a computation in the first approximation, assuming very approximate plastic ice properties, where it is shown that buckling is a limiting factor when the ice thickness is less than 50 cm thick (Lofquist, 1944).

When an ice sheet increases in thickness in a reservoir and the temperature falls, tensile stresses are set up in the ice, giving rise to contraction cracks. When the ice is heated up again, the gap in the crack, if it has not already frozen, will first have to be taken up by the expanding ice before any thrust can be exerted. This phenomenon thus reduces the force exerted by an expanding ice sheet.

Expansion and contraction of lake ice due to temperature changes has long been known to be responsible for the formation of ice ramparts along the shore beaches, or of pressure ridges in the ice itself.

The underlying principle involved in the formation of ice ramparts has been well established (Buckley, 1900). A rapid fall in air temperature causes the ice on a lake to contract, so that tension cracks are developed in the ice sheet. These cracks immediately fill with water, which refreezes, so that the total mass of the ice cover is increased. A subsequent rise in air temperature results in expansion of the ice, so that the total surface area is greater than it was before contraction. This increase in area of the ice cover results in compressive forces against the shore so that shore debris is forced into ridges called "ice ramparts".

Extensive measurements were made of this phenomenon at Wamplers Lake, Michigan, during the winters of 1951-52 and 1952-53 (Wilson *et al*, 1954). The first winter a movement of ice on the shore of 49 cm was measured. The expansion movement of the ice was not normal to the shore but was related to the form of the lake; more expansion was observed along the long axis than along the short one on this oblique lake. In the second winter the lake level was lower and a new rampart was built at a lower level from boulders and gravel that were normally submerged beneath the water of the shore region.

On one occasion the atmospheric temperature dropped from $-1^{\circ}$ to $-13^{\circ}C$ in 12 hours and cracks developed with widths ranging from a few tenths of a cm to 4 cm. The ice thickness at that time was 20 cm and the snow cover was 9 cm. One set of cracks radiated from the central part of the lake and another set roughly followed the shoreline.

Let us note that the formation of a compression ridge was observed only once during these two winters, when the ice was 20 cm thick. The axis of this ridge lay at right angles to the direction of movements of the sheet as recorded by markers. Prior to this time no buckling had taken place, but considerable thrust had developed on the shore.

## 4.2 - IMPACT OF DRIFTING ICE

### 4.2.0 - INTRODUCTION

Most older river bridges of northern countries end their lives by pier movement following scour and ice pressure at breakup time. Very little has been written on this unfortunate event, although it would be quite revealing for the design of these structures. There is no doubt that one of the most important forces that can lead to the collapse of a bridge is the force of moving ice floes but only the failure of larger bridges under ice conditions has been, at times, reported in the literature. Examples are the Niagara Falls View bridge and the Brattleboro bridge (Pratley, 1938; Roads and Bridges, 1945; Kirham, 1927; Eckhard, 1920).

Also reported by Danys (1972) are the failure of many light-houses in the St. Lawrence river and Great Lakes under the impact of ice.

Ice may adversely affect a bridge in a number of ways. Premature snow melt in a river basin may suddenly increase the river flow, which in turn raises the solid ice sheet adhering to the piers partially lifting them out and endangering the stability of their foundation. The piers may be located in such a manner on the river bed as to hinder the passage of ice, and to prime ice jams. These increase the pressure on the piers, and the superstructure may be destroyed if the ice is raised high enough to impinge on it. The water may also find a new channel within or around the ice jam and high-velocity flow may badly scour the approaches or the pier foundation. Finally the moving ice floes at breakup usually exert an appreciable force when hitting the bridge piers. This force may displace a pier enough to cause the collapse of the bridge.

Moving ice may also impinge on other types of hydraulic structures. Such is the case of dam facings or spillway gates and of wharf piles or offshore structures. Much interest is now brought up by the design of Arctic structures where the ice conditions are much different than those of bridge piers in rivers.

Codes have been set up for the computation of the impact force of ice on bridge piers. The 1974 Canadian Standard Association Code allows for a nominal effective pressure on a vertical face pier varying from 6.9 to 27.6 x $10^5$ Pa without making a distinction between the nominal pressure and the actual compressive strength of ice. The U.S.S.R. Code SN 76-66 for river structures at spring break-up recommends effective pressure of 3 to 13.2 x $10^5$ Pa. These codes can be used only for the purpose for which they were prepared and cannot be extrapolated to conditions of cold winter ice, estuarine or Arctic ice; or for any mode of failure other than compression.

## 4.2.1 - MOVEMENT OF ICE FLOES

Both wind and moving water exert a tangential stress on an ice cover. This wind or water stress can be estimated using the equation:

$$\tau = \rho \ Cd \ U^2 \tag{4-13}$$

where: $\tau$ - tangential stress imposed by the wind or water on the ice cover

U - the wind speed or water current measured at a distance d from the respective surface

Cd - the drag coefficient associated with the height at which U is measured

$\rho$ - the density of the air or water.

It is usually assumed that the wind speed or current velocity varies logarithmically with distance from the surface. The speed at distance d can be determined from measurements at distance z using the equation:

$$U = \frac{Uz \ (\ell n \ d - \ell n \ z_o)}{(\ell n \ z - \ell n \ z_o)} \tag{4-14}$$

where $z_o$ is a measure of the roughness of the surface. The drag coefficient depends on the roughness of the surface and is related to $z_o$ by the equation:

$$Cd = k^2 \left[ \ell n \left( \frac{d + z_o}{z_o} \right) \right]^{-2} \tag{4-15}$$

where k is Von Karman's constant (usually taken as 0.4). Measured values of Cd and associated values for $z_o$ are given in Table 4.1 along with a qualitative description of the roughness of the surface.

The values given for the drag coefficients are for a surface of uniform roughness. Ridges are usually a nonuniform feature that must be considered separately. There is little information on the drag coefficient for ridges. One calculation for an assumed ridge density of 7 per kilometer and ridge height greater than 0.6 m gave a value for the drag coefficient for wind of the same order as that for uniformly rough surfaces (i.e. 0.0015) (Banke and Smith, 1973).

| Height (or depth) of measurement m | $C_d$ x $10^3$ | $z_o$ x $10^3$ | Nature of surface | Source |
|---|---|---|---|---|
| 10 (air) | 0.95 | 0.023 | smooth snow covered surface | (1) |
| 10 (air) | 2.61 | 3.98 | gently rolling hummocked | (1) |
| 10 (air) | 3.3 | 9 | rafted | (2) |
| 0.5 (water) | 5 | 1.7 | smooth | (3) |
| 0.5 (water) | 20 | 30 | rough | (4) |
| 0.5 (water) | 47 | 94 | very rough | (4) |

Table 4.1 - Measured drag coefficients, $C_d$, for ice surfaces. (1) Banke and Smith (1973) - measured in Beaufort Sea, Arctic Ocean, Robeson Channel. (2) Langleben (1972) - measured in Beaufort Sea. (3) Michel (1971) - River ice. (4) Johannessen (1970) - measured in Gulf of St. Lawrence and Arctic Ocean.

If the subscript a is used for air and w for water stresses, it can be seen from Table 4.1 that:

$$1.3 \times 10^{-3}\ Ua^2 < \tau_a < 4.4 \times 10^{-3}\ Ua^2$$

where $\tau_a$ is in Pa and Ua in m/sec.

The usual value of $C_d$ for the ice-water interface is:

$$5 \times 10^{-3} < C_d < 2 \times 10^{-2}$$

The corresponding water stress in Pa, is usually between

$$5\ Uw^2 < \tau_w < 20\ Uw^2$$

Once an ice floe has attained an equilibrium position under the action of wind, its velocity in permanent regime is shown on Fig. 4.17. It is a function of concentration of hummocks. For an isolated floe its maximum drift velocity cannot be more than 3% of Uw.

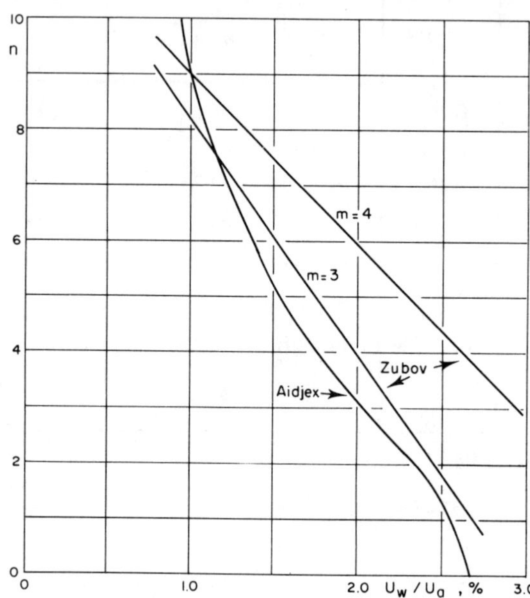


*Fig. 4.17. Drift velocity Uw of ice floes in function of wind velocity $U_A$ and ice concentration n in tenths.*

In transitory regime the drift velocity is not a constant and increases with time. The equation of movement is:

$$(\tau_a - \tau_w)\ A = m\ \frac{dUw}{dt} \tag{4-16}$$

With initial conditions t = 0, Uw = 0 we have the solution:

$$\frac{Uw - K_1}{Uw + K_1} = -\ \exp\ (-2K_1K_2t) \tag{4-17}$$

where

$$K_1{}^2 = \frac{\tau_a}{\tau_w}\ Uw^2 \quad \text{and} \quad K_2 = \frac{Uw^2}{\rho' \tau_w h}$$

With the maximum value of the wind stress and the lesser value of the water resistance we obtain the maximum possible drift of the ice floes. The equilibrium velocity is then:

$$Uw \simeq 0.03 \ U_a \qquad (4\text{-}18)$$

## 4.2.2 - INDENTATION BY A VERTICAL FACE STRUCTURE

### 4.2.2.0 - INTRODUCTION

The destruction of an ice floe by a pier with a vertical edge occurs in different ways, depending mostly on the stored kinetic energy of the floe and its strength. Upon impact with a pier a small ice floe is locally crushed, stops and then continues to move across an adjacent opening. A slightly larger floe generally collapses by splitting while a large ice field is cut without splitting. Finally, under certain conditions the ice will buckle at a certain distance from the face of the pier. During the penetration of a pier into a large ice field, the width of indentation gradually increases and attains a maximum value equal to the total width of the pier, after which it remains constant.

### 4.2.2.1 - GENERAL RELATIONSHIPS

If we consider an ice floe hitting a pier, the energy transmitted to the obstacle is made up of the energy of deformation inside the ice floe and the energy required for its crushing. As the energy of deformation is negligible compared with the energy of crushing for sizeable impacts, the fundamental relationship of mechanics gives:

$$F \, dx = m \, v \, dv \qquad (4\text{-}19)$$

where F is the force exerted on the obstacle for a displacement dx of the floe, and m and v are the mass and instantaneous velocity of the floe.

On the other side, at crushing, we have:

$$F = \sigma_i \; y \; h \tag{4-20}$$

where $\sigma_i$ the crushing strength of ice by indentation, y is the width of the failure zone and h is the thickness of the ice.

For an ice floe, with an initial velocity $V_o$, that is completely stopped by the pier we get, using Eqs (4-19) and (4-20):

$$\int_{x=0}^{x=xo} h \; \sigma_i \; (x) \; y \; dx = \frac{1}{2} \; m \; V_o^{\;2}$$

Because of the geometry of the pier, the term on the left has an upper limit when x = d and y = B where d is the length of the pier nose and B is the total width of the pier.

If the kinetic energy in Eq. (4-21) is adequate to attain this upper limit, the floe will not stop and the maximum force will be given by:

$$H_{max} = \sigma_i \; B \; h \tag{4-22}$$

In the case where the indentation strength of the ice can be taken as independent of the indentor width y, the force required to stop an ice floe of a limited size can be easily computed:

$$Ao = \int_{o}^{x=xo} y \; (x) \; dx = \frac{m \; V_o^{\;2}}{2 \; \sigma_i \; h} \tag{4-23}$$

$$H = \sigma_i \; y_o \; h \tag{4-24}$$

$A_o$ is thus the horizontal crushed surface of the ice floe when it stops as shown on Fig. 4.18 and can be determined for any type of pier shape (Michel *et al.*, 1965). Actually, $y_0$ is the pier width corresponding to the $A_o$ area.

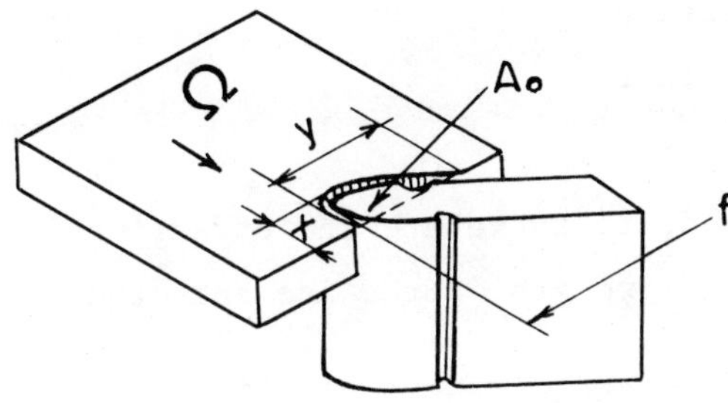


*Fig. 4.18. Crushing of an ice floe against a pier.*

For the case of a triangular-nosed pier of angle 2α attacked by an ice floe along its axis we obtain the Petrunichev (1954) formula for limited size floes:

$$H_{max} = V_o \; h \; \sqrt{2\rho' \sigma_i \; \Omega \; \tan \alpha} \tag{4-25}$$

where
- $H_{max}$ - maximum horizontal force
- $V_o$ - water velocity
- $h$ - ice thickness
- $\rho'$ - density of ice
- $\Omega$ - area of the ice floe
- $\sigma_i$ - indentation strength of ice

### 4.2.2.2 - CRUSHING OF ICE AGAINST A VERTICAL FACE

The relatively simple case of indentation of ice by vertical face structures show that the largest forces are associated with crushing along the contact perimeter in a more or less continuous manner. The force acting on the structure is usually defined by the effective ice pressure, p:

$$F = p\ b\ h \tag{4-26}$$

F - force on the structure

p - effective ice pressure

b - width of the structure

h - thickness of the ice

The main question, then, is to relate the effective ice pressure to the actual crushing strength of the ice and other geometrical factors.

Tests done by Michel and Toussaint (1977) with plates of $S_2$ ice at -10 C, indented with a square edge indentor have shown that the failure mechanism differs considerably in function of the rate of indentation. Three cases have to be considered:

a) Pure ductile behaviour

Under pure ductile behaviour, micro-cracks are formed under the indentor as shown on Fig. 4.19. These cracks are oriented at random to the main shear direction and their length is of the order of magnitude of the crystal diameter. They tend to propagate to the full width of the plate along the longitudinal axis of the crystals. As the indentor progresses, the concentration of these minor cracks increases and a zone of continuous fissured material of whitish appearance is formed right under the indentor. This zone, which we might call the zone of plastification, extends to about 1.6 times the indentor width where yield occurs. It was also noted, in all tests with a ductile behaviour, that very few cracks will appear at a distance beyond 2.5 times the indentor width.

This process of pure ductile behavior appears to be valid with increasing strain rates up to an effective indentation rate of $2.5 \times 10^{-5}$ $s^{-1}$.

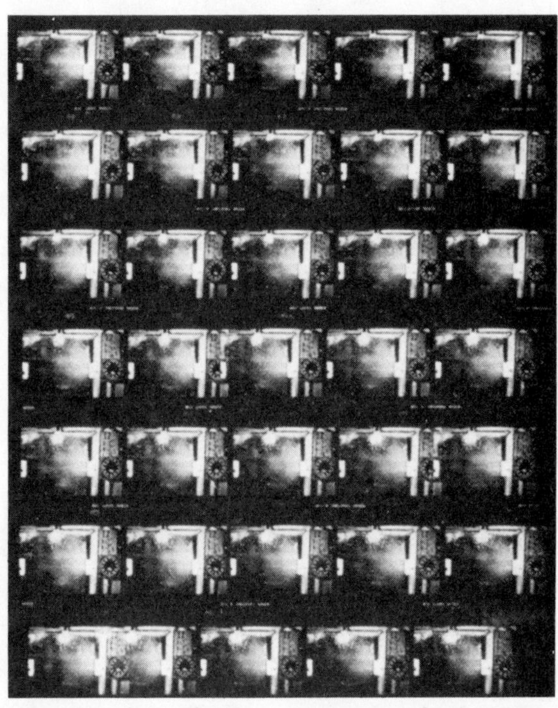

*Fig. 4.19. Formation of micro-cracks in pure ductile indentation. For loading curve see Fig. 4.20.*

The test for that rate is shown on Fig. 4.19. The corresponding recording of load versus displacement is shown on Fig. 4.20. It can be seen that the recording is continuous but that there is a large reduction in effective pressure after the peak has been attained. This reduction is in the order of 40%.

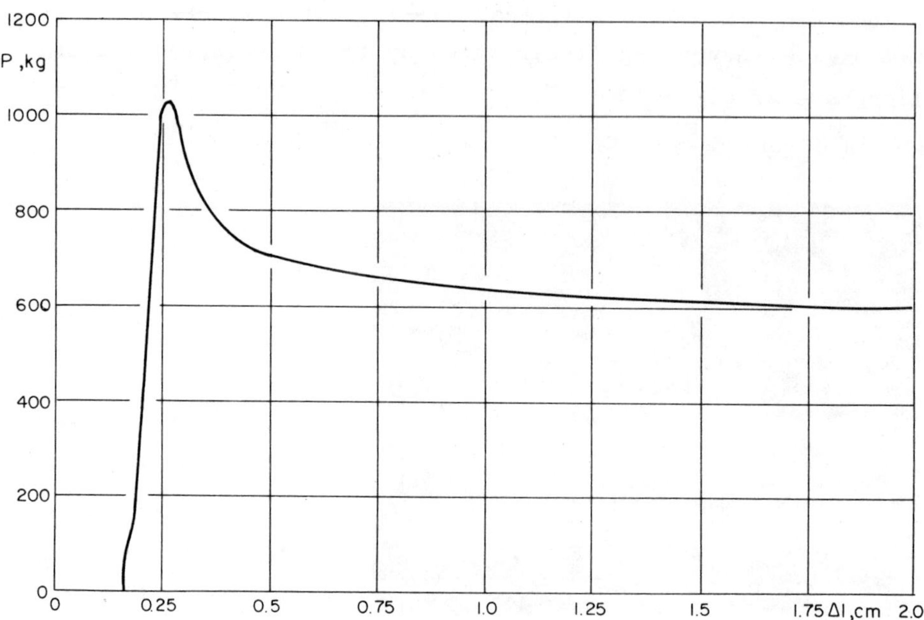


*Fig. 4.20. Loading curve for indentation test in the ductile zone. Load P versus displacement $\Delta \ell$. Indentor width $b = 2.54$, cm, thickness $h = 5.09$ cm, velocity of indentation $v = 2.54 \times 10^{-6}$ m/s.*

We have previously used the expression effective strain rate, which need to be defined in such a way that it will be compatible with measurements of ice strength under uniaxial conditions in the ductile range. As said previously, it was observed in all tests that the area of micro-crack production did not extend beyond 2.5 times the width of the indentor and that this value appears to be a constant whatever be the indentor's width, the thickness of the plate or the rate of indentation in the ductile range. With this observation we have set up a physical model of deformation mechanisms shown in Fig. 4.21. Close to the indentor is the plastification zone where the lines of slippage are shown, followed by a circular area of micro-cracking and ductile behavior of the material. Outside of this area, the ice is deformed very little in the elastic range.

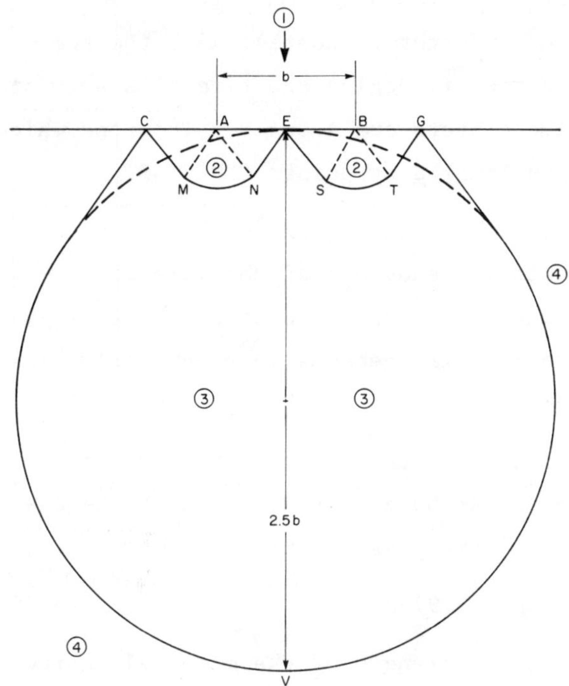


*Fig. 4.21. Physical model of indentation process; (1) indentor, (2) zone of plastic flow, (3) zone of ductile deformation with micro-cracks, (4) zone of elastic deformation without crack.*

From this model and taking into account the biaxial effect we can define the effective strain rate as the rate of deformation in the zone of ductile behavior with the expression:

$$\dot{\varepsilon} = V/4b \qquad (4\text{-}27)$$

where V is the rate of indentation and $\dot{\varepsilon}$ the equivalent uniaxial strain rate in $s^{-1}$.

Fig. 4.21 shows that close to the indentor we may use, accroding to Hill (1956) a zone of perfectly plastic material where the glide lines satisfy the equations of plastic theory. The four equilateral triangles - CMA - ANE - ESB - BTG represent regions where the glide lines are straight and are making a $45^{\circ}$ angle with the plane of the indentor in contact with the semi-infinite material. Both corner triangles are tied with segments of circles whose respective centers are A and B. The zone which deforms plastically is then limited by points CMNESTGS.

Adjacent to this fully plastic zone of large deformation there is a zone of ductile behaviour of the material extending to the limit CVG where the maximum distance is EV = 2.5b. Outside of this limit, the material behaves in a perfectly elastic manner as said previously.

With the plastic theory, the nominal pressure of indentation under the indentor can be computed. With the Huber-Von Mises failure criterion, this pressure is:

$$p = 2.97\ \sigma_o \tag{4-28}$$

where $\sigma_o$ is the uniaxial yield strength of the material in the zone of ductile behaviour.

Let us now compare the computed yield strength of ice obtained from indentation measured directly in uniaxial compression. This is done in Fig. 4.22 where the strength diagram for $S_2$ ice at $-10^{\circ}C$ given in Chapter 2 is reproduced.

It may be observed that the ice strength increases in the ductile range up to a maximum for $\dot{\varepsilon}$ close to $5 \times 10^{-4}\ s^{-1}$ and then stays a constant for higher strain rates.

For the indentation tests, results from Michel and Toussaint (1977), from Frederking and Gold (1975) and some results from Hirayama *et al.*, (1974) were reduced with the value of C = 2.97 and plotted on the same Fig. 4.22.

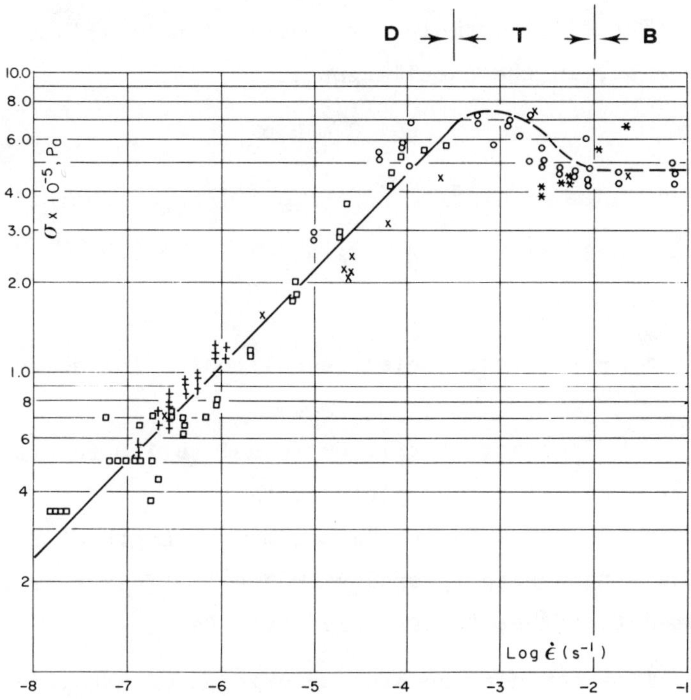


*Fig. 4.22. Universal curve for uniaxial crushing and indentation of $S_2$ ice at -10°C; Michel and Paradis, 1976 (• uniaxial), Carter and Michel, 1971 (o uniaxial), Frederking and Gold 1975 (+ indentation) Hirayama et al., 1974 (* indentation), Michel and Toussaint, 1977 (× indentation). We note the ductile zone (D), the transition zone (T) and the brittle zone (B).*

It can be seen that the representation of Fig. 4.22 is universal both for uniaxial compression and indentation, and that the fit is very good for all tests in the ductile and transition zones. For brittle fracture, the indentation process is however completely different and the theory of ductile behavior does not apply.

The final formula for indentation for plates of $S_2$ ice would then be, in the ductile zone:

$$p = C\ m\ k\ \sigma_o$$
$$\sigma_o = \overline{\sigma}_y\ (\dot{\varepsilon}/\dot{\varepsilon}_o)^{0.32} \qquad (4\text{-}29)$$

valid for: $10^{-8}\ s^{-1} < \dot{\varepsilon} < 5 \times 10^{-4}\ s^{-1}$

with: C - 2.97

m - form coefficient equal to 1.0 for a rectangular indentor

k - contact coefficient equal to 1.0 for full contact and 0.6 after, or with incomplete contact

$\overline{\sigma}_y$ - 7000 k Pa is the maximum strength of $S_2$ ice at $-10^\circ C$ at a strain rate $\dot{\varepsilon}_o$ of $5 \times 10^{-4}\ s^{-1}$. This strength could be computed for other temperatures with Eq. (2-154).

This formula can also be written with Eq. (4-27), which has the same form as that given by Frederking and Gold (1971):

$$p = C'\ m\ k\ \overline{\sigma}_y\ V_o^{0.32}\ b^{-0.32} \qquad (4\text{-}30)$$

where C' is a coefficient depending in this type of ice which is equal to $5.64 \times 10^{-2}$ for $S_2$ ice at $-10^\circ C$.

Hirayama *et al.*, (1974) found experimentally that the dependance of ice pressure on the indentor's width was expressed with an exponent equal to -.32 of the width. Because the indentor's width intervenes directly in the strain rate by Eq. (4-27), the so-called size effect can be explained simply by a fundamental understanding of the creep properties of ice. Seaki *et al.*, (1978), working with sea ice, suggest an exponent value of -0.50, but their data would better be represented with a smaller value like -0.39 as shown on Fig. 4.23.

For $S_2$ sea ice, Isnard (1978) has shown that an equation of the same type could be used. At -10°C, his relation for a flat indentor is:

$$p = 9.66 \times 10^4 \, \dot{\varepsilon}^{0.22} \text{ in k Pa} \qquad (4\text{-}31)$$

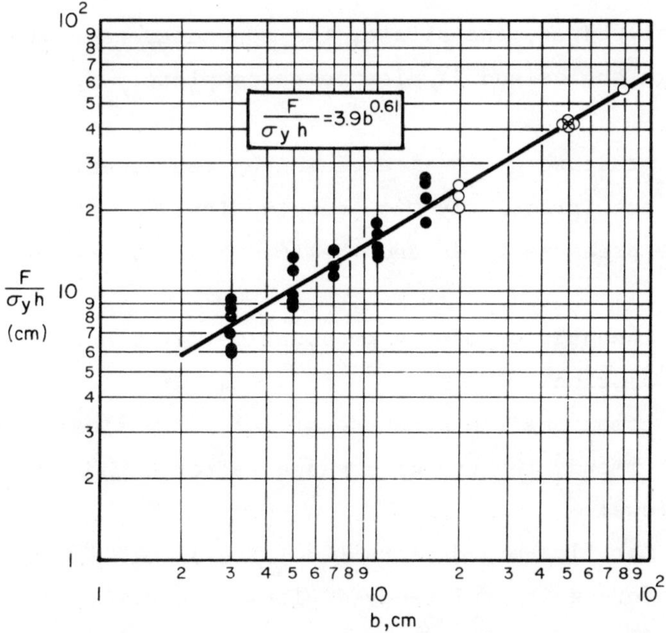


*Fig. 4.23. Tests of indentation by a flat edge pile on sea ice. o Points are with a small rig for thin ice. • Points are for field tests with coastal ice floes. (Seaki et al., 1978).*

The shape of the cutting edge of the indentor influences the strength of ice considerably according to experimental investigations made by Korzhavin (1962). He studied the effect of six different indentor shapes: semicircular, flat and wedge shapes, the latter with wedge angles $2\alpha = 120^{\circ}$, $90^{\circ}$, $75^{\circ}$ and $60^{\circ}$. The tests were made at $0^{\circ}C$ and the rate of indentation was 1.5 cm/min. Altogether 73 tests were made of which 52 were with lateral confinement.

He summarized his experimental results as follows:

$1^{\circ}$) The shape of the crushed zone near the indentor depends on the shape of the indentor. For flat indentors it is semicircular. For wedges, the size of the zone decreases with decreasing wedge angles and it disappears completely for wedges of $2\alpha = 60^{\circ}$.

$2^{\circ}$) The pressure on the ice caused by the indentor increases continuously but irregularly with penetration. At the instant of cracking, the pressure falls and the load has a vibrating character.

$3^{\circ}$) The indentation of ice produces mostly a brittle failure with crack formation.

$4^{\circ}$) For computing the pressure of large ice floes on piers, Korzhavin recommends the use of a shape factor m whose values are given as follows:

a) for flat-edged piers ($2\alpha = 180^{\circ}$), $m = 1.0$.

b) for semicircular-edged piers, $m = 0.90$.

c) for wedge-shaped piers ($60^{\circ} < 2\alpha < 120^{\circ}$).

$$m = 0.85 \sqrt{\sin \alpha} \tag{4-32}$$

More recently, Seaki *et al.*, (1978) have found experimentally, relative values of $m = 1$ for a flat-edged indentor, $m = 0.735$ for a cylindrical pile and $m = 0.66$ for a wedge at $\alpha = 45^{\circ}$. Isnard (1978) finds a coefficient of 0.54 for a wedge with $2\alpha = 102^{\circ}$.

b) Failure in the Transition Zone

For an effective indentation rate of 6.25 x $10^{-5}$ $s^{-1}$ there is an important evolution of the failure mechanism in the Michel and Toussaint tests. Because of the increase in the indentation rate, the stresses concentration at the corners of the indentor are high enough to propagate major cracks through the plate before yielding can occur in the ductile manner. These macro-cracks separate the plate in many sections but they do not attain the physical limits of the constrained plate. On Fig. 4.24, the cracks at the corner make an angle of about 30° with the vertical. Other macro-cracks may also appear with increasing strain rates.

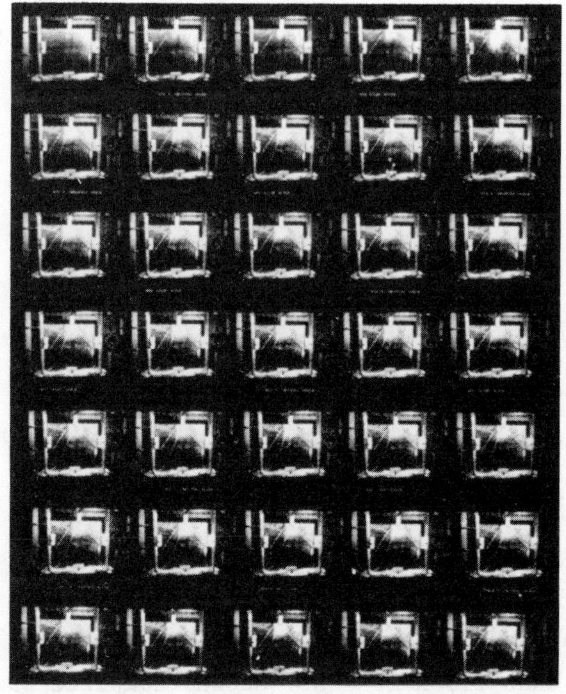

*Fig. 4.24. Macro-crack formation during indentation in the transition zone. For loading curve see Fig. 4.25.*

Because of the fact that these macro-cracks cannot fail the constrained plate as such, the mechanism of micro-crack formation is still repeated under the indentor as for pure ductile behavior and the yield strength continues to increase with strain rates up to a limit around $5 \times 10^{-4}\ s^{-1}$.

Fig. 4.24 shows the failure that gave the maximum pressure during the investigation by Michel *et al.*, (1977) at a rate of $2.43 \times 10^{-3}\ s^{-1}$. The recording of the load is given on Fig. 4.25. It can be seen that because of major cracking within the plate, the recording fluctuates after the first peak and the residual load is only about 25% of the peak.

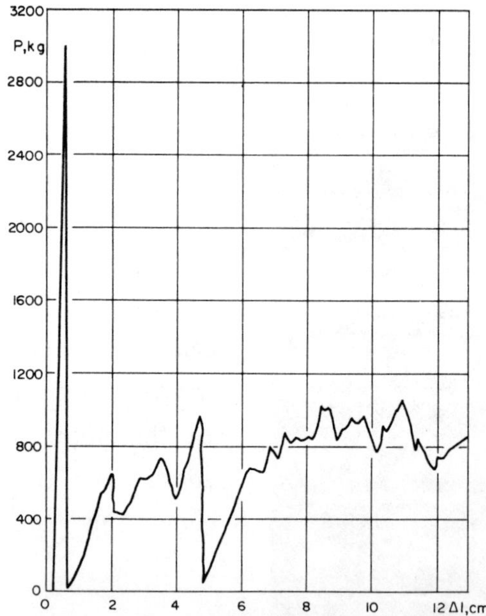


*Fig. 4.25. Loading curve for an indentation test on the transition zone. Load P versus displacement $\Delta l$. Indentor width $b$ = 5.08 cm, ice thickness $h$ = 5.08 cm, velocity of indentation $v$ = $4.93 \times 10^{-4}$ m/s.*

c) Pure brittle fracture

Five tests carried out by Michel and Toussaint (1977) at strain rates higher than $2.3 \times 10^{-2}\ s^{-1}$ showed complete brittle fracture. They were characterized by major crack formation, peeling of the ice and very little micro-crack formation. In fact, no plastification zone could be seen in front of the indentor.

The major crack that formed were similar to those observed during the transition from ductile to brittle behavior. The first micro-cracks appeared at the corners of the indentor making an angle between $30^{o}$ to $40^{o}$ with the vertical. Other major cracks were formed and sometimes a vertical fissure appeared separating the plate in two sections. However, these major cracks did not produce the failure of the plate.

The failure of the plate appeared to be produced by peeling of the ice on both sides of the plate under the indentor. Sometimes the peeling was made over a very short distance under the indentor. Other times, one or two cleavage cracks would appear and extend inside and in the plane of the plate. They would then reappear on the sides to form a semi-circular scale of ice that fell off. This cleavage process is very similar to that observed by Shadrin *et al.*, (1962) and Hirayama *et al.*, (1974), the latter suggesting that they are caused by tension inside the ice plate because of the lateral restraint of the indentor. The remaining ice layers are then finally crushed in very small pieces that fell off the side of the plate in a manner that could probably be analyzed with transverse shear as done by Tryde (1973).

The recording for five tests in that range shows that the cleavage cracks that propagate under the indentor, followed by crushing, are similar to those obtained for conditions under uniaxial testing, A typical test is shown on Fig. 4.26.

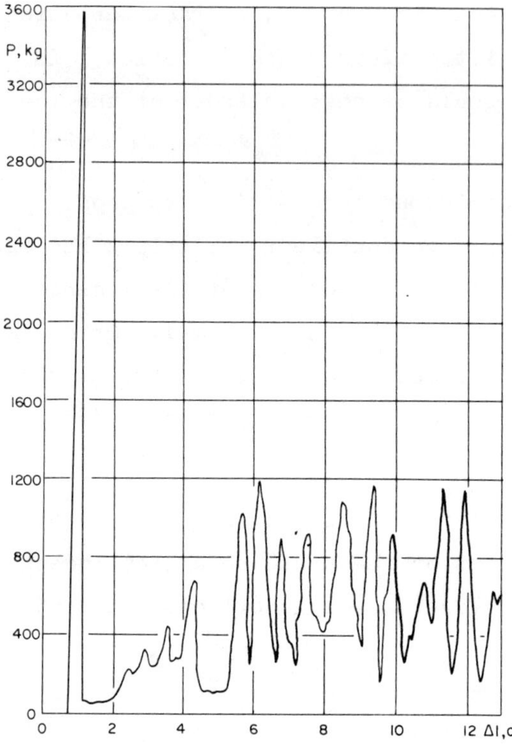


*Fig. 4.26. Loading curve for an indentation test in the brittle zone. Load P versus deplacement Δℓ. Indentor width b = 5.08 cm, ice thickness h = 5.08 cm velocity of indentation v = 3.30 × $10^{-2}$ m/s.*

Isnard (1978) has shown that the pressure for brittle fracture can be determined simply from the theory of pure elastic deformation of the plate with fracture determined by the Tresca criteria. Consider a plate loaded uniformly on a length 2a as shown on Fig. 4.27.

The stresses at point P are given by (Durelli, Phillips and Tsao, 1958):

$$\sigma_x = -\frac{P}{\pi}\left[\text{arc tg}\left(\frac{y}{x-a}\right) - \text{arc tg}\left(\frac{y}{x+a}\right) + \frac{2\,ay\,(x^2-y^2-a^2)}{[(x+a)^2+y^2]\,[(x-a)^2+y^2]}\right] \tag{4.33}$$

$$\sigma_y = -\frac{P}{\pi}\left[\text{arc tg } \frac{y}{x-a} - \text{arc tg}\left(\frac{y}{x+a}\right)\frac{2ay\ (x^2-y^2-a^2)}{[(x+a)^2+y^2]\ [(x-a)^2+y^2]}\right] \tag{4-34}$$

$$\tau_{xy} = -\frac{P}{\pi}\left[\frac{4\ a\ x\ y^2}{[(x+a)^2 + y^2]\ [(x-a)^2 + y^2]}\right] \tag{4-35}$$

with the notation:

$$\text{tg } \alpha_2 = y/(x-a)$$
$$\text{tg } \alpha_1 = y/(x+a)$$

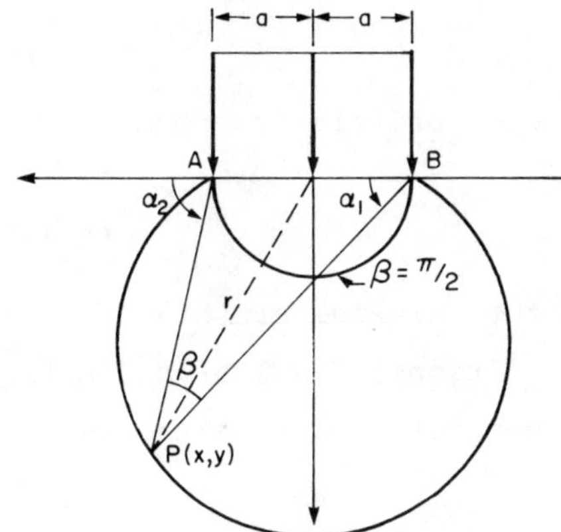


*Fig. 4.27. Edge loading by a flat indentor over a semi-infinite elastic plate.*

Eqs (4-33), (4-34) and (4-35) become:

$$\sigma_x = -\frac{P}{\pi}\,[(\alpha_2 - \alpha_1 + \tfrac{1}{2}\ (\sin 2\alpha_2 - \sin 2\alpha_1)] \tag{4-36}$$

$$\sigma_y = -\frac{P}{\pi}\,[(\alpha_2 - \alpha_1 - \tfrac{1}{2}\ (\sin 2\alpha_2 - \sin 2\alpha_1)] \tag{4-37}$$

$$\tau_{xy} = \frac{P}{2\pi}\,(\cos 2\alpha_2 - \cos 2\alpha_1) \tag{4-38}$$

The principal stresses can readily be obtained from these relations:

$$\sigma_1,\ \sigma_2 = -\frac{P}{\pi}\,[(\alpha_2-\alpha_1) \mp \sin\,(\alpha_2-\alpha_1)] \tag{4-39}$$

With the Tresca criteria:

$$\frac{\sigma_1 - \sigma_2}{2} = \tau_{max} = \frac{P}{\pi} \sin (\alpha_2 - \alpha_1) \qquad (4\text{-}40)$$

The maximum value of $\tau_{max}$ is for $(\alpha_2 - \alpha_1) = \pi/2$, which gives:

$$P = \frac{\pi}{2} (\sigma_1 - \sigma_2) \qquad (4\text{-}41)$$

The experimental results obtained by Isnard with a flat indentor on $S_2$ ice at $-10^{\circ}C$, for the brittle zone, have shown that the indentation coefficient C was sensibly equal to 1.57 given by the elastic theory.

The indentation pressure under brittle fracture can then also be expressed by:

$$p = C\ m\ k\ \sigma' \qquad (4\text{-}33)$$

valid for $\dot{\varepsilon} > 10^{-2}\ s^{-1}$ where $\sigma'$ is the uniaxial crushing strength of the ice under brittle conditions, C = 1.57 and k is of the order of 0.30 for incomplete contact, which is the usual case of practical interest.

Table 4.2 shows the rate of advance of an ice sheet that produces the maximum pressure and the limiting speed that gives the least pressure; the first one at a strain rate of $5 \times 10^{-4}\ s^{-1}$ in the ductile range and the second at $10^{-2}\ s^{-1}$ at the beginning of the pure brittle fracture. In applications with drifting ice, it can be seen that velocities of impact would normally be high enough to produce fracture in the brittle manner. The possibility of the maximum pressure occuring in the ductile range seems to be remote and limited only to the case of small speed movements of large ice fields set up by the wind.

During brittle fracture of an impinging ice sheet, the force oscillates during the crushing process. The frequency of this fluctuation depends primarily on the speed of the floe and the size of the crushed ice which is cleared in front of the indentor, between successive contacts. Observations by Michel and Toussaint (1977) and Hirayama *et al.*, (1974) show that this zone has a width $\delta$ of:

$$0.25 < \delta < 0.5h$$

| b-m | v-m/s $\dot{\varepsilon} = 5 \times 10^{-4}\ s^{-1}$ | v-m/s $\dot{\varepsilon} = 10^{-2}\ s^{-1}$ |
|---|---|---|
| 1 | $2 \times 10^{-3}$ | $4 \times 10^{-2}$ |
| 5 | $1 \times 10^{-2}$ | $5 \times 10^{-2}$ |
| 10 | $2 \times 10^{-2}$ | $1 \times 10^{-1}$ |
| 20 | $4 \times 10^{-2}$ | $2 \times 10^{-1}$ |
| 30 | $6 \times 10^{-2}$ | $3 \times 10^{-1}$ |

Table 4.2 - Critical rate of advance of ice sheet.

Thus the frequency of the oscillation is given by:

$$f = V/\delta \qquad (4\text{-}42)$$

For values of V between 0.3 to 1 m/s and h from 0.6 to 1 m thick, the frequency of the crushing oscillation is between 0.6 and 6 $s^{-1}$. Particular attention must be given if this frequency is close to that of the natural frequency of the vertical structure.

### 4.2.2.3 - BUCKLING OF ICE FIELDS

For larger structures and a semi-infinite ice field the maximum pressure is usually limited by conditions of buckling of the ice at a certain distance from the structure. Sodhi *et al.*, (1977) have made the analysis by the finite element method of buckling of an elastic plate on an elastic foundation.

The basic equation of the plate behavior due to in-plane loading only is, according to Timoshenko and Gere (1961):

$$D \Delta^2 w + kw = N_{xx} \frac{\delta^2 w}{\delta x^2} + 2 N_{xy} \frac{\delta^2 w}{\delta x \delta y} + N_{yy} \frac{\delta^2 w}{\delta y^2} \qquad (4\text{-}43)$$

where: $N_{xx}$, $N_{xy}$, $N_{yy}$ are in-plane stress resultants (force per unit length).

With the boundary conditions for an in-plane load P applied over a width b, the finite element computation is carried on a computer program. The results of the lowest buckling load is plotted in non-dimensional form $P/k\ell^3$ with respect to the $b/\ell$ ratio in Fig. 4.28. The full line is obtained from the computation and it tends to approach the value of the non-dimensional concentrated buckling load ($P/k\ell^3 = 3.32$) as the value of $b/\ell$ approaches zero. For high values of the $b/\ell$ ratio, the curve tends to approach the Hetenyi's (1946) buckling load for a semi-infinite beam on an elastic foundation, which is shown by the dashed line on the figure. The results of the analysis can be approximated with the following formula:

$$\lambda = \frac{P}{k\ell^3} = \frac{b}{\ell} + \frac{3.32}{1+\frac{b}{4\ell}} \qquad (4\text{-}44)$$

where k is the foundation's modulus 1000 $kg/m^3$ and $\ell$ the characteristic length:

$$\ell = \sqrt[4]{\frac{E\ h^3}{12(1-\nu^2)\ k}} \qquad (4\text{-}45)$$

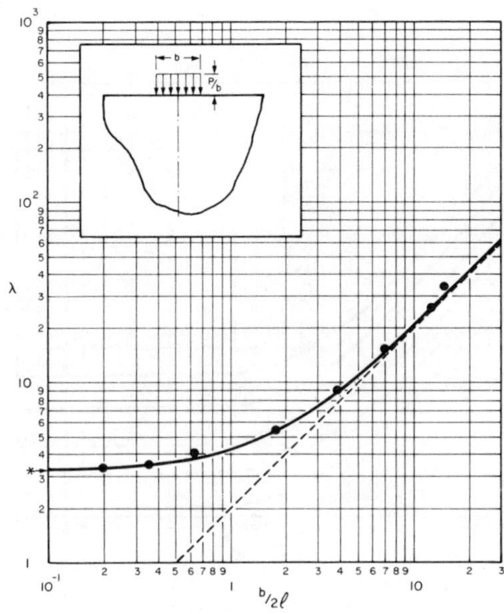


*Fig. 4.28. Buckling load of an ice sheet. The points represent computations made by finite element analysis (Sodhi et al., 1977)*

Numerical computations with values of E = $10^{10}$ Pa $\nu$ = 0.33 and h = 0.5, 1.0 and 2.0 m have been made and the buckling pressure p/b is represented in full lines in Fig. 4.29. Buckling conditions can also be estimated for very slow movements of the ice sheet. The ductile behavior of ice might perhaps be presented by an elastic-plastic response where the elastic modulus is reduced to included the delayed elasticity. Taking a value of E = 2.5 x $10^9$ Pa, the buckling conditions are computed and shown in dashed lines in Fig. 4.29. For large structures, thin ice sheets and slow movement, buckling will control instead of failure by indentation. Each condition of loading has to be computed to determine the lower load that causes failure.

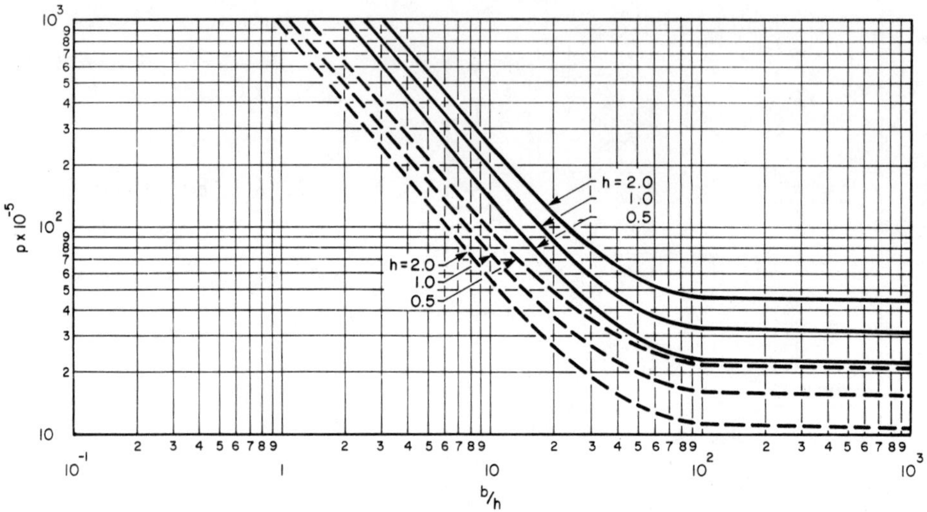


*Fig. 4.29. Limiting buckling pressure. In full line for the elastic condition, in dashed line, the ductile condition.*

## 4.2.2.4 - SPLITTING OF ICE FLOES

We will now consider the case of an intermediate size floe which is not big enough to be stopped by the pier nor crushed but is split along a line of minimum resistance. The minimum resistance may be given either by shear at an angle on both sides of the pier or on the axis of the pier, or by tension cracks forming in the floe.

Shear cracks. Let us consider a floe of size L x $B_o$ x h (Fig. 4.30) whose size is not big enough to attain the critical value that will produce pure crushing. The shape of the failure crack can be taken as a straight line. Contact forces between the pier and the floe can be reduced to two normal forces T and two friction forces F. For equilibrium conditions (Korzhavin, 1962):

$$P - 2\ T \sin \alpha - 2\ F \cos \alpha = 0 \qquad (4\text{-}46)$$

Because we have:

$$F = \mu\ T$$

where $\mu$ is the friction coefficient of ice on the pier surface, we obtain:

$$T = P/2\ [(\sin \alpha + \mu \cos \alpha)] \qquad (4\text{-}47)$$

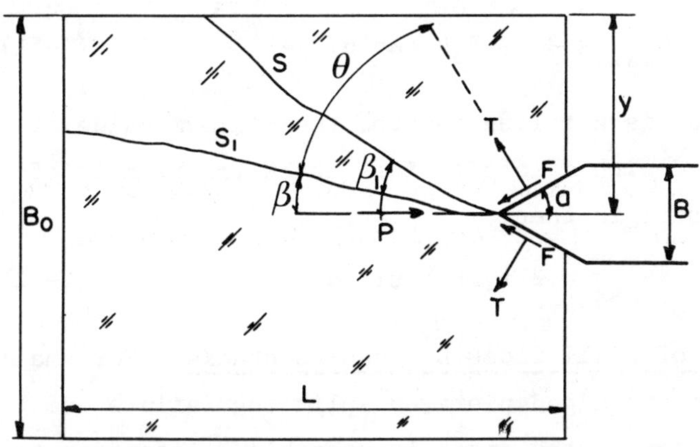


*Fig. 4.30. Splitting of an ice floe by a pier having a vertical edge.*

Assuming the length of the crack $S_1$ we get along the line of shear failure:

$$T\ (\cos \theta + \mu \sin \theta) = hS_1\tau_o \qquad (4\text{-}48)$$

where $\tau_o$ is the average splitting stress in the section corresponding to the shearing strength of the ice. It is evident that such an averaging process is permissible only for relatively small floes. Neglecting friction forces in the first approximation we get:

$$P = \frac{2\ \tau_o\ h\ S_1\ \sin\alpha}{\sin(\alpha + \beta)} \qquad (4\text{-}49)$$

When the crack reaches the side of the floe, we have $S_1 = (B_o/2)\sin\beta$. The force P becomes:

$$P = \frac{\tau_o\ h\ B_o \sin\alpha}{\sin\beta\ \sin(\alpha + \beta)} \qquad (4\text{-}50)$$

The splitting occurs for $P_{min}$, when:

$$\beta = (90^o - \alpha)$$

And the maximum axial force acting on the pier with this lateral splitting is:

$$H_{max} = 2\ \tau_o\ B_o h\ \tan\alpha/2 \qquad (4\text{-}51)$$

If the floe is not long enough, the minimum value of the force is attained with a longitudinal crack in the axis of the pier and this force is then:

$$H_{max} = 2\ \tau_o\ L\ h\ \sin\alpha \qquad (4\text{-}52)$$

Splitting of small floes by tension cracks. For small ice floes and very sharp edged piers, a splitting failure due to normal tensile stresses at the crack is possible. According to Korzhavin the following expression is suggested for the case:

$$\beta = \alpha/2 \qquad H_{max} = n\ L\ h\ \sigma_t \qquad (4\text{-}53)$$

where $\sigma_t$ is the tensile strength of ice and n is a shape factor for the pier varying as a function of the wedge angle according to Table 4.3.

In all cases of floe splitting, the force on the pier is always smaller than the maximum force required to crush the ice by complete indentation through the floe. Thus in the case of large floes the splitting is always partial with the formation of cracks of limited length.

| Pier angle $2\alpha$ | $60^\circ$ | $70^\circ$ | $80^\circ$ | $90^\circ$ | $100^\circ$ | $120^\circ$ |
|---|---|---|---|---|---|---|
| n | 0.25 | 0.29 | 0.33 | 0.38 | 0.43 | 0.53 |

Table 4.3 - Shape factor, n, and pier angle $2\alpha$.

## 4.2.3 - FAILURE OF ICE FLOES ON AN INCLINED STRUCTURE

### 4.2.3.0 - INTRODUCTION

Inclining the cutting edge of a pier usually produces an important decrease in the horizontal component of the pressure and the appearance of a vertical component. This facilitates the work of ice cutting and greatly increases the stability of the pier.

After a floe contacts the inclined edge of a pier, its lower sharp edge is crushed and a reaction appears, as shown in Fig. 4.31 which can be decomposed into two forces V and T, related to each other in the simplest case by:

$$V = T \cos \beta \qquad (4\text{-}54)$$

Under the action of these components the floe can fail in three different ways:

$1^\circ$) by bending in section 1-1

$2^\circ$) by shearing in section 2-2

$3^\circ$) by crushing under the forces T, which are approximated by:

$$H = 2\ T \sin \alpha \qquad (4\text{-}55)$$

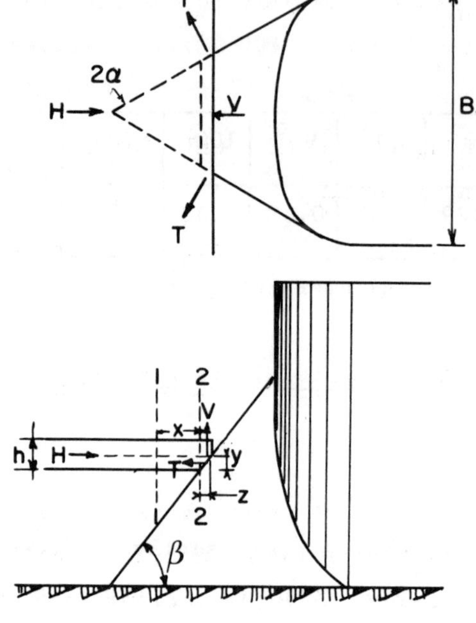


*Fig. 4.31. Failure of ice on the inclined edge of a pier.*

The type of failure that takes place depends on which kind of stress attains its ultimate value first. It can be shown that failure by shearing is the most probable at breakup. As is known, the ice cover of rivers and lakes is often made up of an agglomerate of vertically oriented crystals, separated from one another by narrow layers where impurities can collect. During the spring thaw the resistance of ice to sliding along these crystal faces is much reduced. Observations in nature show that failure by shear occurs in the immediate vicinity of the pier, so that a channel of crushed ice of width B remains behing the pier.

If the process of spring thaw is not yet sufficiently developed, a bending failure of the ice cover is possible. In that case too the fracture occurs in the vicinity of the pier at a distance not exceeding 8 to 6 times the thickness of the ice cover. The width of channel formed by the pier in the ice field is larger than in the previous case, but is still only a little larger than the pier.

Fracture by pure crushing is possible only for rigid piers when the vertical component is small.

Unfortunately no exact analysis of theses cases exist at present and we must rely only on approximations.

### 4.2.3.1 - RELATIONS BETWEEN THE VERTICAL AND HORIZONTAL COMPONENTS

A complete development of the relations between the horizontal and vertical components of the forces acting on an inclined pier has been made by Carter (1977) for various forms of piers. He assumed full contact between the ice and the periphery of the structure.

At the time of impact of the ice on the structure, the pressure exerted by ice at any point can be resolved into a normal stress and a stress tangential to the wall. In the elastic domain, the strains are proportionnal to the deformations. To determine the distribution of normal stresses we will consider an infinitesimal deformation, $e_x$, in the direction of the floe movement. Using symbols defined in Fig. 4.32 we obtain first the displacement normal to any point of the structure.

$$e_n = e_x \cos (r) \cos (s) \qquad (4\text{-}56)$$

and the corresponding normal stress:

$$\sigma_n = C\, e_x \cos (r) \cos (s) \qquad (4\text{-}57)$$

where C is the elastic constant.

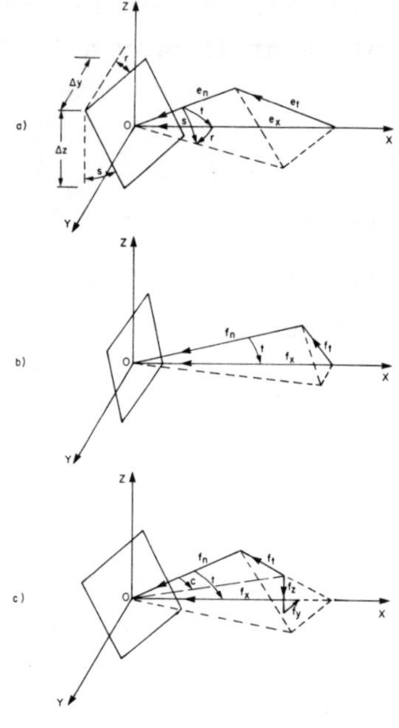

*Fig. 4.32. Definitions of symbols in analysis of forces on an inclined pier. The ice moves in direction* ox. *a) deformations b) friction angle* c *larger than t, c) friction angle* c *smaller than t.*

There is a complication in the determination of the tangential stress. It must be seen that this stress can not be greater than:

$$\tau_{max} = \mu \, \sigma_n \tag{4-58}$$

where μ is the friction coefficient, usually taken between 0.04 and 0.34 (Pounder, 1965), that can also be expressed by:

$$\mu = tg\,(c) \tag{4-59}$$

where c is the friction angle.

Referring to Fig. 4.32, two conditions may exist. When $t \le c$, where t is the angle between the normal to the surface and the direction of motion of ice, we than have:

$$\tau = \sigma_n \, tg\,(t) \tag{4-60}$$

However, if $t \geq c$, then:

$$\tau = \sigma_n \text{ tg } (c) \tag{4-61}$$

In a general manner we can write:

$$\tau = \sigma_n \text{ tg } (t_1) \tag{4-62}$$

where $t_1$ is the smaller of t or c.

In Fig. 4.32 an element of surface can be expressed by:

$$\Delta A = \frac{\Delta y}{\cos (r)} \times \frac{\Delta z}{\cos (s)} \tag{4-63}$$

So that, the normal stress becomes:

$$\sigma_n = C\, e_x \cos (r) \cos (s) \tag{4-64}$$

The equilibrium of forces in the three directions give:

$$f_x = C\, e_x \left[ \cos (r) \cos (s) + \text{tg } (t_1) \sin (t) \right] \Delta y\, \Delta z$$

$$f_y = C\, e_x \left[ \sin (r) \cos (s) - \text{tg } (t_1) \cos (t) \frac{\text{tg } (r)}{\text{tg } (t)} \right] \Delta y\, \Delta z$$

$$f_z = C\, e_x \left[ \sin (s) - \text{tg } (t_1) \frac{\sin (s)}{\text{tg } (t)} \right] \Delta y\, \Delta z \tag{4-65}$$

where $f_x$, $f_y$ and $f_z$ are the elementary forces acting respectively in directions ox, oy and oz.

The total forces acting on the structure are obtained adding the elementary forces:

$$H = C\, e_x \int_{-B/2}^{+B/2} \int_{-h/2}^{+h/2} f_x\, (r, s, t_1, t)\, dy\, dz$$

$$V = C\, e_x \int_{-B/2}^{+B/2} \int_{-h/2}^{+h/2} f_y\, (r, s, t_1, t)\, dy\, dz$$

$$T = C\, e_x \int_{0}^{+B/2} \int_{-h/2}^{+h/2} f_z\, (r, s, t_1, t)\, dy\, dz \tag{4-66}$$

where H - total horizontal force

V - total vertical force

T - total transversal force

$f_x$, $f_y$, $f_z$ - functions defined previously

h - ice thickness

B - width of structure

A computer program can be used to obtain the vertical and horizontal component for any type of pier, ice thickness to width ratio and friction coefficient.

### 4.2.3.2 - SIMPLIFIED ANALYSIS. (KORZHAVIN'S FORMULAS)

Korzhavin (1962) analyzed the various ice failure modes with some simplified hypothesis. A summary of his work is given by Michel (1970).

Failure of large ice floes by shearing. It is assumed that at the moment of shear failure the edge of the floe has already been crushed to a length z (Fig. 4.31). Investigations show that failure occurs at the pier surface when z is small. We then find:

a) The length of the shear surface:

$$bs = m\ p \tag{4-67}$$

where p is the perimeter of the cutting edge of the pier and m is a coefficient of contact.

b) The greatest possible vertical force acting on one-half of the pier:

$$V = \frac{bs\ h\ \tau_o}{2} \tag{4-68}$$

c) The greatest possible horizontal force:

$$T = (1+\mu)\ V \tan \beta' \qquad (4\text{-}69)$$

The coefficient $\mu$ is introduced, to take account of dynamic friction forces between the pier and the ice floe. $\beta'$ is the angle of inclination of the sides of the cutting edge to the horizontal:

$$\tan \beta' = \frac{\tan \beta}{\sin \alpha} \qquad (4\text{-}70)$$

d) The greatest possible horizontal force acting on the whole pier:

$$H_{max} = 2\ T \sin \alpha = 2\ V \tan \beta\ (1+\mu) \qquad (4\text{-}71)$$

For a pier with a wedge-shaped cutting edge the maximum possible length of the shear surface can be found from the relation:

$$bs = \frac{m\ B}{\sin \alpha}$$

and the calculated forces on the whole pier are equal to:

$$V_{max} = \frac{m\ B\ \ h\ \tau_o}{\sin \alpha} \qquad (4\text{-}72)$$

$$H_{max} = \frac{(1+\mu)\ m\ B\ \ h\ \tau_o \tan \beta}{\sin \alpha} \qquad (4\text{-}73)$$

In the case of a semi-circular cutting edge we get:

$$V_{max} = \pi/2\ m\ B\ \ h\ \tau_o \qquad (4\text{-}74)$$

$$H_{max} = \frac{(1+\mu)\pi}{2} m\ B\ \ h\ \tau_o \tan \beta \qquad (4\text{-}75)$$

where in both cases the contact coefficient m can be taken as in the case of ice crushing.

Failure of large ice floes by bending. This case might theoretically be investigated with the theory of a plate lying on an elastic foundation, as the inverse problem of the bearing capacity of an ice sheet. This theory shows that the failure of the ice sheet begins with the formation of radial cracks starting from the load and the collapse occurs when a circumferential crack develops. This latter condition gives the maximum loading condition on a pier. However, observations by Korzhavin do not substantiate the theory for this application.

The plate theory is derived for vertical loading only. Here the horizontal component of the force on the pier combines with the vertical one to modify the stress distribution. The theory is applied for a uniformly distributed load over a circular area. Here we have peripheral loading along different shaped pier noses and observations show that these forms are a prime factor affecting the ice pressure.

In elastic deformation the theory shows that the circumferential crack would appear at a distance on the order of twenty times the thickness of the sheet while observations by Korzhavin show that the dimensions of broken pieces of ice do not exceed 3 to 6 times this thickness, and are frequently smaller. In the theory of the ideal plastic plate the position of the circumferential crack has to be indirectly assumed.

We will thus present here Korzhavin's semi-empirical derivation where the main hypothesis is that the collapse of the ice sheet occurs at a distance of 3 h from the sides of the piers. This distance was also chosen because it is safe for engineering design, since it gives a higher force than could probably act.

When an ice floe contacts a massive pier, indentation occurs first when the lower edge of the floe is crushed. The ice then comes in contact with the side of the pier, a vertical and horizontal reaction is set up, and the ice is ruptured with the formation of at least three radial cracks as shown in Fig. 4.33. These radial cracks are followed by transverse cracks breaking a piece of the floe of form 1 - 2 - 3 - 4. After that, the force on the pier is reduced until a new contact is restored and the same process continues. In this manner, the resulting reaction of the pier on the ice floe is reduces to two forces on each side, V and T (Fig. 4.31).

Again, taking into account the dynamic friction force of ice against the structure μ, we get:

$$H = 2\ T \sin \alpha \tag{4-76}$$

$$H = 2\ (1+\mu)\ V \tan \beta \tag{4-77}$$

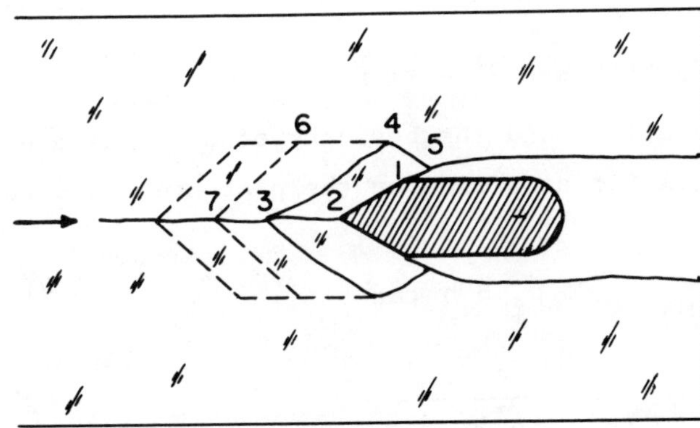


*Fig. 4.33. Progression bending failure of ice on the inclined edge of a pier.*

As the separation of the ice into two strips precedes its breaking, the greatest pressure on the ice occurs at the moment of failure along line 3 - 4. The length of this line can be written:

$$\ell_{34} = \frac{n_o\ B}{2 \sin \alpha} \tag{4-78}$$

where $n_o$ is a coefficient higher than unity taking into account the plan-shape of the pier. Korzhavin considers the strip 2 - 3 - 4 - 5 as a finite element clamped in section 3-4 and acted upon by forces V and T at the other end. The element being short (3 h) and its deformation small before failure, its weight and hydrostatic reaction are neglected in relation to the breaking forces. The bending moment in section 3-4 is then given by:

$$M = Vx - Ty \tag{4-79}$$

Because there is no deep indentation of the ice floe (Fig. 4.31) the forces V and T are not applied at the center of the ice sheet but near its lower surface. Korzhavin arbitrarily proposed to take y = 0.25 h for this point of application. At failure the yield moment of ice for an elastic plate is:

$$Mo = \frac{\sigma_f \; h^2 \; \ell_{34}}{6} \tag{4-80}$$

where $\sigma_f$ is the flexural strength of ice.

With these relations and the basic hypotheses x = 3 h and y = 0.25 h, the total forces acting on the pier are found to be:

$$H_{max} = Co \; \sigma_f \; B \; h \; \tan\beta \tag{4-81}$$

$$V_{max} = \frac{Co \; \sigma_f \; B \; h}{(1+\mu)} \tag{4-82}$$

and

$$Co = 0.73 \; n_o/[12 \sin\alpha - \tan\beta] \tag{4-83}$$

Assuming a similarity between the bending and indentation mechanisms, Korzhavin finds a relation between the coefficient $n_o$ and the angle of the pier $2\alpha$. Adjusting its value so that it will fit experimental data, he finally proposes:

| $2\alpha$ | - | 45° | 60° | 75° | 90° | 105° | 120° |
|---|---|---|---|---|---|---|---|
| $n_o$ | - | 0.94 | 1.18 | 1.42 | 1.68 | 1.98 | 2.0 |

The values of $C_o$ can then be determined from Eq. 4.83 (Table 4.4).

| $\beta$ | $2\alpha$ = 45° | 60° | 75° | 90° | 120° |
|---|---|---|---|---|---|
| 45° | 0.20 | 0.17 | 0.16 | 0.16 | 0.15 |
| 60 | 0.24 | 0.20 | 0.19 | 0.18 | 0.17 |
| 70 | 0.38 | 0.27 | 0.28 | 0.21 | 0.19 |
| 75 | 0.79 | 0.38 | 0.29 | 0.26 | 0.22 |

Table 4.4 - Values of coefficient $C_o$.

It can be seen that the influence of the angle $2\alpha$ becomes important at small values ($2\alpha < 60°$). An increase of $H_{max}$ for small angles $2\alpha$ can be explained by the fact that a sharper pier penetrates farther into the floe and the length of broken pieces is correspondingly increased. If the cutting edge is rounded in plan, some additional cracks appear; Korzhavin then recommends taking the rounded edge as approximately equivalent to a sharp edge as shown in Fig. 4.34 whose angle $\alpha$ is given by the formula:

$$2\alpha = 2\alpha_1 + \frac{4r}{B}(40^\circ - \alpha_1) \qquad (4\text{-}84)$$

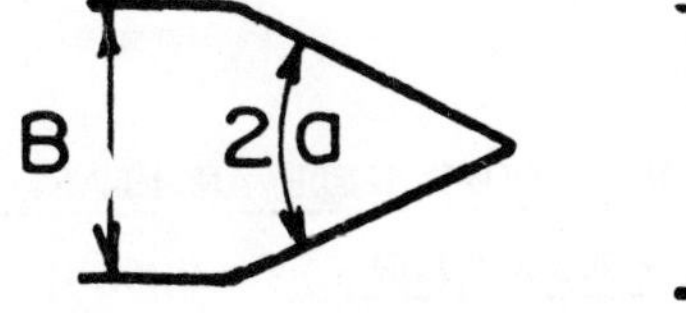

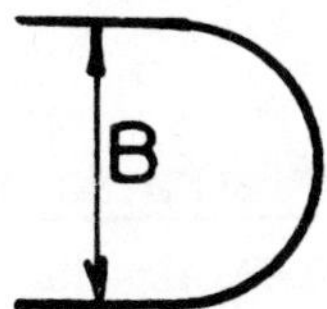

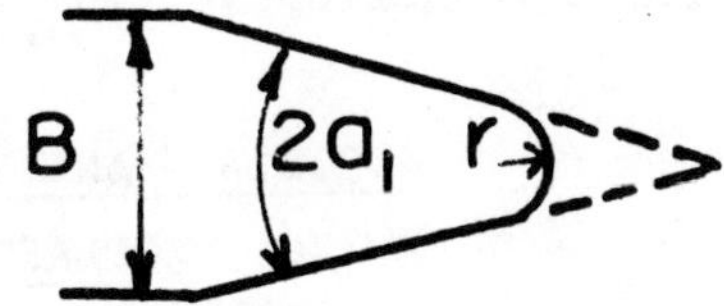


*Fig. 4.34. Shape of pier noses.*

### 4.2.3.3 - CRUSHING FAILURE BEFORE RIDE-UP

Because indentation will occur first on the lower edge of the ice sheet before any movement on the inclined structure, the maximum force can be obtained for the condition of crushing before other ice failure types. For this condition the Marine Aids of the Canadian Coast Guard has used the following formulas for calculation of horizontal and vertical ice forces on cone-shaped light piers (Danys *et al.*, 1977):

$$H = m\, n_1\, B\, h\, \sigma_o \qquad (4\text{-}85)$$

$$V = m\, n_2\, B\, h\, \sigma_o \qquad (4\text{-}86)$$

where m is the shape and contact coefficient; $n_1 = \cos^2 \delta$ and $n_2 = \cos \delta \sin \delta$ are slope coefficients; $\delta$, the slope angle with the vertical; B , width of structure at the design water level; h, the effective thickness of the ice sheet; and $\sigma_o$, is the effective compressive strength of ice.

The $\cos^2 \delta$ term in Eq. (4-85) is explained by the double decomposition of the crushing force and crushing surface in a direction parallel to the ice movement.

The crushing failure for inclined face piers with the reduction factor given by the cosine function appears to apply for piers with an angle with the vertical of less than $15^o$. For angles higher than $30^o$, the failure is caused by bending with a large reduction in impact forces.

### 4.2.3.4 - ANALYSIS OF BENDING ON A CONE WITH AN ELASTO-PLASTIC MATERIAL AND FOUNDATION

For large ride-up of ice on a cone it may be expected that the full ice sheet will either be partly unsupported by the elastic foundation and partly submerged. This condition for an elastic-perfectly plastic foundation response is described in Fig. 4.35.

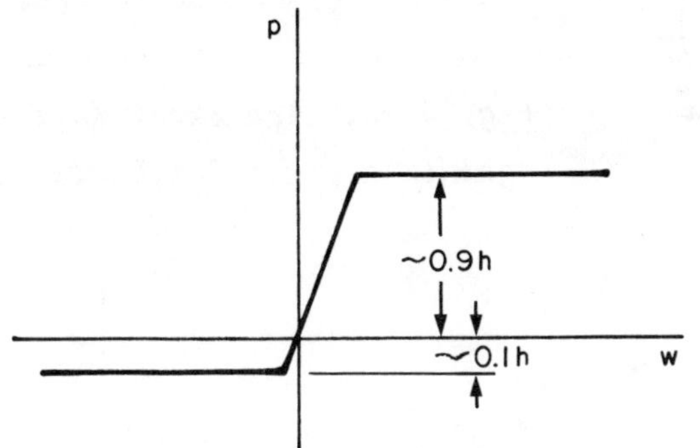


*Fig. 4.35. Elastic-perfectly plastic foundation response.*

The analysis was made by Ralston (1977) for ice behaving also as an elastic-plastic material that would resist an elastic movement up to the moment Mo at failure (yield). It follow the ideas of plastic limit analysis with an upper bound solution for the plasticity problem with methods described by Chen (1977).

The results of this analysis for the case shown on Fig. 4.36 can be expressed in the form:

$$H = [A_1 \sigma_f h^2 + A_2 \rho g h D^2 + A_3 \rho g h (D^2 - D_o^2)] A_4$$

$$V = B_1 H + B_2 \rho g h (D^2 - D_o^2) \tag{4-87}$$

The dimensionless coefficients $A_1$, $A_2$, $A_3$, $A_4$, $B_1$ and $B_2$ are given in Fig. 4.37 as a function of the appropriate parameters. These equations have been found to verify well the experimental results obtained by Croasdale and Edwards (1976) and shown in Fig. 4.38.

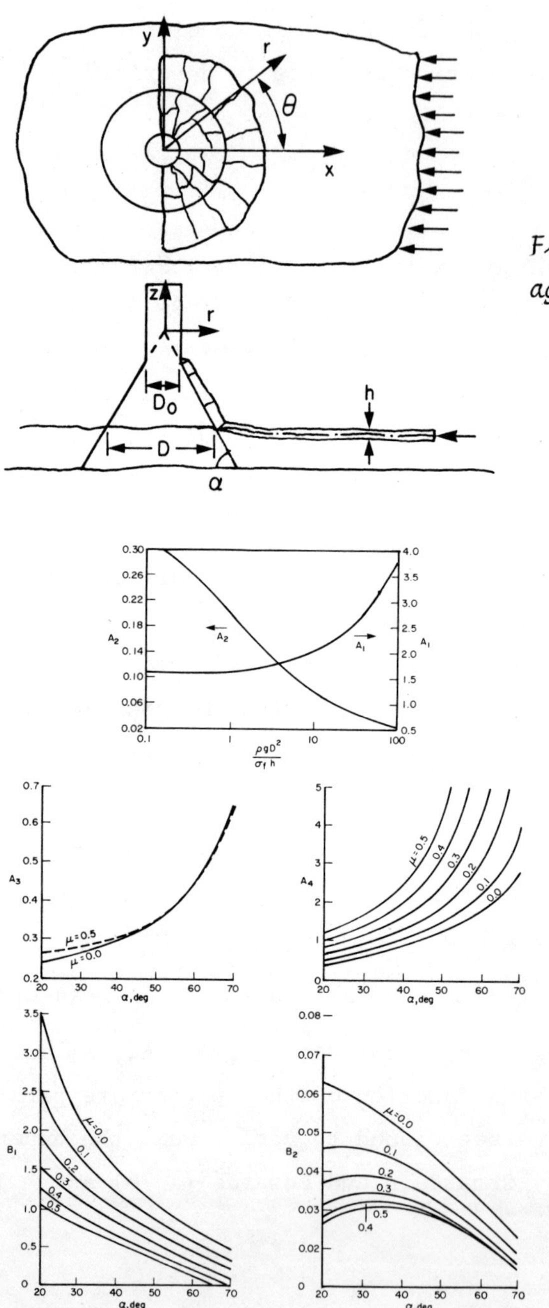


*Fig. 4.36. Ice sheet failure against a conical structure.*

*Fig. 4.37. Ice force coefficients for plastic analysis.*

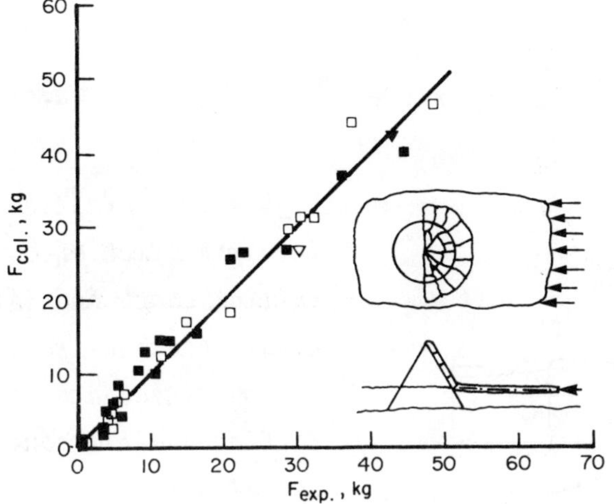


*Fig. 4.38. Comparison of the Ralston (1977) analysis with Edwards and Croasdale's (1976) model test data.*
*• Horizontal forces*
*o Vertical forces*
*▽ High friction tests.*

## 4.2.3.5 - IMPACT OF A MULTIYEAR ICE RIDGE

The failure of multiyear ice ridges moving against a conical structure has been discussed (Croasdale, 1974) with the theory of ice beams on an elastic foundation. The ridge first cracks in the center and then fails by hinge crack formation as shown on Fig. 4.39.

The equation of a rectangular beam on an elastic foundation is (Hetenyi, 1946):

$$EI \frac{d^4y}{dx^4} = -ky + q \tag{4-88}$$

The reaction on the cone is the limit condition, which gives q = 0, for which the solution of Eq. (4-88) is:

$$y = e^{-x/\ell} (C_1 \cos x/\ell + C_2 \sin x/\ell) \tag{4-89}$$

For a concentrated load on an infinite beam the values of $C_1$ and $C_2$ give:

$$y = \frac{P}{2k\ell} e^{-x/\ell} (\cos x/\ell + \sin x/\ell) \tag{4-90}$$

and the moment:

$$M = -\frac{P_o \ell}{4} e^{-x/\ell} (\cos x/\ell + \sin x/\ell) \qquad (4\text{-}91)$$

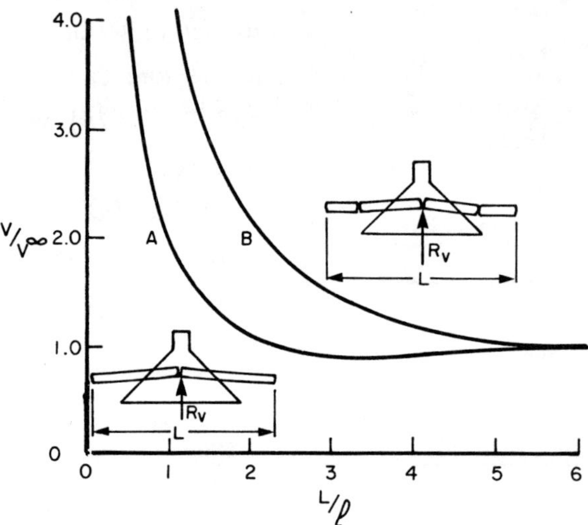


*Fig. 4.39. Elastic interpretation of initial crack and (A) hinge crack (B) forces for multiyear ridges failing against conical structures.*

For x = 0, the maximum moment at failure is attained which gives a vertical force $V_1^\infty$ for the formation of the initial crack at the center:

$$V_1^\infty = 4 \frac{\sigma_f I}{\ell z_b} \qquad (4\text{-}92)$$

where $z_b$ is the distance of the extreme upper fiber of the beam from the neutral axis, and I is the moment of inertia.

In the case of a semi-infinite beam loaded at the end, the maximum moment occurs at a distance of $x/\ell = \pi/4$ from the edge. The computation then gives:

$$V_2^\infty = \frac{6.20 \ \sigma_f I}{\ell z_b'} \qquad (4\text{-}93)$$

where $z_b'$ is here the distance of the extreme lower fiber of the beam from the neutral axis.

For ridges of finite length the vertical force exerted by the structure has to be much higher than for ridges of infinite length. A computation by Ralston (1977) using a method of superposition of loads and moments give the results of $V/V\infty$ shown in Fig. 4.39 for the two cases of loading.

#### 4.2.3.6 - DIMENSIONAL ANALYSIS

The preceding analysis of the force exerted by ice on inclined structures are very simple mathematical models that give only the first crushing load and ignore the subsequent variable forces, the weight of ice upon the inclined surface, its movement around the structure, abnormal ice configurations and many other factors.

Dimensional analysis applied to both model and full-scale data has recently tended to supplement mechanical analysis in studies of the ice force problem, in relation both to structures and to ships. Relevant references include Lavrov (1969), Michel (1970), Edwards and Wheaton (1972), and Schwarz *et al.*, (1973). At the time of writing, only the general nature of expected functional relationships has been sketched out: the extensive experimental data needed for development of generalized design charts are not freely available, although a considerable amount of unpublished data is known to be held by private organizations involved in Arctic engineering.

In the case of steady motion against inclined faces, characteristic parameters are as follows: H = horizontal force developed; h = sheet thickness, B = pier width or tower diameter at ice level; V = sheet velocity; $\sigma_o$, $\tau_o$, $\sigma_f$ = ice strengths in compression, shear, bending; $\dot{\varepsilon}$ = corresponding test strain rates (may require three components); E = Young's modulus of ice; g = gravitational acceleration; $\rho$ = water density, $\rho'$ = ice density; and $\beta$ = angle of face.

These require arrangement into 10 dimensionless groups; for example (Edwards and Wheaton, 1972).

$$\frac{H}{\rho g\ h^3} = F\left(\frac{V}{gh}, \frac{E}{\rho g\ h}, \frac{\sigma_f}{E}, \frac{\sigma_o}{\sigma_f}, \alpha, \beta, \psi, \mu\right) \tag{4-94}$$

where $\alpha$, $\beta$, $\psi$ are angles characterizing the geometry of the structure. Even if several of these groups can be regarded as virtually constant or as of secondary importance, the number remaining is suffisant to make interpretation of data quite difficult. The functional relationship with the minimum number of significant combinations, can probably be represented, for a given structure geometry, by:

$$\frac{H}{\rho g\ B\ h^2} = F\left(\frac{V}{\sqrt{gh}}, \frac{\sigma_f}{\rho g\ B}, \frac{\ell}{h}, \frac{B}{h}, \frac{\sigma_o}{\sigma_f}, \mu\right) \tag{4-95}$$

where B has been introduced, and the characteristic length $\ell$ is used instead of E.

For high Froude numbers, the ice behaves in a pure brittle manner, and the velocity term can be dropped. Croasdale and Edwards (1976) have found, in that case, a good relationship for conical structures, of the form:

$$\frac{H}{\rho g\ B\ h^2} = C_o + C_1 \frac{\sigma_f}{\rho g\ B} \tag{4-96}$$

## 4.2.4 - FIELD VERIFICATIONS

In general, three methods have been used to obtain full-scale measurements and estimates. These are (i) analyses of cases of failure or damage and of existing structure performance; (ii) indirect estimates from observed mass and deceleration of ice floes impacting on a structure - the "kinematic" method; and (iii) direct measurements using force or pressure transducers.

Table 5 lists the principal features of investigations reviewed by Neill (1976). Readers should refer to the original documents for details of test installations and techniques.

Description of the principal results of each investigation follow. We cite extensively the publication by Neill:

### Siberia, 1933-1961

"Investigations were reported by Korzhavin (1962) and summarized by Michel (1970).

Photogrammetry was used to estimate ice-floe mass and deceleration while striking bridge piers and spillway buttresses in large rivers and reservoirs, during spring break-up. Both vertical and inclined structures were monitored. Values of $B_o/h$ ranged from 4.5 to 19.

Korzhavin's force values are remarkably low: calculated effective pressures do not exceed (160 $kN/m^2$). Later observations suggest the reason is that the technique yields only at time-average value over the duration of an impact episode. Short-duration forces fluctuate rapidly and may be several times greater than time-average forces (Neill, 1972). It is also possible that observation periods did not include critical conditions when ice strengths were high. No related strength data were reported.

Cook Inlet, Alaska, 1963-1969

Investigations were reported by Peyton (1966, 1968a, b) and by Blenkarn (1970).

A number of drilling platforms on vertical cylindrical legs, subject to moving sea ice including ridge formations, were instrumented at different times by strain gauges and load cells. Maximum forces reported from ridge impact were up to 963 kN/m of pile diameter. Maximum effective pressure from steady crushing of a uniform sheet was apparently about 1100 kN/m$^2$.

Table 4.5 summarizes some features and results of the program. As described by Blenkarn, the ice was first-year sea ice, generally not more than 0.6 m thick. Tide and wind move floes and sheets, some of very large extent. Pressure ridges may project 3 m or more below water. Velocities are usually from 1.2 to 2.1 m/s.

Most of Blenkarn's data were presented as plots of force per unit width vs. ice thickness. Actual pile diameters were not quoted in most cases. Force and pressure values quoted for different seasons are not directly comparable in all cases, but the general pattern is evident. To quote from Blenkarn's conclusion directly:

"1$^{\circ}$) Effective structural loading ice pressure in Cook Inlet is less than 860 kN/m$^2$.

2$^{\circ}$) Ice pressure does not appear to be strongly influenced by ice floe temperatures.

3$^{\circ}$) Winters for which force measurements were made are not much less severe than historical extreme seasons.

$4^{o}$) Including allowance for dynamics, forces imposed by pressure ridges are two to three times the forces imposed by uniform floes.

$5^{o}$) Maximum measured intensity of pressure ridge loading is in the range of 875 to 1020 kN/m of leg diameter."

The terminology requires some comment. "Structural loading ice pressure" (1) apparently means sustained effective pressure from uniform sheets crushing against piles. "Allowance for dynamics" (4) appears to include some structural response effects, so that it does not seem quite correct to write of "forces imposed by pressure ridges" in this connection, or of "maximum measured intensity" (5). Peyton (1968a, b) suggests that most of the data were affected by structural response and did not reflect imposed forces correctly."

It is useful to compare the data with predictions of formulas and codes. Data by Peyton (1968a, b) suggest that compressive strengths for high loading rates were in the order of 1200 $kN/m^2$.

Formula (4-33) gives for fracture in the brittle range

$$p = C\ m\ k\ \sigma'$$

with C = 1.57, k = 0.50, m = 0.9, the effective pressure is $p = 0.7\ \sigma'$, which is of the order of 832 K Pa. This is in the range of effective pressures reported at Cook Inlet.

Both Peyton and Blenkarn gave considerable attention to the force oscillations which were characteristic of their records. Peyton (1968a, b), in reference to records from a test pile used in 1963-1964, writes of "sharp, ratcheting force oscillations caused by very slowly moving ice... the frequency of

| Location | Dates | Types of Structure | Types of ice action | Method of force determination |
|---|---|---|---|---|
| Siberia | 1933-1961 | Bridge piers and spillway buttresses | Moving river ice in spring | Photogrammetric estimation of ice-floe mass and deceleration |
| | Korzhavin (1962) - Michel (1970) | | | |
| Cook Inlet, Alaska | 1963-1969 | Drilling platforms | Moving sea ice including ridges | Strain gauges on structural members; test piles with load cells |
| | Peyton (1966, 1968a, b) - Blenkarn (1970) | | | |
| Alberta | 1967 | River bridge piers | Moving river ice in spring | Pier nose and pile supported by load cells; evaluation of past performance |
| | Sanden and Neill (1968) | | | |
| St-Lawrence River | 1970 | Navigation light piers and towers | Moving river and estuary ice | Evaluation of performance and failures; test pier with load panels |
| | Atkinson et al. (1971) - Danys (1972) | | | |
| Eider River estuary, Germany | 1967-1969 | Pile on bridge pier | Thin moving estuary ice | Array of pressure cells covering contact area |
| | Schwarz (1970) | | | |
| Baltic Sea | 1966-1969 | Lighthouses | Low pressures sures by pack ice | Analysis of damage and failure |
| | Reinius et al. (1971) - Bergdahl (1972) | | | |
| Beaufort Sea, Arctic Canada | 1970 | Test piles | Passive pressure | Jacking of piles into solid coastal sheet |
| | Croasdale (1974) | | | |

Table 4.5 - Principal features of full-scale investigations reviewed by Neill (1976).

oscillation is about one cycle per second." This remark applies only to part of the record shown, which taken as a whole indicates a spectrum of frequencies ranging mainly from 1 to 10 cycles/s. After testing a small-scale pile in the laboratory, Peyton concluded that the frequency cited, although it corresponded closely to the natural frequency of the platform, originated basically from ice failure mechanisms and that there was a great possibility for forced resonant oscillation of the structure.

This interpretation corresponds to the value of the frequency of oscillations given by Eq. (4-42) for the fracture of an ice sheet.

Alberta Rivers, 1967-Present

Investigations were reported by Sanden and Neill (1968), and Neill (1970,1972, 1976) is cited here:

"The data consist mainly of force measurements during spring break-up on two special bridge piers, one vertical and the other inclined. Analyses were also made of a number of old piers. Results from the two sites will be discussed separately.

Vertical Pile, Pembina River

In this case a difficulty of interpretation arises because of a peculiarity of many of the recorded signals (Fig. 4.40). Most events were short-duration impacts of a second or less duration, and many exhibit sharp "spikes" of virtually zero duration. It appears likely that these spikes represent structural response to sudden impact. The spike values can have little significance for pier because they contain no significant energy, but they might be significant for thin structural members.

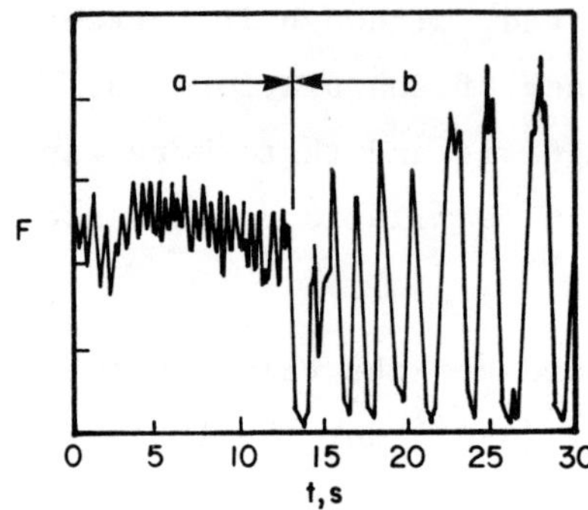


*Fig. 4.40. Examples of ice force records. Alberta test structures, a) crushing failure, b) bending failure (Neill, 1972).*

The following points emerge from examination of the results.

(*i*) Discounting spikes the maximum effective pressure is of the order of 1800 kN/m$^2$.

(*ii*) Field strength tests on 100 mm x 100 mm x 200 mm specimens a few days before break-up in 1972 and 1974 yielded unconfined compressive strengths ranging from less than 690 to 3100 kN/m$^2$ in the horizontal direction, depending on the location sampled, the ice type, the prevailing weather, and the depth in the sheet.

(*iii*) In situ ram tests were conducted in 1972 at the same time as the unconfined compression tests, using a special apparatus to push a small pile into the ice (Haynes *et al.*, 1975), and thereby measuring effective pressures at a B /h ratio of approximately 0.35. Pressures ranged from 1240 to 3240 kN/m$^2$ in a period of a few days before break-up, with an average of approximately 2400 kN/m$^2$."

(*iv*) On the basis of previously discussed analyses, a reasonable design formula for this situation, based on formula (4-32) might be:

$$p = C\ m\ k\ \sigma' \tag{4-33}$$

with C = 1.57, k = 0.5, m = 0.9 we get:

$$p = 0.7\ \sigma'$$

with a value of p = 1800 K Pa, this corresponds to a strength value $\sigma'$ = 2600 K Pa, which is very typical of the crushing strength at melting point of $S_2$ ice, as can be seen in Chapter 2. It is in the range of the results for the strength of the small ice samples of good quality ice and corresponds also to the pressure of the ram tests.

## Inclined Pier, Athabasca River

The following comments and conclusions by Neill (1976) emerge form examination of the data:

(*i*) "Ice failure against the pier (inclination 23$^\circ$ from vertical) occurs sometimes by crushing and sometimes by bending and splitting. Peak instantaneous forces for each mode are similar, but time-average forces over periods of a second or more seem to be greater for crushing failure (Fig. 4.40). Crushing generally seems to be associated with lower velocities, although this is not clear from the tabulated data.

(ii) No definite effect of B /h ratio or of ice velocity on calculated pressures is apparent.

(iii) The maximum effective pressures at the Athabaska site are 30% lower than those of the Pembina site. The effect of the pile inclination would reduce the pressure by only 15% with formula (4-85) because of the term $\sin^2 67^\circ = 0.85$. On the other hand, Korzhavin's formula for a bending failure gives:

$$p = Co \tan \beta \, \sigma_f$$

with $\beta = 67^\circ$ and a rounded edge the value of Co is about 0.35, the value of $\sigma_f$ at melting is $\sigma_f \simeq 0.6\, \sigma'$ (Chapter 2), so that:

$$p = .49\, \sigma'$$

which gives a reduction which is about 30% of that given for the vertical faced pier and corresponds better to the measured data.

## Forced Oscillations

Some statistics on forced oscillations at both sites, derived from extensive records of 1971, were quoted by Neill (1972). For three well-documented impact events at the Pembina site, time-average forces estimated by Korzhavin's "kinematic" method agreed closely with the average values of the forces signal over the duration of impact, whereas recorded peak instantaneous forces were up to six times greater. At the Athabasca site, peak forces were up to four times greater then maximum one-second means. These Athabasca results should be interpreted with some caution, however, because the 1971 peaks were associated with bending-splitting failure, whereas data from other years suggest the possibility of sustained high forces from continuous crushing.

Typical oscillograms and a spectral density plot for the Alberta sited are shown in Fig. 4.41. There is a marked resemblance to the Cook Inlet data.

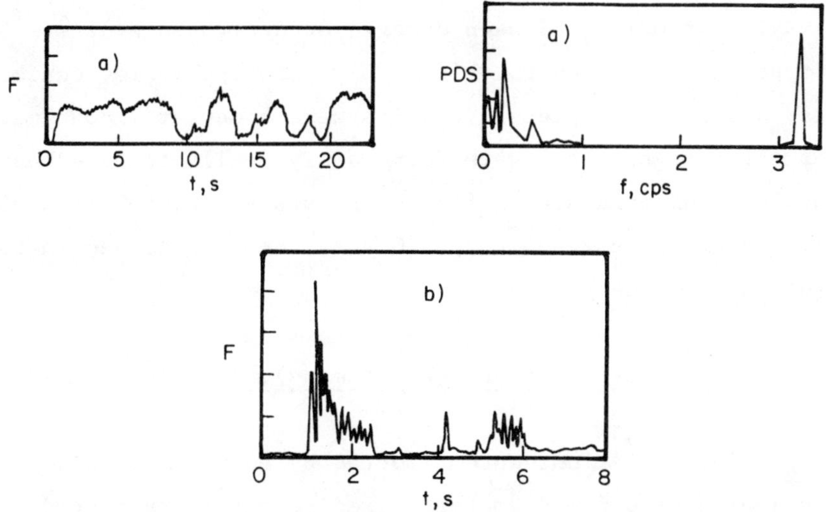


*Fig. 4.41. Typical oscillogram and power spectral density (PSD). a) inclined pier b) vertical pile with impact from separate floes.*

Analyses of Old Piers

A number of old bridge piers, several of which had been known to withstand severe ice runs over periods of up to 60 years, were analyzed to estimate ultimate strengths and associated failure forces (Sanden and Neill, 1968). It appears that several of those should have failed under sustained ice pressures of 1720 $kNm^{-2}$ or less. Failures actually experienced were all ascribable to ice erosion of timber crib piers, or to buckling of steel framed piers or superstructures.

## Upper St.Lawrence River System, 1958-1968

Investigations were reported by Danys (1972). Analyses were conducted on 22 offshore light piers in the St.Lawrence River waterway, 7 of which had been damaged or destroyed by ice. Structure types included square blocks with sloping walls, conical towers, and inclined piers. Results were somewhat inconclusive: all of the damaged piers were found analytically to have been incapable of withstanding effective pressures greater than 690 $kNm^{-2}$. No piers designed for 2070 $kNm^{-2}$ or greater had in fact been damaged by ice action.

## Eider River Estuary (W. Germany), 1967-1969

Investigations were reported by Schwarz (1970). A vertical pile of 0.6 m diameter, fixed to a temporary bridge pier in tidal flats, was fitted with an encircling shield consisting of 50 separate pressure-measuring plates, so that both total force and its areal distribution could be determined. Ice thickness varied from 9 to 15 cm. Critical ice action arose from sustained crushing over periods of several seconds.

Data analysis was concerned mainly with (i)relationship between effective pressure and ice strength; (ii) relationship between pressures averaged over different sizes of area; and (iii) statistical distributions of instantaneous pressure values. The original report included frequency distributions of ice pressure and cube strength for five separate episodes (see Fig. 4.42) and the power spectral density analysis.

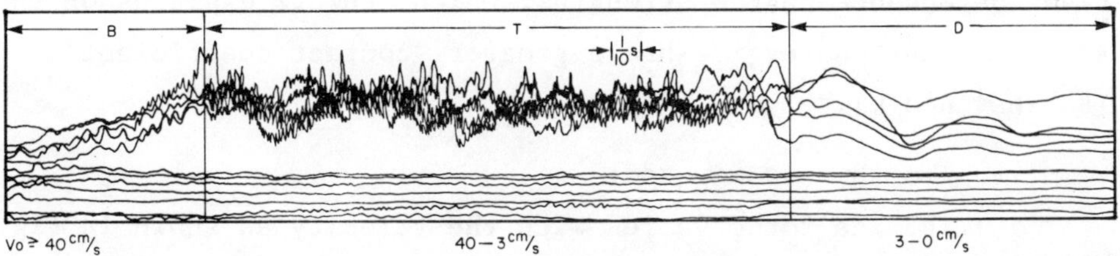


*Fig. 4.42. Examples of ice pressure records on 5 adjacent cells. Eider Estuary test pile (Schwartz, 1970).*

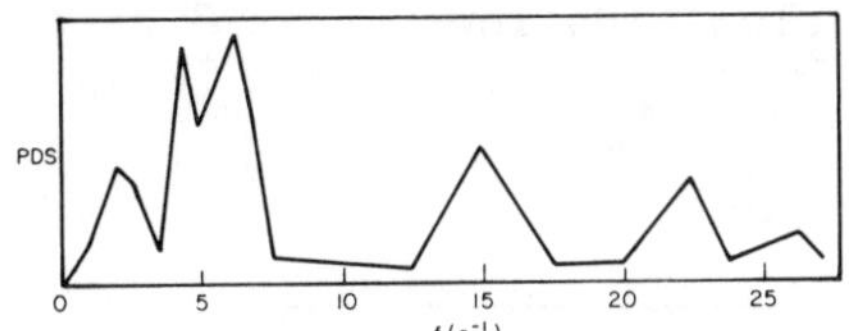


*Fig. 4.43. Power spectral density from single cell record. Eider Estuary test pile (Schwartz, 1970).*

Some results of the study have been summarized by Neill (1976) as follows:

"$1^o$) Maximum effective pressure averaged over the whole projected contact area was approximately 760 $kNm^{-2}$. Corresponding cube strengths were apparently from 1720 to 2420 $kNm^{-2}$ at $-1^oC$. For all episodes analyzed, reported cube strength was several times greater than average effective pressure.

$2^o$) Because of incomplete timewise correlation of pressure maxima on adjacent elements of area, maximum instantaneous pressures on small areas were greater than averages over larger areas. Pressure maxima on an area of 25 $cm^2$ were roughly twice those over the total contact area of 840 $cm^2$.

$3^o$) Higher effective pressures were associated with thicker ice, that is with smaller B/h ratios.

4°) Measured pressures do not appear to have shown any very clear dependence on ice strengths, which Schwarz explains on the argument that weaker ice has a greater "contact coefficient" (Korzhavin's k)."

To this we may add:

5°) The ice force varies with the velocity as shown in Fig. 4.42. There is a striking similarity with the ice behaviour and that discussed in Section 4.1.2. For high velocities over 40 cm $s^{-1}$, the ice fails brittlely and the maximum effective pressure is reduced. The maximum pressures occur in the transition zone for velocities between 3 cm $s^{-1}$ and 40 cm $s^{-1}$. Typical ductile behaviour is observed for velocities lower than 3 cm $s^{-1}$, with reduced pressure.

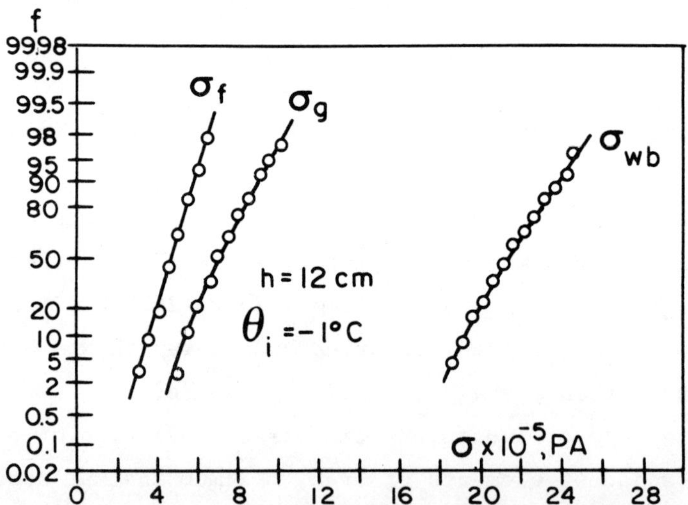


*Fig. 4.44. Typical frequency plot of ice strength and pressure:* $\sigma_f$ *average pressure on area of 225 cm*$^2$, $\sigma_g$ *effective pressure over pile diameter,* $\sigma_{wb}$, *strength of 10 cm cubes (Schwarz, 1970).*

A striking peculiarity of Schwarz's results in relation to previously discussed experimental and analytical data is their low ratio of effective pressure to ice strength, even recognizing that the cube strength reported is likely to be higher than the prism strength used by some investigators. For example, Fig. 4.44 indicates a ratio of approximately 0.25, whereas other results discussed previously would indicate a value of close to 0.7. Schwarz explained the low ratios by postulating a special reduction coefficient to allow for incomplete correlation of pressure maxima. Neill (1976) discussing these results, suggested that the higher frequencies that were observed reflect structural response rather than inherent ice fracture frequencies. This may influence results. Another explanation might be that the strength tests were made at strain rates corresponding to the maximum yield strength of ice, with ice cubes instead of ice cylinders.

Baltic Sea, 1966-1969

Investigations were reported by Reinius *et al.*, (1971) and Bergdahl (1972). Analyses were conducted of one lighthouse which was broken in Northern Sweden and of another in Southern Finland which was pushed horizontally before completion of foundation grouting. Both structures consisted of cylindrical towers on top of flat conical caissons.

The inferred failure forces per unit of tower diameter are considerably higher than measured in Cook Inlet or Alberta rivers. In the Finnish case, ice drift and pile up occured following several weeks of temperatures from -10 to -26$^{0}$C. At such temperatures the ice strength might be expected to be considerably higher than encountered in these other locations, so that on this basis the unit pressure estimate of 2400 to 3200 kN/m$^2$ is quite credible.

### Beaufort Sea, 1970

Investigations were reported by Croasdale (1974). A series of "nutcracker" tests were conducted on Arctic coastal ice near the Mackenzie Delta, by jacking apart two steel piles frozen into the ice sheet, until the ice failed by crushing. Because thickened ice collars had formed around the piles prior to jacking, reported effective pressures at failure (referred to in the reference as "measured crushing strengths") were calculated using an estimated effective thickness somewhat greater than the undisturbed sheet thickness. A few uniaxial compressive tests were made on samples cut from the sheet.

Except for uncertainty over the effect of initial freeze-in, the nutcracker data should be expected to check quite well with results of Section 4.1.3, since no question of a contact coefficient or statistical coefficient arises as in the case of active ice pressure from moving sheets.

The results are plotted on Fig. 4.45 in function of strain rates (V/D). It can be seen that they have exactly the same form as those of Fig. 4.22, the maximum pressure increasing in the ductile range up to a real strain rate of the order of $10^{-4}$ $s^{-1}$ and then decreasing in the brittle range. Because the contact coefficient should be equal to 1, the effective pressure for this condition should be equal from Eq. (4-22) to 0.9 x 2.97 $\sigma_o$, which seems very plausible for this case.

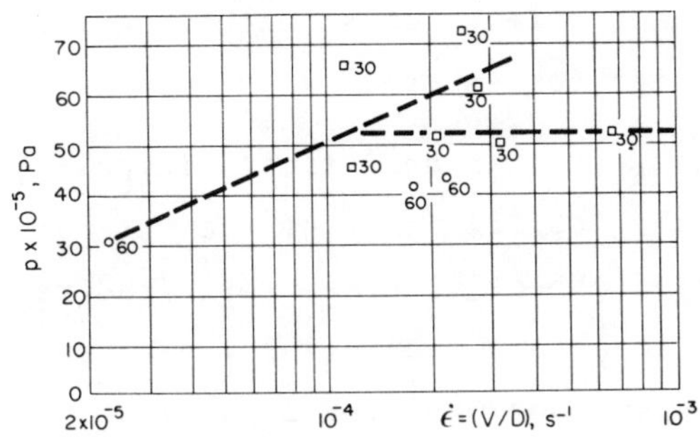


*Fig. 4.45. Results of Arctic ice crushing tests. Maximum pressure p in function of strain rate ε. The number indicates the diameter of test piles in inches.*

## Summary

Some of the most significant results and implications of the full-scale studies have been summarized as follows by Neill (1976).

1 - "Ice forces are highly erratic and oscillatory in nature, exhibiting a wide range of oscillographic patterns and frequencies. In some measurement programs, readings were affected by structural response and are not true pictures of the force input, but in others this objection has little validity. Frequencies in the range of say 0.5 to 5 cycles/s seem to account for much of the signal variance, but until more widespread analysis of records is undertaken it might be unwise to generalize. Depending on the characteristics of the structure, the oscillatory characteristics and instantaneous peak values of the force input may or may not be important. Thus for given environmental conditions, the appropriate ice strength or effective pressure to be used for design may depend on the mass and elastic characteristics of the structure itself. It is possible, for example, to envisage a statically adequate frame structure collapsing under ice action while a massive but statically weaker solid structure, subjected

to identical action, is not overstressed. This point is important when reliance is placed on experience of structure performance for estimation of safe design forces.

2 - The principal weakness, from an analytical viewpoint, in most sets of field data is the insufficiency of ice strength data for comparison with measured effective pressures. The reasons for this are compelling: to obtain meaningful ice strengths at time of ice movement in a large body of water is a hazardous and formidable undertaking, given the point-to-point variations in quality in an ice sheet, the sample-to-sample variations in strength at a given point, and the care that must be taken to obtain consistent strength data (expecially in compression) from a brittle temperature-sensitive material of large crystal size and variable crystal orientation. Nevertheless the problem seems to merit priority in further field studies.

3 - All of the full-scale data discussed are from single piles and piers. Combined forces on multiple-pile structures such as wharves and ocean platforms seem to merit model-scale investigation, since design assumptions as to simultaneous force components on various elements of the structure appear to be particularly arbitrary at present. It is clear that the probability of a given force magnitude occurring simultaneously on a number of identical piles is less than the probability of a given pressure average on a large contact area is less than on a small area - as the Eider River investigation showed."

To this we may add that:

$1^o$) For vertical pile tests, the field measurements show that basic formula 4-29 could be used with good confidence to predict the force exerted on the structure

$$p = C\ m\ k\ \sigma_o \qquad (4\text{-}29)$$

In this formula the value of $\sigma_o$ is that obtained from uniaxial tests of the same ice at the same temperature under the same strain rate, and which could be obtained from the strength diagrams of Chapter 2. The influence of the strain rate is extremely important, as it explains the failure mode, the effect of the so-called scale ratio B /h which intervenes in the computation of the strain rate, and the values of the indentation coefficient and residual contact coefficient k. For very large structures, the strength $\sigma_o$ will decrease with the computed decreasing strain rate, down to a lower limit which has not been investigated. The residual contact coefficient k is very small in the brittle and transition zone and more accurate information is needed on its value for the time average effective pressure as well as the residual peak effective pressures. The value of the form coefficient m has not been clearly established for various failure modes.

$2^o$) The only full-scale data for inclined piers are from the Athabasca River when the inclination is $23^o$ from the vertical. The measurements show a reduction in effective pressure which will give more support to Korzhavin's formula (4-81) than to formula using the crushing strength of the ice (4-85). Korzhavin's parameters in his formula are entirely empirical and some solid experimental evidence is needed before it could be used with full confidence. As such it shows a large decrease in horizontal force. For example, with average values of the parameters, a $45^o$ angle reduces this force by a factor in the order of 10, whereas the factor given by Eq. (4-85) is only 2.

### 4.2.5 - ABRASIVE ACTION OF MOVING ICE

When an ice pack is moving either during breakup or

during wintertime, the floes may come in close contact with wharves, piers, retaining walls or dams and may cause abrasion of the structures at the water level. Deep cuts in a wall, of the order of 0.3 m, have brought about the collapse of a wharf (Korzhavin, 1962). Piles have been cut in less than five or six hours (Kirkham, 1927). During a spring breakup, concrete walls were locally eroded to a depth of about 2 to 8 cm (Telechev *et al.*, 1961).

The erosion of river shores by moving ice is well known and may be much more severe in northern countries than scouring action by water alone. Even rock banks are extensively eroded by ice. One of the main drawbacks of shore protection with boulders, artificial blocks, sand bags, etc... is their poor resistance to ice abrasion. This action is particularly severe on concave river banks and an interesting solution has been used in Minneapolis on the Mississipi River. A log boom was attached at the upstream end of a bend and allowed to swing free; it stabilized at a certain distance from the bank because of the spiral motion of the flow. Ice floes tending to hit the bank were deflected by the boom, whose inertia, together with the inertia of the water behind it, easily guided the ice out, avoiding the bank.

The maximum local pressure that ice can exert while moving along a longitudinal structure corresponds to local crushing of the ice. The normal component to the structure face could be computed from the previous indentation formula:

$$\sigma_n = k\,\sigma_o \tag{4-97}$$

The maximum pressure over a large width will usually be limited by buckling and can be obtained from Section 4.4.2.3.

The longitudinal pressure of the moving ice is the dynamic friction force:

$$\tau = \mu \sigma_n \qquad (4\text{-}98)$$

with $\mu$ usually taken as 0.1.

These are in effect localized point loading and should not be applied simultaneously over the whole wall surface. Because of their repetitive character, abrasion results in the long run. That is why it is recommended that the surfaces of structures subjected to severe ice conditions of that type be hardened either by using special surfacing concrete or by adding extra protection such as granite or steel revetments.

As an indication, the ice breaker design codes suggest local ice pressures for design up to 85 x $10^5$ Pa for the bow plating. No failure have been reported in the plating of ships designed to this pressure.

## 4.3 - ICE PRESSURE BY UNCONSOLIDATED ACCUMULATIONS

### 4.3.0 - INTRODUCTION

The problem of the determination of the thrust exerted by a broken-up ice field on an hydraulic work has been studied since 1935 by many investigators in the U.S.S.R. (Petrunitchev and Mamaieff, 1941; Proskuriakoff, 1941; and Zubov, 1935). Latyshenkoff (1946) found experimentally that the force exerted on the structure did not increase indefinitely with the length of the cover but attained a constant value after a length of 3 to 4 times the width of the field.

Beccat and Michel (1959) used the Janseen theory of grain elevators to show that there was a limit to the thrust on a boom holding ice floes. This limit is attained when the reaction of the shores of the river becomes equal to the increment of the total external forces acting on the accumulation.

In (1968) Michel used the Caquot theory of grain elevators to study the stability of unconsolidated ice jams and the thrust they exert on retaining structures. This is the work that will be presented here.

## 4.3.1 - EQUILIBRIUM OF AN ICE ACCUMULATION

Let us consider an unconsolidated ice accumulation of constant thickness h in a rectangular channel of width B with uniform flow (Fig. 4.46).

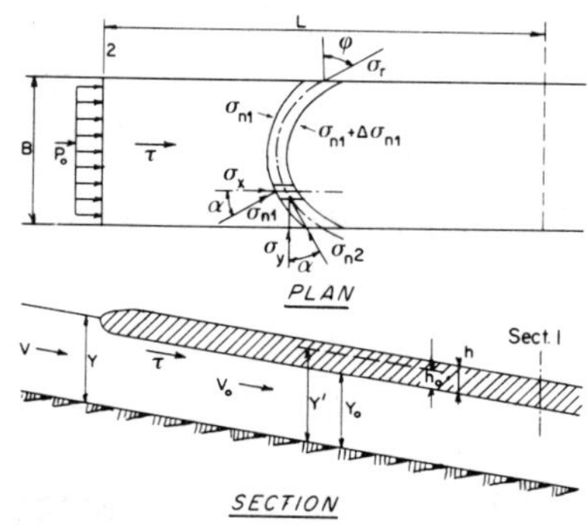


*Fig. 4.46. General stress distribution in an unconsolidated ice cover.*

The flow and the wind exert a hydrodynamic force on the frontal edge of the cover of $p_o$ per unit width, at a short distance from this edge. They also produce a tangential force $\tau$ per unit area in the direction of the flow.

The problem of the distribution of the stresses in such an ice accumulation is then similar to the one of a two-dimensional grain elevator with a top load. Both materials are granular where the unit force $\tau$ takes the place of the weight of the grain. The stress distribution in a grain elevator has been studied for a long time. A theory that verifies well the measurements is that of Caquot *et al.*, (1956), which assumes that part of the total thrust is tramsmitted to the edges by an arching action of the material. Because of the constant loading, this system consists of parabolic arches that correspond to the directions of the principal stresses in the material.

If we consider an element of this system limited by axial planes, one of the principal stresses is $\sigma_{n1}$, which is a constant all along the arch, and the other is the normal stress $\sigma_y$, in the y direction. From considerations of the conjugate stress relationships at equilibrium in granular media we have for "filling" conditions:

$$\sigma_{n2} = K_\alpha \sigma_x \tag{4-99}$$

where $K_\alpha = \tan (45^\circ - \alpha/2)$ is the coefficient of active thrust for an element of obliquity $\alpha$. The limiting condition of equilibrium will be attained for the highest value of $\alpha$ at the wall, where $\alpha = \psi$, $\psi$ being the angle of friction of the material with the wall. We then have:

$$\sigma_r = \sigma_{n1} \cos \psi \, K\psi \tag{4-100}$$

It is now possible to write the equation of equilibrium along the x axis for an arch element of the system:

$$\tau \, B \, dx - d \, \sigma_{n1} \, B \, h - 2 \, \sigma_r \sin \psi \, h \, dx = 0 \tag{4-101}$$

with Eq. (4-100), this gives:

$$\frac{d \, \sigma_{n1}}{dx} + \frac{\sin 2 \, \psi \, K\psi}{B} \sigma_{n1} - \frac{\tau}{h} = 0 \tag{4-102}$$

whose solution is, with $\sigma_{nl}\ h = p_o$ for $x = 0$:

$$T = \sigma_{nl}\ Bh = p_o\ B \exp\left[-\xi\ x/B\right] + \frac{\tau\ B^2}{\xi}\left[1 - \exp\left(-\xi\ x/B\right)\right] \tag{4-103}$$

where T is the total thrust exerted on an arch at a distance x from the edge, and:

$$\xi = \sin 2\psi \tan^2 (45^{\circ} - \psi/2)$$

If the angle of friction of ice on the shores varies from $15^{\circ}$ to $30^{\circ}$ the coefficient $\xi$ varies very little. By taking a value $\xi = 0.3$ the error is less than 4% in that range. This is very useful in the case of ice floes where this angle has not been measured but most probably lies in that range, as for many other granular materials.

With these numerical values we can now write the final formulas:

$$T = T\infty\ (1 - \alpha \exp(-0.3\ x/B))$$

$$T\infty = 3.3\ \tau\ B^2 \tag{4-104}$$

$$\alpha = 1 - p_o/3.3\ \tau\ B$$

Computation of the hydrodynamic force on the frontal edge, $p_o$. Let us consider the frontal edge of an ice cover where the ice pieces have taken a hydrodynamic configuration and where we neglect the friction forces, which are taken into account in other terms of the formula. We then have the conditions shown in Fig. 4.47. At a certain distance from the edge, normal buoyancy of the cover is attained. Considering hydrostatic pressure distribubion in sections (1) and (2):

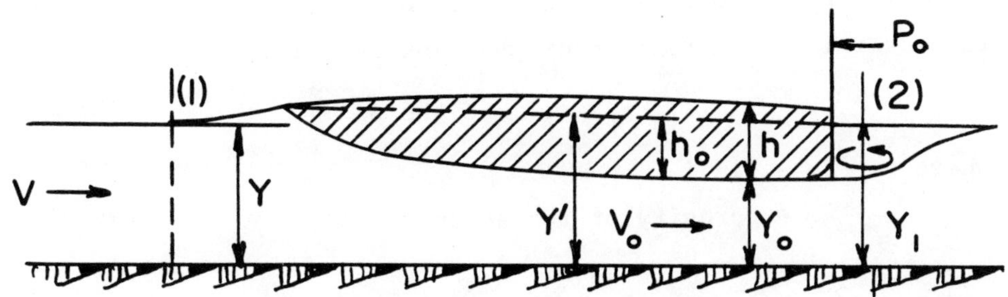


*Fig. 4.47. Hydrodynamic force on frontal edge.*

From energy conservation:

$$Y + \frac{V^2}{2g} = Y_1 + \frac{Vo^2}{2g} \qquad (4\text{-}105)$$

Because of continuity:

$$VY = Vo\ Yo \qquad (4\text{-}106)$$

The momentum theorem gives, between sections (1) and (2):

$$po = \frac{\gamma\ Y^2}{2} - \frac{\gamma\ Y_1{}^2}{2} + \rho\ Y\ V^2 - \rho\ Y\ V\ Vo \qquad (4\text{-}107)$$

And finally, neglecting terms of second order of smallness:

$$p_o = \gamma\ Y\ (1 - Yo/Y)^2\ Vo^2/2g \qquad (4\text{-}108)$$

Computation of the tangential force $\tau$ on the cover. The total tangential force on the cover is made up of the component of the weight of the cover $\tau_g$ and of the water friction force $\tau_{wl}$ under the cover.

Weight of the cover. The component of the weight of the cover in the direction of the flow is (Fig. 4.48):

$$\tau_g = \gamma(1 - e)\ ho\ So \qquad (4\text{-}109)$$

where So - sin i

e - porosity of the accumulation.

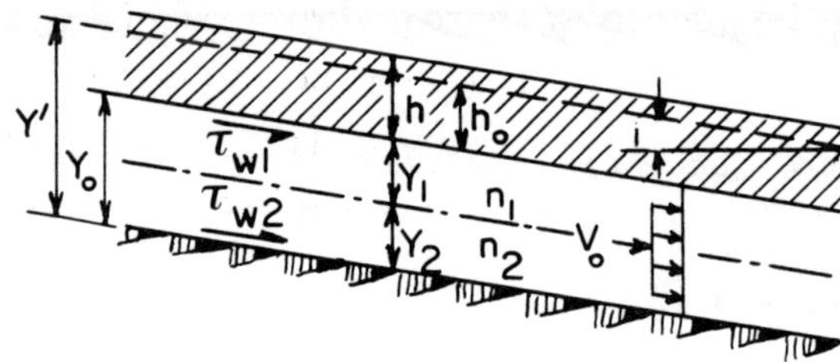


Fig. 4.48. Stress distribution under an ice cover.

Water friction force. The boundary stress distribution can be obtained by the Torok-Sabaneev hypothesis (Fig. 4.48).

$$\begin{aligned} \tau_\omega &= \gamma\ Y_0\ So \\ \tau_{\omega 1} &= \gamma\ Y_1\ So \\ \tau_{\omega 2} &= \gamma\ Y_2\ So \\ \tau_\omega &= \tau_{\omega 1} + \tau_{\omega 2} \end{aligned} \qquad (4\text{-}110)$$

where $\tau_\omega$ is the total friction force of the flow, $\tau_{\omega 1}$ under the cover and $\tau_{\omega 2}$ on the bottom.

With the use of the Chézy-Manning equation:

$$So = \frac{Vo^2\ no^2}{2.22\ (Y_0/2)^{4/3}} = \frac{Vo^2\ n_1{}^2}{2.22\ Y_1{}^{4/3}} = \frac{Vo^2\ n_2{}^2}{2.22\ Y_2{}^{4/3}} \qquad (4\text{-}111)$$

we then have:

$$\begin{aligned} 2\ (no)^{3/2} &= n_1{}^{3/2} + n_2{}^{3/2} \\ Y_1 &= Y_0\ n_1{}^{3/2} / 2\ n_o{}^{3/2} \\ Y_2 &= Y_0\ n_2{}^{3/2} / 2\ n_o{}^{3/2} \end{aligned} \qquad (4\text{-}112)$$

The friction force caused by the flow under the cover is then:

$$\tau_{w1} = \frac{\gamma\, Y_o\, S_o\, n_1^{3/2}}{2\, n_o^{3/2}} = \frac{\gamma\, V_o^2\, n_1^{3/2}\, n_o^{1/2}}{1.76\, Y_o^{1/3}} \qquad (4\text{-}113)$$

Tests carried out in flumes with model wooden logs by Kennedy (1958) are shown on Fig. 4.49. They checked very well Eq. 4-104 with a value of $\alpha$ = 1.0 which is shown as a continuous curve. Because of model length limitations Kennedy was not able to get points for higher values of L/B.

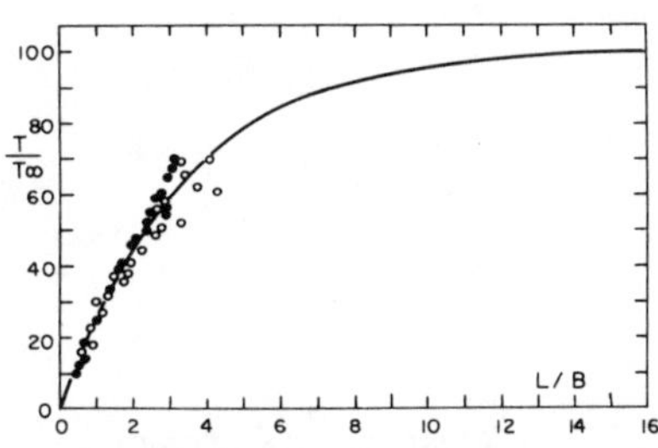


*Fig. 4.49. Forces on a boom caused by a model log jam (Kennedy, 1958).*

Delagrave (1966) made experimental studies in a 2 m wide flume with polyethylene pieces simulating ice floes. These tests were made for thick covers (h/Y in the range of 0.2 to 0.3). It was found that the value of $p_o$ was important here and that Eq. (4-104) represented the phenomenon fairly well in all cases (Fig. 4.50).

Some investigations were made in a small channel, 1.6 m wide, by Latyshenkov (1946). The ice accumulation consisted of a single layer of ice floes and it was found that the force attained a maximum when L ≃ (2.5 to 3.0) B. For practical purposes the maximum force was given by the empirical relation:

$$T\infty = 1.72\ \tau\ B^2 + p_o\ B \qquad (4\text{-}114)$$

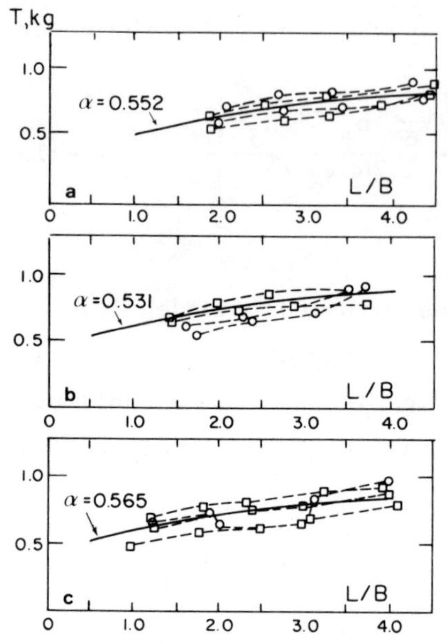


*Fig. 4.50. Thrust measured on a scale model with simulated ice (Delagrave, 1966).*

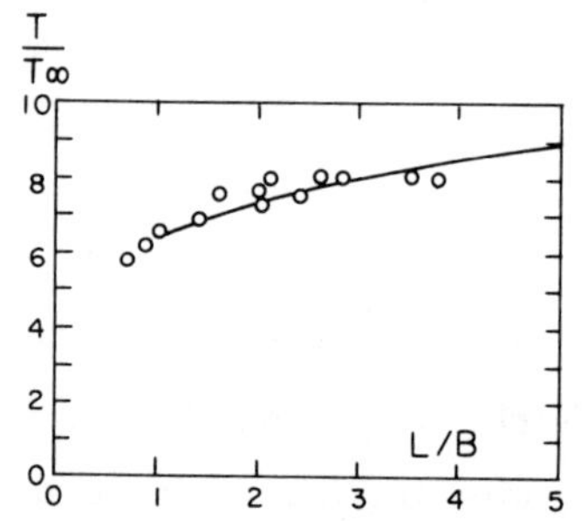


*Fig. 4.51. Results of tests with ice floes (Latyshenkov, 1946).*

It can be suspected that the value of $p_o B$ was also very high in those tests as the ratio h/Y varied from 0.15 to 0.22. Moreover, the hydrodynamic thrust on the edge is higher for a single floe than for an accumulation of ice pieces of the same thickness. Eq. (4-114) checks with Eq. (4-104) if $p_o B$ is taken equal to $1.88\ \tau\ B^2$. In that case the results of the testing and the theory give a very good fit as shown in Fig. 4.38, if we take a limiting value of 0.8 T∞ to measure the very slowly increasing force starting from a length of 3L/B.

## 4.3.2 - STABILITY OF AN IDEALIZED ICE JAM

There is a physical limit to the thickening of an ice cover under the effect of the hydraulic forces. If the discharge is slowly raised under an ice accumulation the cover thickens by shoves until the increase in the hydraulic thrust because of the reduction of the flow area gets higher than the resistance of the cover to this thrust.

Let us consider a section through the cover normal to an imaginary arch (Fig. 4.52). The resistance of the cover developed under the acting principal stress $\sigma_{n1}$ is:

$$p_1 = \frac{1}{K\phi}\,(h - ho)\,\gamma'\,(1 - e) = \frac{(\gamma - \gamma')\,(1 - e)\,ho}{K\phi}$$

$$\sigma_{n1}\,h = \frac{p_1\,h_o\,\gamma}{2\gamma'} \qquad (4\text{-}115)$$

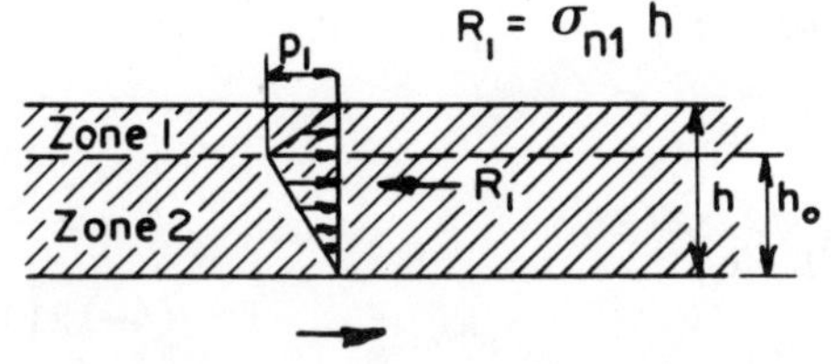


*Fig. 4.52. Stress distribution inside an accumulation of ice pieces.*

The reaction of the shore is given by Eq. (4-100):

$$R = \sigma_r h = \sigma_{n1}\cos\psi\,h = \frac{1}{2}\cos\psi\,\frac{K\psi}{K\phi}\,\frac{(\gamma - \gamma')\,\gamma}{\gamma'}\,(1 - e)\,ho^2 \qquad (4\text{-}116)$$

$K\psi$ and $K\phi$ being respectively the coefficient of active resistance of the accumulation to the thrust at the shore and inside the accumulation.

Because of the stress distribution in the cover (Fig. 4.52) the reaction R at infinity is given by:

$$\tau B = 2 R \sin \psi \qquad (4\text{-}117)$$

In the case of hydraulic thrust only, the value of $\tau$ is given by Eqs (4-112) and (4-116):

$$\tau = 1.14 \gamma \left[ (1 - e) ho + \frac{1}{2} \left( \frac{n_1}{n_o} \right)^{3/2} Yo \right] \frac{Q^2 n_o^2}{B^2 Yo^{10/3}} \qquad (4\text{-}118)$$

If we write:

$$\xi = \sqrt{\frac{2.28 \gamma' n_o^2 K\phi}{\sin 2\psi K\psi (1 - e) (\gamma - \gamma') Yo^{1/3}}} \qquad (4\text{-}119)$$

The conditions of equilibrium of the cover with Eqs (4-116), (4-117), (4-118) and (4-119) become:

$$\frac{[(1 - e) ho + 1/2 (n_1/n_o)^{3/2} Yo]}{ho^2 Yo^3} = \frac{B}{\xi^2 Q^2} \qquad (4\text{-}120)$$

This can be simplified with:

$$Y' = Yo + ho$$

$$\beta = 1/2 (n_1/n_o)^{3/2}$$

$$h' = ho/Y' \qquad (4\text{-}121)$$

$$y' = (Y')^4 B/\xi^2 Q^2$$

and we get:

$$y' = \frac{\beta + (1 - e - \beta) h'}{(1 - h')^3 (h')^2} \qquad (4\text{-}122)$$

This function has a minimum that can be obtained by differentiation. Because $(1 - e - \beta) h'$ is always smaller than $\beta$, this minimum is given very closely by:

$$h' = ho/Y' \simeq 0.4 \qquad y' \simeq 31\ \beta \tag{4-123}$$

These expressions represent the limit of stability of an ice jam. In the case where most of the roughness comes from underneath the ice jam, β = 1, and the condition of limit equilibrium of the jam is:

$$\frac{Y'\ \sqrt[4]{B}}{\sqrt{\xi Q}} = 2.36 \tag{4-124}$$

This equation of equilibrium of an ice accumulation (Eq. 4-122) is shown in Fig. 4.53 and the results of tests carried on with polyethylene blocks of density 0.92, porosity 0.4 and $\phi$ = 30 are shown on the same graph. It can be seen that the general criterion for the stability of a jam is then given by Eq. (4-124).

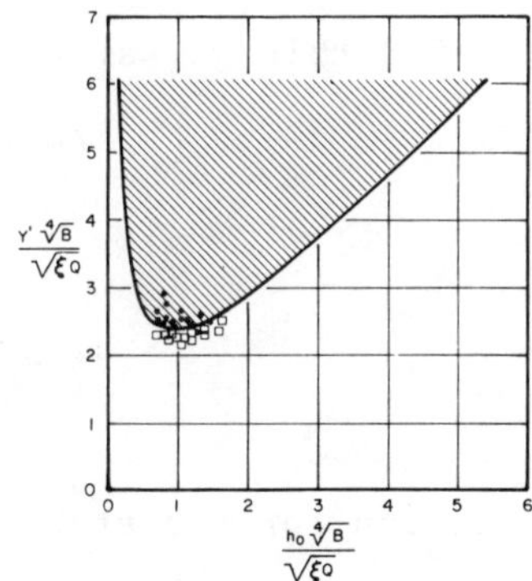


*Fig. 4.53. Stability of static ice jams. The grey zone is the zone of stability. Dark points (in equilibrium). Open circles (not in equilibrium).*

The maximum discharge before shoving is:

$$Q_{max} = \frac{0.037}{n} (Y')^{2.16}\ B^{0.5} \tag{4-125}$$

or

$$F_{rc} = \frac{V_{cr}}{\sqrt{g\ Y'}} = \frac{0.037}{\sqrt{g}\ n} (Y')^{0.66}\ B^{-0.5} \tag{4-126}$$

where

$$n = \sqrt{n_1{}^{3/2}\ n_o{}^{1/2}}$$

## 4.3.3 - MAXIMUM THRUST OF UNCONSOLIDATED ICE ACCUMULATIONS

### Hydraulic thrust

The limiting tangential stress for the condition of instability is obtained from (4-116), (4-117) and (4-123):

$$\tau\ B = 0.08 \sin 2\ \frac{K\psi}{K\phi}\ \frac{(\gamma - \gamma')\ \gamma}{\gamma'}\ (1 - e)\ h^2 \qquad (4\text{-}127)$$

The maximum thrust that can be exerted on a boom, is according to (4-104):

$$T\infty = .288 \sin 2\psi\ \frac{K\psi}{K\phi}\ \frac{(\gamma - \gamma')\ \gamma}{\gamma'}\ (1 - e)\ h^2\ B \qquad (4\text{-}128)$$

With max. $\psi = \phi = 30^o$, min. $e = 0.3$ this gives:

$$T\infty = 16\ h^2\ B \qquad (4\text{-}129)$$

where $T\infty$ is in kg, h and B in m.

### Thrust caused by the wind

Because of this hydraulic limitation for the stability of the accumulation, it might happen that the wind will exert a higher thrust when the discharge is very low and the cover might thicken right down to the bottom of the river.

Let us consider an extreme 33 m/sec wind at standard 10 m above the ground. The tangential stress it exerts on the surface of the ice field is from Eq. (4-13).

$$\tau_a \simeq 4.9 \text{ in Pa} \qquad (4\text{-}130)$$

and the maximum thrust on the structure from (4-104):

$$T\infty = 1.76\ B^2 \tag{4-131}$$

but the cover cannot thicken more than $y_o = h$, because the bottom would then take a great part of the load. So there is also another limitation given by (4-116) and (4-117):

$$T\infty \leq 100\ h^2\ B \tag{4-132}$$

## General representation

The maximum possible thrust on a hydraulic structure may be readily obtained from either formulas (4-129), (4-130) or (4-131).

Because each of these conditions limits the others, it is better to use a general graphical representation for all the cases considered. This is shown in Fig. 4.54 where the thrust per unit width is plotted against the water depth. The width of river is shown with horizontal lines.

In zone A the ice accumulation is grounded, the thrust is given by formula (4-132) which is the upper parabola on the graph. In zone B the flow discharge is small and the cover is in equilibrium with the wind only; the thrust being given by formula (4-131). In zone C, the ice field is subject to hydraulic conditions only. If the discharge increases to its value for limiting stability, the thrust is given by the lower parabola on the graph corresponding to formula (4-129).

Paradoxically, the maximum thrust can occur in most cases with a strong wind and practically no water flow. In those cases the cover might increase in thickness until it reaches the bottom of the river.

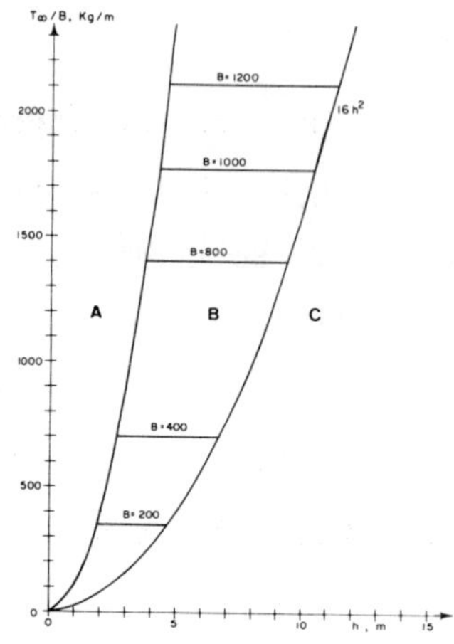


*Fig. 4.54. General diagram of thrust exerted by an unconsolidated ice cover on an hydraulic structure (Michel, 1968).*

## 4.3.4 - ICE PILE-UPS ON STRUCTURES

A moving ice field may, under the action of water currents or wind, hit the shore of a lake or river and any engineering work located there. It will then induce important ice accumulations while slowing down, and dissipating its energy by breaking and accumulating small ice pieces on its edge. Accumulations of large size have repeatedly been reported in the literature and one close to 30 m high has been observed by Estifeev (1958).

These accumulations may have harmful effects on installation and works merely because of their bulk but the total force they exert is usually considerably smaller than the initial impact of ice on the structure. When at rest the force they exert may be estimated by simulating the accumulation of a granular soil. The horizontal pressure at any point above the water surface is given by Michel (1970) as shown on Fig. 4.55.

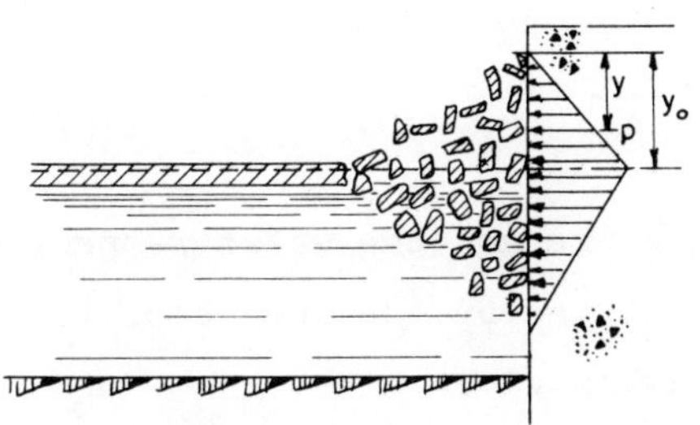


*Fig. 4.55. Horizontal pressure caused by an ice accumulation against a vertical wall.*

$$p = K\phi\ \gamma' (1 - e)\ y \text{ for } y < y_o \qquad (4\text{-}133)$$

where $K\phi = \tan^2 (45^o - \frac{\phi}{2})$, the coefficient of active pressure of the ice pieces

$y$ - height of ice on top of the considered point

$\phi$ - angle of internal friction of the ice pieces

$\gamma'$ - unit weight of ice

$e$ - porosity of the ice accumulation (ratio of voids to total volume).

Of interest to this problem is that of the equilibrium of floating or grounded ice ridges.

The geometries of the two cases are as shown in Fig. 4.56. Then according to Parmerter and Coon (1972) if the sail and keel heights of floating ridge are related such that:

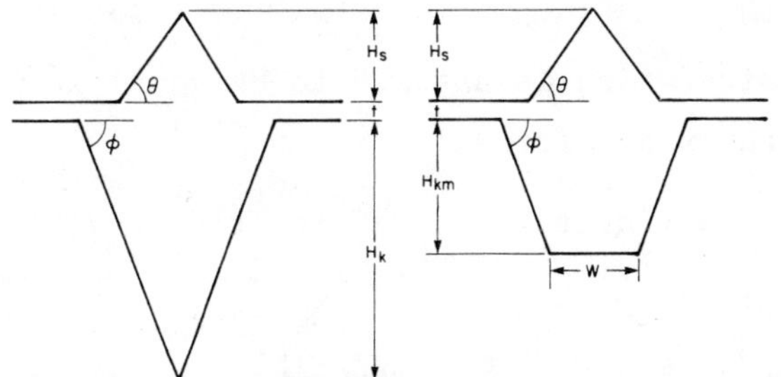


*Fig. 4.56. Cross-sectional profiles of idealized floating and grounded ridges.*

$$\frac{H_k}{H_s} = \frac{\rho'}{\rho - \rho'} \qquad (4\text{-}134)$$

where $\rho'$ and $\rho$ are the density of the ice and the water respectively, then the effective stress exerted by the ice sheet to generate the stored potential energy is given by:

$$\sigma_f = \frac{\rho' g H_s}{2} \qquad (4\text{-}135)$$

where g is the acceleration due to gravity.

Eq. (4-134) implies that along the vertical symmetry plane of the ridge, the ice is locally in isostatic equilibrium. However, if the keel extends further from the center than the sail as indicated in Fig. 4.56 there is a net buoyant force on the ice sheet. When this force becomes large enough, the ice sheet is failed at the edge of ridge which then grows in extent rather than height. This is the height limiting mechanism for floating ridges.

For the grounded ridge Kry (1977) gives the following development. With the geometry and symbols defined in Fig. 4.56, the potential energy per unit length of ridge E, is given by:

$$E = (1 - e)\,(\rho - \rho')\,g\left[\frac{\rho' H_s^{\,3}}{3\,(\rho - \rho')\tan\theta} + \frac{H^3_{km}}{3\tan\phi} + \frac{WH^2_{km}}{2}\right] \qquad (4\text{-}136)$$

where e is the ridge porosity which is assumed to be equal in the sail and keel during the ridge formation.

The volume of ice per unit length of the ridge, V, is given by:

$$V = (1 - e)\left[\frac{H_s^{\,2}}{\tan\theta} + \frac{H^2_{km}}{\tan\phi} + WH_{km}\right] \qquad (4\text{-}137)$$

This volume of ice per unit ridge length must equal the product of the ice sheet thickness and the total extent of the sheet movement U which generated the rubble.

Both the energy and volume per unit ridge length are functions of the sail height $H_s$. The width W is also a function of $H_s$ and with the assumption that as the ridge grows, the horizontal extents of the sail and keel increase at the same rate, then:

$$\frac{dW}{dH_s} = \frac{2}{\tan \theta} \tag{4-138}$$

Thus the effective stress in the ice sheet applied to create the grounded ridge is given from Eqs (4-136), (4-137) and (4-138):

$$\sigma_g = \frac{1}{t}\frac{dE}{dH_s} \cdot \frac{dH_s}{dU}$$

$$= \sigma_f \left[ \left(\frac{\rho - \rho'}{\rho}\right)\left(\frac{H_{km}}{H_s}\right)^2 + 1 \right] \Big/ \left(\frac{H_{km}}{H_s} + 1\right) \tag{4-139}$$

Eq (4-139) states that $\sigma_g < \sigma_f$ since once the ridge grounds:

$$\frac{H_{km}}{H_s} < \frac{\rho'}{\rho - \rho'} \tag{4-140}$$

Therefore, once the ridge grounds, the force required to continue increasing the gravitational potential energy for a given sail height is decreased.

Rubble fields increase the total horizontal loads experienced by artificial islands mainly because they increase the effective cross-section against which a moving ice sheet must fail.

## 4.4 - VERTICAL FORCES EXERTED BY ICE

### 4.4.0 - INTRODUCTION

Among the forces exerted by ice on hydraulic structures, one of the least known, and perhaps most often neglected in design practice, is the vertical force transmitted to a structure by an ice sheet during water level fluctuations.

The adhesive strength of ice on various construction materials is high and considerable moments and forces have to be resisted by the walls of structures following a variation in water level. With a rise in the level, piers, caissons, pile groups, well casings and even rectilinear wharf sheetings may be partly lifted off their foundations and damaged. Observations in northern Sweden have shown that even large caissons can be lifted by the ice and one was raised more than 20 cm (Lofquist, 1944). Damage can also be inflicted on revetments, gates or protruding parts of dams and intake works frozen in the ice sheet, following the usual lowering of the reservoir level during winter operation.

The theoretical determination of the vertical force exerted by an ice sheet is a difficult task for a number of reasons.

The first difficulty arises because water level fluctuations occur frequently during the winter. The initial failure produced in a thin sheet becomes the weak point where subsequent failures will be located. This time-dependent phenomenon defies analysis and is partidularly apparent in tidal waters where high ice caps build up on the upright surfaces of structures. Fortunately these caps do not usually have extensive horizontal projections and the computation can be carried out as with the original boundary if the maximum thickness of the solid ice sheet that can be formed with a stable water level can be predicted.

Another difficulty is the complexity of the analysis of the basic behavior of a floating ice sheet subject to very slowly applied loads, if the viscoelastic properties of the ice have to be taken into account. Many approximate assumptions have to be made to get usable answers. One is to use the theory of elasticity with brittle fracture of ice. The other method is to determine the maximum loading conditions for the ultimate strength of ice on the basis of the ductile behaviour. In this latter connection it might be said that analyses pertaining to the similar problem of the bearing capacity of an ice sheet are most useful and have been applied to this case.

## 4.4.1 - STRENGTH OF ADHESION OF ICE TO STRUCTURES

In some cases the maximum vertical force that can be transmitted to a structure might be limited by the strength of adhesion of ice to the material the structure is made of. Unfortunately there are very few available data on this basic mechanical property of ice and there is a lot of scatter in the experimental results because of insufficient control of the uniformity of the structure of the ice and of other relevant factors.

It has been shown (Jellinek, 1957) that the adhesive strength of ice is a linear function of the temperature up to a point where it becomes larger than the internal shear or cohesive strength of the ice itself. The sharp transition between the adhesive and the cohesive break occurs at -13°C for strong snow ice between two stainless steel plates.

| | Pa x 10^5 |
|---|---|
| Metals: | |
| Mild steel | 8.6 |
| Stainless steel | 8.2 |
| Copper | 8.9 |
| Aluminum | 6.4 |
| Nickel | 6.0 |
| Zinc | 6.4 |
| Galvanized steel | 7.9 |
| Woods: | |
| Douglas fir | 3.2 |
| Courbaril | 4.6 |
| Lignum vitae | 5.7 |
| Plastics: | |
| Polymethyl methacrylate | 2.9 |
| Cellulose acetate | 1.8 |
| Glass: | 4.6 |
| Rubber: | 1.4 to 10.7 |
| Paints: | |
| Gray deck paint (Navy) | 5.7 |
| Haze gray paint (Navy) | 7.1 |
| Resin films: | |
| Acrylics | 5.7 to 9.3 |
| Alkyds | 5.4 to 6.8 |
| Epoxys | 5.7 to 9.3 |
| Polytetrafluoroethylenes | 4.3 to 5.0 |
| Polytrifluorochloroethylene | 64 |
| Phenolics | 2.1 to 7.9 |
| Silicones | 2.9 to 6.8 |
| Urethanes | 6.8 to 9.3 |
| Vinylidenes | 3.6 to 5.7 |
| Vinyls | 6.4 to 8.2 |

Table 4.6 - Adhesion of ice to various materials and coatings (Freiberger and Lacks, 1961).

Among the results of experimentation for artificial ice are those shown in Table 4.6 which were obtained for rather high rates of loading, on the order of $3 \times 10^4$ Pa/sec, and results depend strongly on the rate. It must be noted that in most cases the strength of adhesion of ice to construction materials is as high as the shear strength of the river or lake ice itself, which is usually taken to be from 4 to $10 \times 10^5$ Pa, as discussed in Chapter 2.

## 4.4.2 - VERTICAL FORCE ON A STRAIGHT WALL

This case can be simulated by a strip load action on an infinite plate on an elastic foundation (Michel, 1970). If a number of equal, central concentrated loads are applied in line on a floating ice sheet, the analysis is simplified by ignoring longitudinal bending and using an equivalent strip load of 2p per unit length.

The differential equation of the deflection of this long beam is:

$$\frac{d^4w}{dx^4} = \frac{q - kw}{D} \tag{4-141}$$

where, as previously, q is the applied load, k the foundation modulus, and D the flexural rigidity of the plate.

In this case of loading, 2p per unit length, the solution is:

$$w = \frac{P}{k\ell} \exp\left[-x/\ell\sqrt{2}\right] \sin\left(x/\ell\sqrt{2} + \pi/4\right) \tag{4-142}$$

For the elastic state the maximum positive transverse bending moment $M_r$, per unit length, under the center of the load is:

$$M_r = p\ell/\sqrt{2} \tag{4-143}$$

The maximum negative transverse moment occurs at a distance of about $x = 2.2\ell$ from the load and is, approximately, $M = -0.14$ pl. The maximum deflection is:

$$w_o = \frac{p}{k\ell\sqrt{2}} \tag{4-144}$$

and the width of the deflection band is about $x = 3.3\ \ell$. The yield load at which the first longitudinal crack develops at the bottom of the ice sheet below the load can be estimated from Eq. (4-143). A strip load has to be more than $3\ell$ from the edge of large sheets to satisfy the above equations and develop the estimated yield load.

Chankin (1961) has proposed a formula very like Eq. (4-143) for computing the maximum force acting on a straight wall. He found this formula to be valid for a variation of water level of from 0.75 to 2.5 cm/hr, ice thickness 1 m, with a value of E of $4.28 \times 10^9$ Pa as obtained from measurements in an artificial reservoir.

In the case of a very rapid rise of water level $\Delta Y$, Lofquist (1944) proposed to use the same formula when the ice is considered to be an elastic material. We then have:

$$p = \sqrt{2}\ k\ \ell\ \Delta Y \tag{4-145}$$

where $p$ - load on the structure, kg/m

$k$ - foundation modulus, 1000 $kg/m^3$

$\Delta Y$ - water rise, m

$\ell$ - radius of relative stiffness, m.

This gives the unit load directly as a function of the variation in water level. In contrast with the case of the isolated load on an ice sheet, the first failure, directly in contact with the wall, is the final one as the wall cannot transmit any more load to the sheet thereafter. With a value of E of $4.28 \times 10^9$ Pa, the deflection $w_o$ at failure is 6.8 cm for a 0.3 m thick sheet of $14 \times 10^5$ Pa bending strength and 12 cm for a 1 m thick sheet. These values of the water level fluctuation can be attained quite rapidly in a river.

The maximum load can also be obtained for the ductile plate. The elastic yield moment of ice $M_b$ should, however, correspond to the plastic bending moment $M_o$. The only difference may be the change of $M_b$ to $M_o$ and the maximum possible vertical force is given by the same Eq. (4-143) which gives here a higher value for p.

$$p = \frac{Mo \sqrt{2}}{\ell} \qquad (4\text{-}146)$$

A discussion by Gold (1977) of this problem shows that the ice will crack along the wall if the rate of change in water level is sufficiently large. If not the ice will behave in a ductile manner without cracking. This implies that for a cyclical change in water level of sufficient height, such as might occur by tidal action, the ice at the wall will be continually broken free of the structure until it is thick enough to bring the maximum strain rate into the ductile range. The deflections required, however, will exceed the thickness of the cover, h, for a time. This has the effect of increasing the thickness near the wall by surface flooding and of developing a "bustle" on the structure. Continuous vertical movement of the ice near the structure may also prevent it from freezing to it even when the thickness exceeds that required for ductile behaviour. As a result, the maximum vertical force that can be developed on the structure becomes indeterminate, and it is necessary to assume an appropriate extreme condition for design. This extreme condition may include taking into account the weight of the "bustle".

### 4.4.3 - LIFTING FORCE ON ISOLATED STRUCTURES

The case of a rising ice sheet attached to a pier or a pile can be ideally represented by the similar one of a concentrated load applied to a large floating sheet of uniform thickness and discussed extensively in Chapter 3.

This approximation should be reasonable for a pile of radius, a, appreciably smaller than the characteristic length of the ice cover, $\ell$. The deflection of the cover at the pile for this condition, assumed equal to the change in water level, is given by Eq. (3-67), with $\alpha = a/\ell$:

$$w_p = \Delta Y = \frac{P\ (1 + \alpha\ \mathrm{ker}'\alpha)}{\pi\ k\ \alpha^2\ \ell^2} \qquad (3\text{-}67)$$

With the circumferential shear stress on the pile $q_p$:

$$q_p = P/2\ \pi\ \alpha\ \ell h$$

the equation becomes:

$$w_p = \Delta Y = \frac{2\ q_p\ h\ (1 + \alpha\ \mathrm{ker}'\alpha)}{k\ \alpha\ \ell} \qquad (4\text{-}147)$$

For $\alpha < 0.15$:

$$1 + \alpha\ \mathrm{ker}'\alpha = \pi\ \alpha^2/8$$

and

$$w_p = \frac{\pi\ h\ \alpha\ q_p}{4\ k\ \ell} \qquad (4\text{-}148)$$

For a long straight wall:

$$w_w = \frac{h\, q_w}{\sqrt{2}\, k\, \ell} \tag{4-149}$$

where $q_w$ - shear force exerted on the wall.

For the same change in water level, i.e., $w_p = w_w$,

$$\frac{q_w}{q_p} = 1.11\ \alpha \tag{4-150}$$

and for $\alpha < 0.1$:

$$\frac{q_w}{q_p} < 0.11$$

That is, the shear stress imposed on the pile for the elastic case is about 10 times that imposed on the wall, as shown by Frederking and Gold, (1971). It is clear that the deformation conditions are much more severe near the pile. This is borne out by field observations and laboratory testing which indicate that for the pile, failure often occurs by shearing at the pile - ice interface, whereas for the wall it is associated with bending and the formation of a crack.

If the creep behavior near the pile can be described by Eq. (2-154) with exponent, n, equal to 3, then the ratio of the vertical displacement at the radial distance r = a and r = b for a pile subject to an uplift shear stress $q_p$, is equal to $\left(\frac{a}{b}\right)^2$ (Frederking, 1974). This shows that the vertical shear displacement occurs relatively close to the pile. The vertical displacement rate of a pile of radius, a, relative to the ice surface sufficiently far away not to be affected by the vertical shear stress is given by:

$$\frac{dY}{dt} = \frac{\dot{\gamma}_c\, a}{(n - 1)} \left(\frac{\tau_a}{\tau_c}\right)^n \qquad (4\text{-}151)$$

Frederking found from tests on "piles" 5 and 10 cm in diameter that $\tau_c$ was about 72 $kN/m^2$ and $\dot{\gamma}_c$ about $4.5 \times 10^{-8}\ s^{-1}$. The maximum shear stress applied was 185 $kN/m^2$, which caused a vertical displacement rate of $1.2 \times 10^{-5}$ mm/s.

If a crack forms at the bottom of an ice cover, it will fill with water which can subsequently freeze. When the water level returns to the average value a thrust must develop due to the increased length of the lower surface. This problem is, in principle, the same as that of thermal thrust. The amount by which the lower surface increases in size will depend on the strain relief associated with the formation of the crack, and the degree of restraint offered by the wall to the vertical movement of the cover. The problem differs from that of the thermal thrust in that the increase in length is localized. If the ice cover is very large, the compressive strain induced would be correspondingly small. If the cover is confined as, say in a reservoir or between two bridge piers, the compressive force induced might be large enough to cause damage. Local damage might occur due to the bending moments included in the restrained situation.

The case which has been considered up to now is that corresponding to a load distributed over the area of a small circle. The case of a load distributed over the area of a rectangle may be obtained by methods developed for simply supported plates (Woinowsky-Krieger, 1953). The equivalent of a square area, in particular, is a circle of radius r = 0.57a, a being the length of the side of the square. The effect of any group of concentrated loads on the deflection of the infinitely large plate can be calculated somewhat by summing up the deflections produced by each load separately (Meyerhof, 1962).

Acutally, the reactions are not distributed over the area of the structure but along its periphery. It is thus difficult to derive the stress distribution theoretically. For construction of a rectangular form it may be safe to replace the shape by a form with straight sides and semicircular ends. Then the lifting force on the straight sides can be computed in the same manner as for a long straight wall.

For a group of piles of rectangular overall form, the U.S.S.R. standard recommends that the total uplift force be taken as for a circle of radius $r = \sqrt{ab}$, where a and b are the sides of the rectangle.

## REFERENCES

Atkinson, C.H., Cronin, D.L.R., Danys, J.V. (1971) - "Measurement of ice forces against a lighpier" 1st Conf. Port and Ocean Eng. under Arctic Conditions, Trondheim, Norw.

Banke, E.G., Smith, S.D. (1973) - "Wind stress on Arctic sea ice" J. Geophys. Res. 78, 7871-7883.

Barnes, H.T. (1928) - "Ice engineering" Montreal: Renouf Publishing Co., 364 p.

Beccat, R., Michel, B. (1959) - "Thrust Exerted on a Retaining Structure by Unconsolidated Ice Covers" Proc. 8th Congress AIRH.

Bergdahl, L. (1972) - "Two lighthouses damaged by ice" Int. Assoc. Hydraul. Res., Proc. 2nd Ice Symp., Lenningrad, U.S.S.R.

Blenkarn, K.A. (1970) - "Measurement and analysis of ice forces on Cook Inlet structures" Offshore Technol. Conf., Houston Tex.

Buckley (1900) -"Ice Ramparts" Wisconsin Academy of Science Transactions, Vol. 13, p. 141-162.

Butkovich, T.R. (1957) - "Thermal expansion of ice" Journal of Applied Physics, Vol. 30, No. 3, p. 350-353.

Caquot, A., Kérisel, J. (1956) - "Traité de Mécanique des Sols" Gauthier-Villars.

Carter, D. (1977) - "Impact des glaçons sur les ouvrages maritimes" Proc. Third National Hydrotechnical Conference, Can. Soc. Civil Eng., p. 815-833.

Carter, D., Michel, B. (1971) - "Lois et mécanismes de l'apparente fracture fragile de la glace de rivière et de lac" Ice Mechanics, Lab., Univ. Laval, Report S-22.

Chankin, P.A. (1961) - "On the stability of concrete revetments on slopes submitted to ice forces" Gidroteknicheskoe Stroitel'stvo, No. 3.

Chen, W.F. (1975) - "Limit analysis and soil plasticity" Elsevier Scientific Pub., New York.

Cook, H.J. (1941) - "Protecting a reservoir against ice" Waterworks and Sewerage, Dec., p. 549.

Croasdale, K.R. (1974) - "Crushing strength of Arctic ice" Symp. on Beaufort Sea Coastal and Shelf Res., Arctic Inst. North Am., Montreal, Que., Canada, p. 377-399.

Croasdale, K.R., Edwards, R.Y. (1976) - "Indentation tests to investigate ice pressures" Reprint to Symp. on App. Glac., Int. Glac. Soc., Cambridge, England.

Danys, J.V. (1972) - "Effect of ice forces on some isolated structures in the St. Lawrence River" Int. Assoc. Hydraul. Res., Proc. 2nd Ice Symp., Leningrad, U.S.S.R.

Danys, J.V., Bercha, F.G., Carter, D. (1977) - "Influence of friction on ice forces acting against sloped surfaces" J. Glaciol. 19, No. 81, p.

Delagrave, M. (1966) - "Etude expérimentale sur modèle, des forces exercées par un champ de glace morcelée" Thesis M.Sc., Civil Eng. Dept., Laval University.

Dorsey, N.E. (1940) - "Properties of ordinary water substances" New York: Reinhold.

Drouin, M. (1968) - "Static ice force on extended structures" N.R.C. Conference on Ice Pressure, Tech. Memorandum No. 92, p. 95-108.

Drouin, M., Michel, B. (1971) - "Les poussées d'origine thermique exercée par les couverts de glace sur les structures hydrauliques" Rapport S-23, Faculté des Sciences, Génie Civil, Université Laval, Québec, Qué., Canada.

Durelli, A.J., Philipps, E.A., Tsao, Ch. (1958) - "Introduction to the theoretical and experimental analysis of stress and strain" McGraw Hill.

Eckhard, G.F. (1920) - "Large highway bridge wrecked by pressure of cake ice" Engineering News, Vol. 84, p. 902-904.

Edwards, R.Y., Wheaton, J.W. (1972) - "Experimental determination of ice impact loads on marine vehicles" Int. Assoc. Hydraul. Res., Proc. 2nd Ice Symp., Leningrad, U.S.S.R.

Engineering Record (1899) - "Failure of a Minneapolis dam by ice pressure" Vol. 39. p. 542-543.

Engineering News (1912) - "Failure of dam from ice pressure" Vol. 65, p. 681.

Estifeev, A.M. (1958) - "Frazil control at power plants" (text in Russian) Moscow, 180p.

Flinn, A.E. (1928) - "Arch dam investigations - Ice pressures" Proceedings ASCE, Vol. 54. No.3, p. 268-269.

Frederking, R., Gold, L.W. (1971) - "Ice forces on an isolated circular pile" 1st Conf. Port and Ocean Eng. under Arctic Conditions. Trondheim, Norw.

Frederding, R. (1974) - "Downdrag loads developed by a floating ice cover: field experiments" Can. Geotech. J., 11, p. 339-347.

Frederking, R., Gold, L.W. (1975) - "Experimental study of edge loading of ice plates" Can. Geotech. J. Vol. 12, pp 456-464.

Freiberger, A., Lacks, H. (1961) - "Ice-phobic coatings for deicing naval vessels" In Proceedings of The Fifth Navy Sciences Symposium, p. 234-237.

Gold, L.W. (1977) - "Ice pressures and bearing capacity" Chapter 10 of 'Geotechnical Engineering for Cold Regions' Edited by D. Anderson.

Haynes, F.D., Nevel, D.E., Farrell, D.R. (1975) - "Ice force measurements on the Pembina River" 2nd Can. Hydrotech. Conf., Burlington, Ont.

Hetenyi, M. (1946) - "Beams on elastic foundations" The University of Michigan Press.

Hill, H.M. (1935) - "Field measurements of ice pressure at Hastings Lock and Dam" Military Engineer, Vol. 27, p. 119-122.

Hill, R. (1956) - "The mathematical theory of plasticity" Oxford University Press, 353p.

Hirayama, K., Schwarz, J., Wu, H. (1974) - "An investigation of ice forces on vertical structures, Iowa Inst. Hydraul. Res. Rep No. 158.

Isnard, J.L. (1978) - "Contribution à l'étude de l'indentation de plaques de glace" Thèse M.Sc., Université Laval.

Jellinek, H.H.G. (1957) - "Adhesive properties of ice" USA SIPRE Research Report 38.

Johannessen, O.M. (1970) - "Note on some vertical profiles below ice floes in the Gulf of St. Lawrence and near the North Pole" J. Geophys. Rs. 75, pp2857-2861.

Kennedy, R.J. (1958) - "Forces Involves in Pulpwood Holding Grounds" Eng. J. Canada, January.

Kirkham, I.E. (1927) - "Five Missouri River highway bridges" Engineering News Record, Vol. 99, No. 19.

Korzhavin, K.N. (1962) - "Action of ice on engineering structures" Novosibirsk, Akad. Nauk. SSSR, 202p.

Kry, P.R. (1977) - "Ice rubble fields in the vicinity of artificial islands" Proc. POAC Conference 1977, p. 200-212.

Langleben, M.P. (1972) - "A study of the roughness parameters of sea ice from wind profiles" J. Geophys. Res. 77, pp. 5935-5944.

Latyshenkov, A.M. (1946) - "Investigations of Ice Booms" Hydrotechnic Structures, No. 15:4, U.S.S.R.

Lavrov, V.V. (1969) - "Deformation and strength of ice" Hydromet. Publ. House, Leningrad, U.S.S.R. Transl. by Israel Prog. for Sci. Transl., Jerusalem, Isr.

Lofquist, B. (1944) - "Lifting force and bearing capacity of an ice sheet" Teknish Tidskrift No. 25, Stockholm. National Research Council of Canada, Technical Translation 164, 1951.

Lofquist, B. (1954) - "Studies of the effect of temperature variation - Ice pressure against dams" Transactions ASCE, Paper 2656.

Meyerhof, G.G. (1962) - "Bearing capacity of floating ice sheets" Transactions of the American Society of Civil Engineering, Vol. 127, part 1, No. 3327, p. 524-557.

Michel, B., Drouin, M. (1965) - "Impact of an ice floe on an obstacle" (In French) In Proceedings International Association for Hydraulic Research, Vol. 5, p. 60-63.

Michel, B. (1968) - "Thrust exerted by an unconsolidated ice cover on a boom" Proc. Conference on ice pressure against structures, Nat. Research Council Tech. Mem. No. 92, p. 163-171.

Michel, B. (1970) - "Ice pressures in engineering structures" Monograph III - Blb, Cold Reg. Res. and Eng. Lab., U.S. Army Corps of Eng., Hanover, N.H., U.S.A.

Michel, B. (1971) - "Winter regime of lakes and rivers" CRREL Monograph III-Bla, Cold Reg. Res. Eng. Lab., U.S. Army Corps of Eng., Hanover, N.H., U.S.A.

Michel, B., Toussaint, N. (1977) - "Mechanisms and theory of indentation of ice plates" J. Glaciology 19, No. 81, p. 285-301.

Michel, B., Paradis, M. (1976) - "Analyse statistique des lois du fluage secondaire de la glace de rivière et de lac" Ice Mechanics Lab., Univ. Laval, Report GCS-76-02.

Monfore, G.E. (1952) - "Ice pressure against dams: experimental investigations by the Bureau of Reclamation" Proceedings ASCE, Vol. 78, Technical Separate No. 162.

Monfore, G.E., Taylor, F.W. (1949) - "The problem of an expanding ice sheet" In Proceedings of the Western Snow Conference, 16th Meeting, p. 30-46.

Neill, C.R. (1970) - "Ice pressure on bridge piers in Alberta" Canada Int. Assoc, Hydraul. Res., Proc. 1st Ice Symp., Reykjavik, Iceland.

Neill, C.R. (1972) - "Force fluctuations during ice-floe impact on piers" Int. Assoc. Hydraul. Res., Proc, 2nd Ice Symp. Leningrad, U.S.S.R.

Neill, C.R. (1976) - "Dynamic forces on piers and piles" An assessment of design guidelines in the light of recent research. Can. J. Civ. Eng. Vol. 3, pp. 305-341.

Parmerter, R.R., Coon, M.D. (1972) - "Model Pressure Ridge Formation in Sea Ice" Journal of Geophysical Research, Vol. 77, No. 33, pp. 6565-6575.

Petrunitchev, N.N., Mamaieff, N.M. (1941) - "An Example of the Computation of the Forces Exerted by the Flow velocity on an Ice Field" Arctic Problems, U.S.S.R. No. 5.

Petrunichev, N.N. (1954) - "Dynamics of ice pressure on hydraulic structures" In Ledotermicheskie voprosy v Gidroenergetike. Ed. by D.N. Bibikov, Moscow: Gosudarstveneo Energeticheskoe Izdatel'stvo, p. 17-46.

Peyton, H.R. (1966) - "Sea ice strength" Univ. Alaska Geophys. Inst., Rep. 182.

Peyton, H.R. (1968b) - "Sea ice forces" Nat. Res. Counc. Can. Tech. Memo. 92.

Peyton, H.R. (1968a) - "Ice and marine structures" In Ocean Industry Gulf Publ. Co., March September and December.

Pounder, E.R. (1965) - "Physics of ice" New York: Pergamon Press, 151p.

Pratley, P.L. (1938) - "Collapse of Falls View bridge" Engineering Journal, Vol. 21, No. 8, p. 375-381.

Proskuriakoff, B.V. (1941) - "On the Analysis of the Thrust Exerted by an Ice Field" Arctic Problems, U.S.S.R. No. 5.

Ralston, T.D. (1977) - "Ice force design considerations for conical offshore structures" Proc. POAC Conference 77, pp. 741-752.

Reinius, E., Haggard, S., Ernstsons, E. (1971) - "Experience of offshore lighthouses in Sweden" 1st Conf. Port and Ocean Eng. under Arctic Conditions, Trondheim, Norw.

Roads and Bridges - News of the month (1945) - "Ice pressure on Bow River damages bridges. Vol. 83, No. 4, p. 86.

Rose, E. (1947) - "Thrust exerted by expanding ice sheet" Transactions of the American Society of Civil Engineers, Vol. 12, p. 871-900.

Saeki, H., Hamanaka, K., Ozaki, A. (1977) - "Experimental study of the ice forces on a pile" Proc. POAC Conference 77, pp. 695-706.

Sanden, E.J., Neill, C.R. (1968) - "Determination of actual forces on bridge piers due to moving ice" Proc. Toronto Conv., Can. Good Roads Assoc. (now Roads Transp. Assoc. Can), Proc. Toronto Conv.

Schwarz, J. (1970) - "Treibeisdruck auf Pfahle" Mitt. Franzius-Inst., Tech. Univ. Hanover, Ger. Heft 34.

Schwarz, J., Hirayama, K. (1973) - "Experimental study of ice force on piles and the corresponding ice deformation" Iowa Inst. Hydraul. Res., Univ. Iowa.

Schwarz, J., Hirayama, K., Wu, H. (1973) - "Model technique for the investigation of ice forces on structures" 2nd Conf. Port and Ocean Eng. under Arctic Conditions, Reykjavik, Iceland.

Shadrin, G.S., Panfilov, D.F. (1962) - "Dynamics of pressure of ice on hydraulic structures" Corps of Engineers, U.S. Army, Cold Regions Res. Lab., Hanover, N.H. CRREL Draft Transl. (348).

Shenehan, F.C., and others (1931) - "Ice as affecting power plants" Final Reports of committee of power division, Transactions of the American Society of Civil Engineers, Vol. 95, p. 1134-1150.

Sodhi, D.S., Hamza, H.E. (1977) - "Buckling analysis of a semi-infinite ice-sheet" Proc. POAC Conference 1977, pp. 593-607.

Taylor, F.W. (1945) - "Temperature changes in an ice sheet with the lower surface in contact with water at freezing temperature for various conditions of exposure on upper surface" Memorandum to R.E. Glover, Nov. 29, Bureau of Reclamation, Denver, Colorado.

Telechev, V.I., Pinigrin, M.I., Tolokno, V.V. (1961) - "Flow of ice through Mamakavskoi hydroelectric works" Hydraulic Constructions, No. 7.

Timoshenko, S.P., Gere, J.M. (1961) - "Theory of elastic stability" McGraw Hill, 2nd Ed.

Tryde, P. (1973) - "Forces exerted on structures by ice floes" Proc. XXIIIrd International Navigation Congress, Ottawa, Subject 4, p. 31-44.

USSR Building Standard (1967) - "Instruction for determining ice loads on river structures" (SN76-66), State Com. of the Council of Ministry for Constr. (GOSSTROI, USSR), Tech. Trans. 1663, Nat. Res. Counc. Can., Ottawa, Canada 1973.

Willmot, J.G. (1952) - "Measurement of ice thrust on dams" Ontario Hydro Research News. No. 3, p. 23-25.

Wilson, J.T., Zumberge, J.H., Marshall, E.W. (1954) - "A study of ice on an inland lake" USA SIPRE Report 5.

Woinowsky-Krieger, S. (1953) - "Uber die Biegung von Platten durch Einzellasten mit rechteckiger Aufstandsfläche" Ingenieru-Archiv, Vol. 21, p. 331-338.

Zubov, N.N. (1935) - "Considerations on the Movement of Ice Floes Under the Influence of the Wind" Marine Research U.S.S.R. Vol. 21.

# CHAPTER 5

# ICEBREAKERS

---

## 5.0 - INTRODUCTION

Icebreakers operate either under arctic conditions or in ice formed in a temperate climate. These conditions are widely different. Arctic ice is usually several years old, several meters thick, solid, and very hard because of the low prevailing temperatures. The ice in the temperate zones is only a few months old and usually not more than 1 m thick. However, it has the tendency to raft and form pack ice of extensive dimensions. In rivers there are often accumulations of frazil and slush ice right down to the bottom of the navigation channel.

Icebreakers are best defined by their primary functions: to break solid ice, to maneuver in heavy concentrations of pack ice, and to clear channels through which other ships can pass in safety. Ice-strengthened ships, in contrast, are normally cargo transport ships strengthened for use in ice. Heavily reinforced hulls, a sloping forefoot and a ratio of horsepower to displacement of more than unity are characteristics common to both types. The icebreaker is usually distinguished by more powerful engines, higher ratio of beam to length, smaller cargo capacity and such distinctive features as heeling tanks. Icebreakers designed for use in interior waters, as in the Gulf of St. Lawrence,may possess one or two forward propellers which are not used in polar icebreakers.

The first ships modified for breaking ice were tugs and harbor vessels with reinforced bows and increased power. They cleared ice and assisted the passage of cargo ships in and near harbors. The first icebreaker, in the true sense, was Eisbrecher 1, built in Hamburg in 1871 for service on the river Elbe. In 1890 the Finnish Murtaja represented the greatest advance in the European type of icebreakers. It was able to break 0.8 m thick solid ice with no snow cover, but was helpless in slush and snow-covered ice. The full-lined bow pushed the snow and slush forward with the result that the ship was forced to a stop. These vessels, which got into difficulty in pack ice, were able to force their way through by backing into the ice. The natural consequence was that icebreakers were built with bow propellers. Thus in 1888 and 1893 the ferryboats St. Ignace and St. Maria were built with both stern and bow propellers. The primary action of the bow propeller was to wash water and broken ice away from the forward end of the ship and thus reduce the friction between the ice and the bow. Furthermore, the ship could break into pack ice by alternately running backward and forward on the bow propeller, so that water and the ship were forced into the pack ice. The work of one bow propeller is, however, asymmetrical because of the unbalanced torque reaction and in modern inland icebreakers there are two of them, rotating in opposite directions.

The first polar icebreaker, the Russian Ermak, was completed in Newcastle, England, in 1899. With the advance of shipbuilding technology, the design, power and size of icebreakers developed steadily. By the outbreak of World War II, they were in common use in the Baltic, the Gulf of St. Lawrence and, though less frequently, in the Russian and Canadian Arctic. Military demands during the war stimulated the building of the Russian Stalin class and the United States Wind class icebreakers, the Argentine General San Martil, the Canadian Labrador, John A. MacDonald and Louis St-Laurent, the Japanese Fuji, and the Russian Moskva class. The Russian Lenin with a hitherto unprecedented nominal horsepower of 44,000 was the world's first nuclear-powered surface ship. All these ships are primarily polar icebreakers. The development of other icebreakers has similarly been rapid, particularly for use in the Baltic. The first icebreaking activity in the U.S. was on the western rivers and Great Lakes in support of waterborne commerce.

The discovery of crude oil along the North Slope of Alaska and in the Canadian archepelago raised the question of the feasibility of commercial seaborne traffic across several thousand miles of ice-covered waters and consequently initiated interest in the selection of optimum hull forms and power levels for icebreaking ships. In 1969, the bow of the tanker Manhattan was modified to give icebreaking capability and the power of the ship was increased. The ship was able to cross the North-West passage in the Canadian Arctic islands at the beginning of the winter, to demonstrate for the first time in this experimental trip, the technical feasibility of commercial ice-transiting ships in the Arctic.

In the summer of 1977, the Soviet atomic icebreaker Arktika cut its way through the thick ice of the Polar Pack to reach the North Pole, the first surface ship ever to reach the top of the world.

## 5.1 - ICE CONDITIONS AND MODES OF ICEBREAKING

It is very unusual to observe an ice field of continuous, uniform ice in the path of an icebreaking ship and this can be considered only as a reference condition for the design of the ship. Even, in some cases, this may not correspond to the worst condition that hinders the ship's movement, and the design may have to be made for those other particular conditions.

The first ice condition that a ship will encounter is that of an unconsolidated ice cover. There are however quite a number of degrees of resistance to movement of such a cover.

When the water surface is not completely covered with individual ice floes it is usual to define the ice as a fraction in tenths of the surface cover. When the ice cover is five-tenths or less, observation has shown that the operation of a ship is virtually unaffected. This is almost as if the ship were in open water.

The effect of increasing the ice cover between 5/10 to 10/10 is to add progressively to the ice resistance. Depending on the sizes and other characteristics of the ice floes making the pack the retardation effect will vary. This is what is usually called the resistance caused by broken or pack ice and is usually much smaller than that caused by a continuous, solid ice cover.

However, if the ice floes continue to accumulate underneath a 10/10 cover because of water velocities, wind or repeated ship passages; an ice floe jam or a pressured ice pack may be formed. Such conditions are the worst of oppose the movement of ships in certain waterways like the St. Lawrence navigation system. Even the mighty icebreaker Louis St-Laurent, that goes through Polar ice, cannot navigate through a jam under pressure,

6 to 10 m thick, that forms occasionnaly in the St. Lawrence river at the Quebec bridge. Five to eight Canadian icebreakers have to be called to destroy that jam, as no icebreaker has yet been designed specifically, for such purposes.

The usual condition for design of an icebreaking ship is that of navigation in a continuous uniform ice field. Although the ice in a uniform field may vary in thickness and strength; this condition is considered as representative of navigation in large expanses of water like the Baltic sea, the Arctic sea, the Gulf of St. Lawrence, etc...

In continuous ice fields there are two conditions that affect seriously the performance of an icebreaking ship. They are the presence of ice ridges and that of pressure ice. The ridges may vary considerably in bulk and in strength. Old reconsolidated ridges will present a considerable resistance to a ship and it might have to go through them by ramming. Little is known on the actual performance of ships in this type of configuration. The movement of a ship is also seriously hindered in a pressure ice field. A few measurements have shown a major increase in resistance under such conditions, and no particular design takes this effect specifically into account.

In a uniform ice field there are essentially two modes of icebreaking; by continuous icebreaking or by ramming if the ship does not have enough power.

We will examine the mechanics of these various modes of icebreaking.

## 5.2 - SHIP RESISTANCE IN UNCONSOLIDATED ICE

When a ship moves through a surface layer of unconsolidated ice, the ice floes coming in contact with the hull will acquire a certain velocity and move aside to allow passage of the ship. Depending on the concentration and sizes of these ice floes the motion may extend well beyond the ship periphery because of the interaction of adjacent floes. The ship does not tend to break this ice, but only to push it aside. Thus it can be expected that the effect of such interaction on the resistance to ship motion will be a direct function of ice concentration. As said previously, if the concentration is less than 5/10, the broken ice does not represent any significant obstacle to the ship motion. To take this concentration into effect, German *et al.*,(1975) have proposed a linear relationship:

$$V_p = V_o + (V - V_o)\left(\frac{m - 0.5}{0.5}\right) \qquad (5\text{-}1)$$

where: $V_p$ - ship speed in partial cover

$V$ - ship speed in solid uniform unconsolidated cover (no pressure) of the same thickness as the prevailing thickness of the ice floes

$V_o$ - ship speed in open water

$m$ - fraction of ice cover (over 0.5).

One hundred percent surface cover may sometimes produce the greatest difficulties for an icebreaking ship. Ice floes, snow, slush and frazil may form thick accumulations and produce ice jams in rivers. Ice floes and brash ice may accumulate into thick ice packs in the sea. A navigation channel may also become clogged due to repeated passage of ships through the same portion of an ice field. Much of the broken ice tends to remain in the channel after the passage of a ship and the water surface exposed in the channel will freeze at a very fast rate. This refrozen surface is broken up by the passage of the next ship and adds to the material in the channel. Eventually, the ice mixture in the channel can reach thickness of up to 4 m. The condition of growth of this broken ice has been studied by Michel *et al.*, (1971). Such an ice mixture is very similar to the brash ice found in infinite fields due to wind action. The major difference is that the ice pieces are confined within the boundaries of the channel.

It is in pack ice and ice jams that the bow propellers of icebreakers are exceptionnally useful. They produce water lubrication between the bow of the vessel and the ice. Furthermore, they break the ice into smaller pieces and the wake of the icebreaker becomes clearer. There is also a possibility of working on the ice by running in reverse on the bow machinery, while during the forward motion of the stern propeller the water will be thrown forward. The bow propellers should be inward rotating so that the wash of the bow is more powerful and effective on both sides of the vessel. Observations have shown that the heaviest mass of water is then thrown against the bow at approximately a quarter of the length from the stem. The stern propeller, however, should be outward rotating so as to throw the ice pieces outwards and give a clearer track.

Another advantage of two bow propellers is the better maneuverability they give the vessel. At low speed and in heavy ice the action of the rudder is weak, but by suitable regulation of revolutions and the direction of rotation of the propellers, one can make the vessel move at will, even directly sideways. But the bow-propeller system has drawbacks. The costs are higher than with stern propellers only and the water and ice resistance are higher. The total propulsion efficiency is also smaller for the same power. For polar operation where thick hard ice is encountered, icebreakers are not designed with bow propellers because of the decrease in ramming or icebreaking ability.

In heavily rafted light field-ice, heavy slush, and frazil ice, the action of ramming of an icebreaker is not really efficient because the ice just packs closer and forms a stronger cushion after each successive charge. One practice on the St. Lawrence river has been to use the propellers as drills to dislodge the ice pieces from the accumulation. This action is slow but positive and sometimes takes place under conditions where the ice jam is so high that part of it falls on the deck. A ship with two forward propellers is then quite an asset. Otherwise the ship proceeds astern with its stern propellers acting as drills. All Canadian icebreakers (Watson, 1959) are fitted with underwater gear which will stand up to the impact of breaking ice astern and propellers which will in effect drill drift ice piled as high as 6 to 9 m.

The determination of an analytical expression to describe the resistance to ship motion of a high concentration of unconsolidated ice is difficult because of the complex nature of the mechanisms involved.

The resistive forces due to this unconsolidated ice field may by classed into four groups:

$$R = R_s + R_i + R_f + R_w \tag{5-2}$$

where: $R$ - sum of all ice forces acting against the ship movement

$R_s$ - resisting forces arising from the buoyancy of the ice pieces, that are submerged by the ship movement

$R_i$ - resistive forces arising from imparting a velocity to the lateral ice pieces as the ship moves through them

$R_f$ - resistive forces due to ice friction on the hull, not counting the water resistance under the hull

$R_w$ - resistive forces in open water.

Resistive forces due to ice friction, as such, are usually believed to be intimately associated with the second force in that case, and are not usually treated separately, (Lewis and Edwards, 1970; Milano, 1973).

The expression for the resistive forces attributable to the buoyancy of the ice pieces and inertia of these pieces may be expressed in the following manner (Lewis and Edwards, 1970).

Imagine a block of ice of width, B, unit length (in direction of travel), and depth, h. The force required to submerge the block of ice to the waterline is:

$$F_s = \alpha \rho \; g \; Bh \tag{5-3}$$

where $\alpha = (1 - \rho'/\rho)$, $\rho'$ and $\rho$ are the density of the ice and water, respectively. The ship beam is B and h the ice thickness in m. The work required to push the block under the unbroken channel is:

$$W_s = \alpha \rho \; g \; Bh^2 \tag{5-4}$$

thus the mean resistance over the unit distance is:

$$R_s = \frac{W_s}{1} = \alpha \rho \; g \; Bh^2 \tag{5-5}$$

The forces required to tip the broken ice and overcome frictional forces can be handled in a similar manner and added to Eq. (5-5). The total of all these forces may be assumed to be of the form:

$$R_s = C_s \rho \; g \; Bh^2 \tag{5-6}$$

The unknown coefficient in this equation accounts for the effects of the hull form, friction between the ice pieces and the hull, friction between the broken ice pieces and the undersurface of the unbroken ice cover, and other unknown factors.

Resistive forces attributable to extracting momentum from the ship and imparting it to the broken ice pieces are handled as follows. Imagine a piece of ice of width, $B/2$, unit length (in the direction of motion), and thickness, h. The piece of ice is motionless before being contacted by the bow of the ship. After the ship has traveled a unit distance in the direction of motion at a speed, V, the piece of ice whose mass is $\rho' Bh/2$ has a velocity proportional to V. The constant of proportionality would depend on the hull form at the point of contact. The energy transmitted from the ship to this piece of ice during this movement is:

$$W_i = \frac{1}{2} \rho' \frac{B}{2} h (C_i' V)^2 \qquad (5\text{-}7)$$

Thus the mean resistance over the unit distance is, with a change in constant:

$$R_i = \frac{W_i}{1} = C_i \rho BhV^2 \qquad (5\text{-}8)$$

With expressions (5-6) and (5-8), and neglecting the open water resistance, the total resistance to motion is given by:

$$R = C_s \rho gBh^2 + C_i \rho BhV^2 \qquad (5\text{-}9)$$

Two dimensionless numbers can be used to represent this equation, as used by Edwards *et al.*, (1972):

$$r = \frac{R}{\rho gBh^2} \qquad (5\text{-}10)$$

r - reduced dimensionless resistance

$$Fr = V/\sqrt{gh} \qquad (5\text{-}11)$$

Fr - Froude number for the ice.

With these new parameters, Eq. (5-9) becomes:

$$r = C_s + C_i F_R^2 \tag{5-12}$$

This is the expression used by Edwards *et al.*, (1973) and Levine *et al.*, (1974), to represent model tests in brash (or mush) ice for different ships. For an iron ore bulk carrier of 245,000 dwt the dimensionless coefficients were found to be $C_s = 1.29$ and $C_i = 1.365$. For the Canadian Icebreaker CCGS Norman McLeod Rogers', $C_s = 0.60$ and $C_i = 0.83$. For the new Canadian "R" class type icebreaker $C_s = 0.60$ and $C_i = 0.57$. The results for these last ships obtained from model tests are shown in Fig. 5.1.

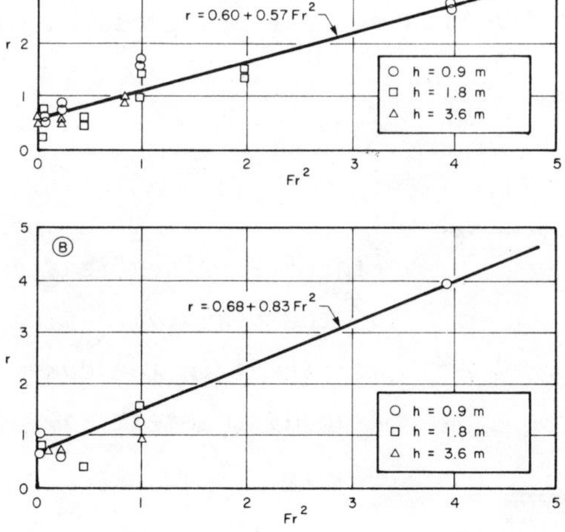


*Fig. 5.1. Dimensionless towing resistance versus dimensionless velocity squared for R class icebreaker (A) and N. McLeod Rogers (B). Obtained from model tests (Edwards et al., 1973).*

An older empirical equation had been used for ship resistance in ice concentrations of almost 100%, based on measurements on the U.S.S.R. Stalin and Kapitan Belousov (Buzuev and

Ryblin, 1961):

$$R = 8.6\ L^{0.5}\ B^{0.25}\ [0.4 + 8\ (V/\sqrt{gL})^{1.25}] \qquad (5\text{-}13)$$

where R is in metric tons, L and B the ship length and beam in meters.

The resistance of the accumulations under pressure, in heavily rafted ice, frazil and ice jams has not been quantitatively investigated at present.

## 5.3 - THE MECHANICS OF CONTINUOUS ICEBREAKING IN UNIFORM SOLID ICE

To understand the mechanics of icebreaking in the continuous mode it is first necessary to study the behavior of the ice cover when it is penetrated by an icebreaker. Fig. 5.2 shows a typical continuous mode ice breakage pattern produced by an icebreaker.

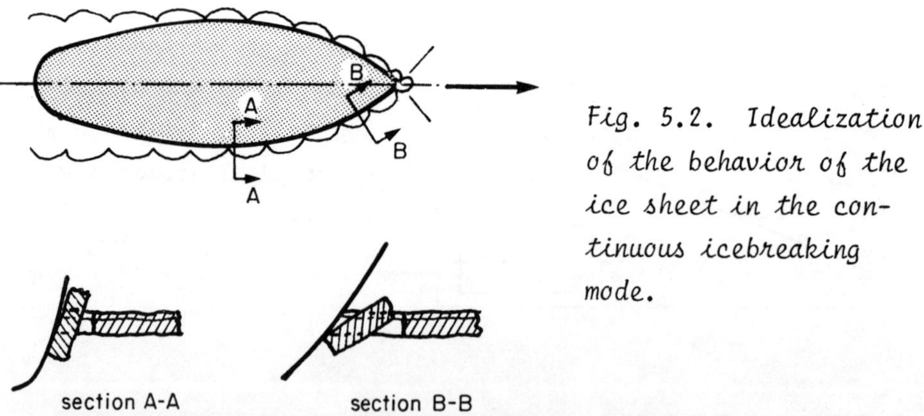


*Fig. 5.2. Idealization of the behavior of the ice sheet in the continuous icebreaking mode.*

The following is the description of the icebreaking process by Vance (1975). "As the ship progresses through an ice sheet with continuous motion, it starts to ride up on the ice until a sufficient downward force is developed to crack the ice sheet. This force is developed at the stem and along the cheeks of the vessel. In very thin ice, the major portion of the breaking is done at the stem with very little pitch angle developed by the vessel. As the ice gets thicker, the bow cheeks assume a more dominant role in the breaking process. As the vessel runs up on the ice, there is some local crushing of the ice sheet (this is actually a small part of the breaking process and can be neglected) and extensive bending of the ice sheet. The bending, whether it be at the stem or at the cheeks, causes a breaking pattern that initiates radial cracks in the under portion of the ice sheet causing the formation of wedges; these are forced down until circumferential cracks are formed in the upper fibers as shown in Fig. 5.2. The ice ultimately fails in bending and the ship settles down in the water and progresses ahead to repeat the cycle again. The frequency of the cycle depends on the thickness of the ice, the speed of the vessel and the natural pressure the ice is experiencing. As the ship progress through the ice field, the broken pieces of the ice impact with the side of the vessel, rubs against the side of each other, are upturned and submerged".

The broken ice pieces have the form of half-moons (cusps) and behave in different ways depending on their position (Lewis *et al.*, 1970). Those near the centerline of the ship and bow are simultaneously submerged and pass underneath the hull of the ship, to reappear in the broken channel after the ship moves by. Some cusps on the side are being pushed under the unbroken ice on the side. The cusps near the beam are simply up-ended and locked into place by the remainder of the ice field. As the icebreaker moves forward, the wedge-shape areas formed by two cusps are loaded by the hull until failure occurs and new cusps are formed. This formation of cusps continuously occurs and speads from the centerline towards the outer extremities of the vessel. The resulting channel left by the icebreaker is only somewhat larger than the beam, and scalloped.

Depending on the form of the ship and the ice thickness, more or less ice will be left in the open channel in the lee of the vessel. A blunt edge ship will break the ice in smaller pieces and push more on the sides. An icebreaker with a small bow angle will produce larger ice slabs that will be forced down under the body of the ship and reappear in the open channel, behind the ship.

The total resistance of a vessel progressing in the continuous icebreaking mode may be expressed by:

$$R = Rb + Rs + Ri + Rf + Rw \qquad (5\text{-}14)$$

where R, Rs, Ri and Rw have been defined for Eq. (5-2) and:

Rb - resistive force needed to break the ice

Rf - resistive force caused by friction of ice pieces on the hull of the vessel or by ice pressure.

The expression for the force needed to break the ice can be obtained by considering the broken cusps at the bow as a serie of adjacent cantilever beams of total width B to be broken in front of the ship (Kashteljan *et al.*, 1978; Enkvist, 1972: Vance 1975). The resistance can then be seen as the average work required per unit length of travel. This work $W_b$ to break the ice will be of the form:

$$W_b = Pdw \qquad (5\text{-}15)$$

where w is the vertical deflection at the contact of the ice pieces with the ship where the load P is applied.

The vertical load P for failure of a cantiliver beam of width B has the form (Hetenyi, 1946):

$$P = C_1 \frac{\sigma_f \ Bh^2}{\ell} \qquad (5\text{-}16)$$

where $C_1$ is a constant, $\sigma_f$ is the flexural strength of the ice and $\ell$ the characteristic length defined by Eq. (3-51) in Chapter 3.

The deflection at the top of such a beam is given by an expression of the form (Hetenyi, 1946):

$$w = C_2 \frac{\sigma_f \ell^2}{Eh} \qquad (5\text{-}17)$$

where $C_2$ is a constant and E the equivalent Young's modulus of the ice.

Eqs (5-16) and (5-17) in Eq. (5-15) give an expression of the form:

$$W_b = C_3 \ \sigma_f^2 \ \frac{B\ell h}{E} \qquad (5\text{-}18)$$

If the resistance is given by the average work spent in breaking the ice over a length $C_4\ell$ (where $C_4$ is also a constant) between successive failures, we will finally have:

$$Rb = \frac{1}{2} \frac{W_b}{C_4\ell} = C_b \ \sigma_f \ Bh \qquad (5\text{-}19)$$

with

$$C_b = \frac{1C_3}{2C_4} \frac{\sigma_f}{E} \qquad (5\text{-}20)$$

where Cb is the constant characterizing the resistive force for fracture of the ice.

The solid friction force of ice pieces under the hull of a vessel can be determined by considering Fig. 5.3. If the ice pieces are thick enough, the ship will submerge slabs of ice that will move underneath the hull. These pieces will be held one against the other from the bow of the ship and be unable to accelerate with the ship movement. Thus the ship will move over a continuous layer of motionless ice in the fore section up to the maximum draft of the ship. After this section, separation will occur and the ice pieces will not follow the hull so closely. The same process will occur on the side of the ship when the ice field is under pressure.

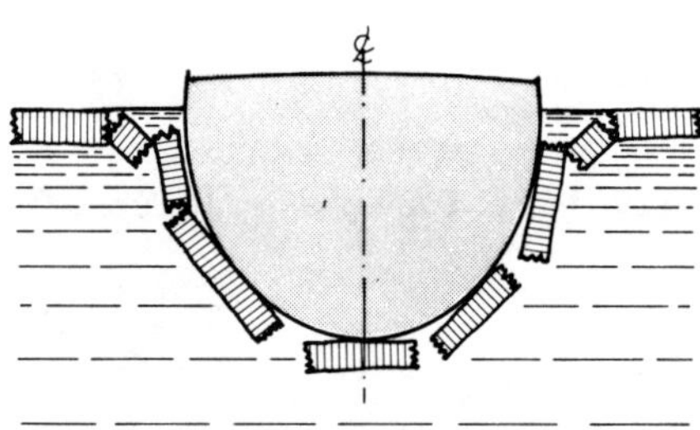

*Fig. 5.3. Movement of fore end of ship over broken thick ice in the continuous icebreaking mode.*

A film of water will lubricate the hull between the ice pieces. Laminar flow may develop and the tangential stress may be given by the equation of a moving plate over a fixed boundary:

$$\tau = \frac{\mu V}{d} \tag{5-21}$$

where $\tau$ is the tangential stress on the hull, V, the velocity, $\mu$ the dynamic viscosity of water and d the distance between the hull and the ice slab.

It is obvious from physical considerations, that the thicker the ice the smaller the gap d between the hull and the ice pieces. If we admit a relation that best fits the experimental data:

$$d \propto h^{-3/2} \tag{5-22}$$

Then we may express the resistive force of ice friction on the hull, if L is the ship's length, by:

$$Rf = C_5 \; BL \; \tau \tag{5-23}$$

where $C_5$ is a coefficient taking into account the part of the foresection of the hull which is ice-covered. With Eqs (5-21) and (5-22) in (5-23) we may finally get the friction force, with a change in the numerical coefficient, in the form:

$$Rf = Cf \; \rho \sqrt{g} \; B \; V \; h^{3/2} \tag{5-24}$$

Cf being the coefficient of ice friction for a hull of given geometry.

The open water resistance can be easily estimated, but it is usually negligible before the resistive forces caused by ice.

The final expression for the resistance of an icebreaker in uniform motion is then with Eqs (5-14), (5-19), (5-6), (5-8) and (5-24).

$$R = Cb \; \sigma_f \; Bh + Cs \; \rho g \; Bh^2 + Ci \; \rho B \; h \; V^2 + Cf \; \rho \sqrt{g} \; B \; V \; h^{3/2} \tag{5-25}$$

With dimensionless resistance and Froude number Fr as originally used by Edwards *et al.*, (1972):

$$r = \frac{R}{\rho g B h^2} \tag{5-10}$$

$$Fr = V/\sqrt{gh} \tag{5-11}$$

We obtain:

$$r = Cb \frac{\sigma_f}{\rho gh} + Cs + Ci\ Fr^2 + Cf\ Fr \qquad (5\text{-}26)$$

This can be taken as the more general form of the equation of resistance of an icebreaking vessel. For unconsolidated ice $\sigma_f = 0$ and $Cf = 0$ and Eq. (5-26) reduces to Eq. (5-12). For a ship with a sharp bow that does not push any ice under the hull, $Cf = 0$ and the resistance is a function of the square of the velocity. For existing icebreakers, it has been found, at this time, that the resistance, for thicker ice, has a linear dependance on velocity. This means that most of the ice goes under the hull and the term $Ci\ Fr^2$ is small compare to the $Cf\ Fr$ term. In thinner ice however, the broken pieces are too small to produce a continuous ice layer under the hull and the quadratic term reappears. For most icebreaking ships, the full Eq. (5-26) would applied with more of less importance on the $Fr$ or $Fr^2$ term depending on the mode of icebreaking and movement of the broken pieces.

Up to now it has been found that the empirical Kasteljan (1968) equation would best fit all results for icebreakers. This equation is:

$$R = [0.004\ \sigma_f\ Bh + 3.25\ \gamma\ Bh^2]\ \mu_o + 0.25\ \frac{B^{1.65}\ Vh}{\eta_2} \qquad (5\text{-}27)$$

where the $\mu_o$ and $\eta_2$ are refered to, as the Kastlejan's coefficients, R is in metric tons, $\sigma_f$ in metric tons/m$^2$ and $\rho$ in metric tons/m$^3$. An equation which has been extensively used with success by Edwards *et al.*, (1976) corresponds to Eqs (5-56) and (5-27):

$$r = Co + Cf\ Fr \qquad (5\text{-}28)$$

From Eq. (5-26) the value of Co would be:

$$Co = Cs + Cb \frac{\sigma_f}{\rho gh} \tag{5-29}$$

Eq. (5-27) has exactly the form of Eq. (5-26) without the Ci term. Fig. 5.4 shows all present data points for the resistance of icebreakers which have been tested in continuous motion. It can be seen that the linear regression is good for each one and also for the bulk of the results, using Eq. (5-28). Indeed the values of the constants Co and Cf will vary for each icebreaker as shown in Fig. 5.5, depending on actual strength of ice and form of the icebreaker. We have computed these values with a best fit linear regression for each vessel and the results are shown in the following Table 5.1.

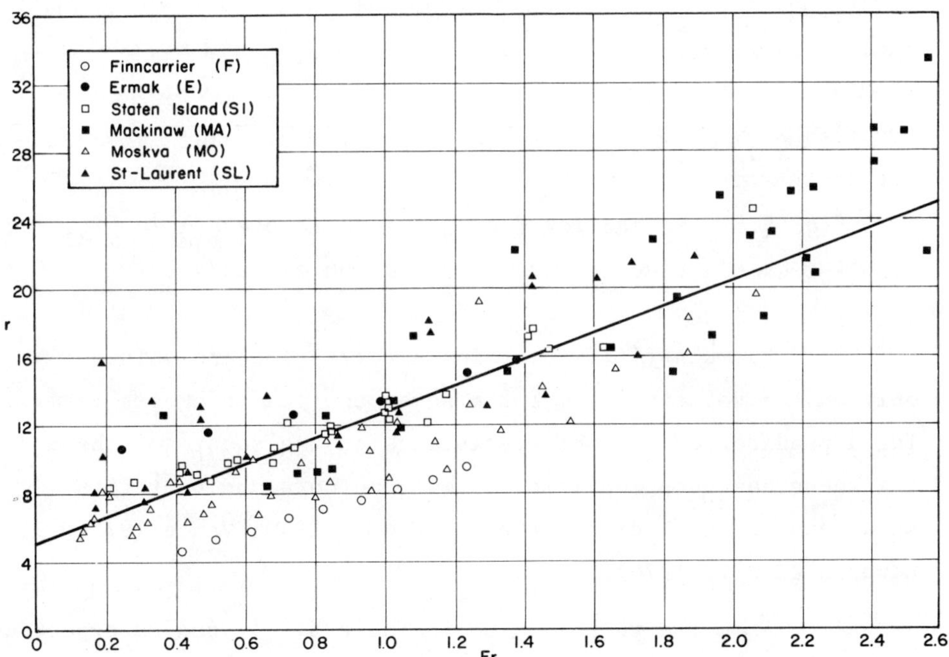


*Fig. 5.4. Data points for field tests on icebreakers resistance. The full line is the regression line on all points.*

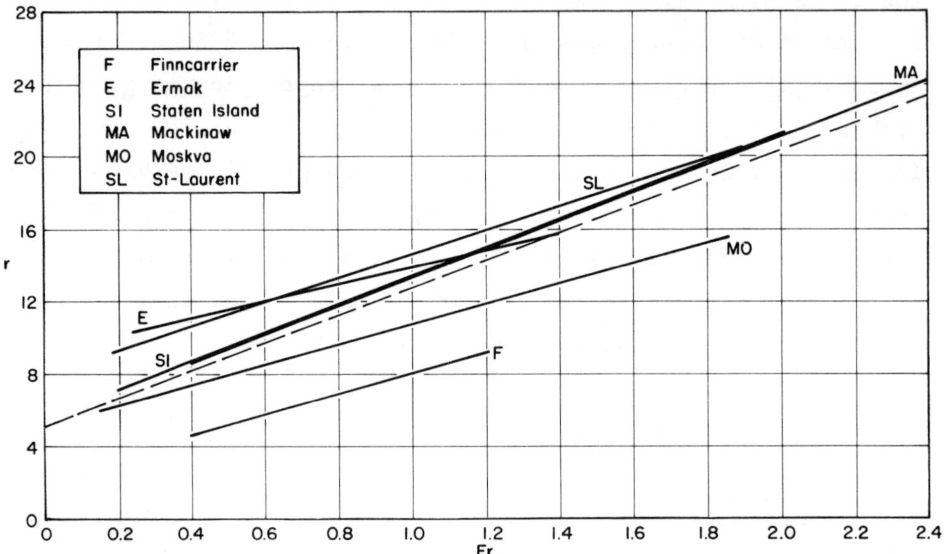


*Fig. 5.5. Regression line for each icebreaker in Fig. 5.4. The dashed line is the regression on all data points.*

Formula (5-29) contains a regression analysis not only on the velocity term but also on the ice thickness term. The dependance of the ship resistance on the square of the ice thickness has been confirmed by many authors and best fits known experimental data. (Lewis *et al.*, 1950; Vance 1975; Edwards *et al.*, 1976).

From the values of Co in Table 5.1 it can be expected that the bow of the icebreaker will have an important influence on its icebreaking resistance. This effect has been studied as early as 1900 by Runeberg and also by White (1965).

These derivations contains certain hypothesis on the friction effect, so we will consider the diagram of Fig. 5.6 where the plane of one cheek of the icebreaker is shown with a bow angle $\alpha$ taken perpendicular to the plane of the bow, and a spread angle $\beta$.

| | Co | Cf |
|---|---|---|
| All icebreakers | 5.14 | 7.63 |
| Louis St-Laurent | 7.94 | 6.61 |
| U.S. Mackinaw | 5.51 | 7.84 |
| U.S. Staten Island | 5.50 | 7.80 |
| U.S.S.R. Ermak | 9.28 | 4.59 |
| U.S.S.R. Moskva | 3.92 | 8.22 |
| Ferry Finncarrier | 2.35 | 5.74 |

Table 5.1 - Resistance coefficients for icebreakers.

From geometrical consideration, we have:

$$\Delta n = \Delta x \sin \alpha \qquad \Delta t = \Delta s \cos \alpha \tag{5-30}$$

$$\cos \gamma = \cos \beta / \sin \alpha \tag{5-31}$$

$$\sin \delta = \sin \gamma / \tan \alpha \tag{5-32}$$

The stresses $\sigma_n$ and $\tau$ normal and tangential to the bow plane are:

$$\sigma_n = E \, \Delta n \tag{5-33}$$

$$\tau = \mu \, E \Delta n \tag{5-34}$$

where E is the equivalent Young's modulus and $\mu$ the solid friction force of the ice during the crushing process.

The corresponding normal ($\Delta N$) and tangential ($\Delta T$)

forces for an area $\Delta A$ on the plane, are:

$$\Delta N = E \Delta n \Delta A \quad (5\text{-}35)$$

$$\Delta T = \mu E \Delta n \Delta A \quad (5\text{-}36)$$

The horizontal forces $\Delta H$ projected along the x axis are:

$$\Delta H = \Delta N \sin \alpha + \Delta T \cos \alpha \quad (5\text{-}37)$$

The vertical forces $\Delta V$ along the z axis are:

$$\Delta V = \Delta N \sin \gamma - \Delta T \sin \delta \quad (5\text{-}38)$$

with the values of $\gamma$ and $\delta$ given in Eqs (5-31) and (5-32) we finally obtain the ratio of horizontal force to vertical force:

$$\frac{\Delta H}{\Delta V} = \left(\frac{\sin \alpha + \mu \cos \alpha}{1 - \mu \tan \alpha}\right) \frac{\sin \alpha}{\sqrt{\sin^2 \alpha - \cos^2 \beta}} \quad (5\text{-}39)$$

To obtain a given vertical force, the horizontal thrust $\Delta H$ is lowest for $\alpha = 0$ and $\beta = \pi/2$, that is for a needdlelike bow, which is indeed impossible to build. The more inclined are the bow cheeks, the more effective they are for breaking the ice by bending.

The effect of internal ice pressure on a ship's resistance to motion is important. Mookhoek *et al.*, (1971) points out that side pressure had a significant effect on the performance of the Manhattan. It is evident that such pressure would increase the frictional resistance along the length of the vessel. Tests done on the Canadian icebreaker "Labrador" (German and Lawrence, 1975) whose results are shown on Fig. 5.6, show that ice pressure increases the resistance by a factor of approximately 3 or greater, for the thickness prevailing in these tests (about 30 cm). The pressure was such that the track behind the ship had either completely closed or had closed approximately 85% within one hour after penetration.

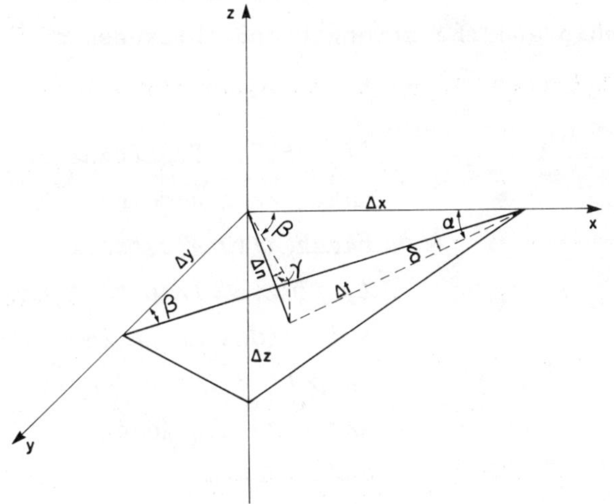


*Fig. 5.6. Sketch of one icebreaker cheek with notation used in analysis to obtain the vertical force component.*

## 5.4 - ICEBREAKING BY RAMMING

An icebreaker resorts to the process of "ramming" when the thrust available from its propellers is not sufficient to overcome the resistance generated in the interaction between the hull and the ice cover.

In ice which is very thick with respect to the limiting continuous-mode thickness discussed previously, the motion of the ship will be more accurately described by the velocity versus time curve shown in Fig. 5.7. (Lewis *et al.*, 1970). The ship will actually ride up on the ice, trimming until sufficient downward force is developed to cause failure of the ice sheet. Upon failure, the trim will decrease gradually as the broken material is accelerated down under the ship. The velocity will continue to decrease since the resistance associated with accelerating and submerging the large broken slabs of ice is greater than the propeller thrust. This discontinuous deceleration process will continue for a length of time which depends upon the mass and impact velocity of the ship and the strength and thickness of the ice.

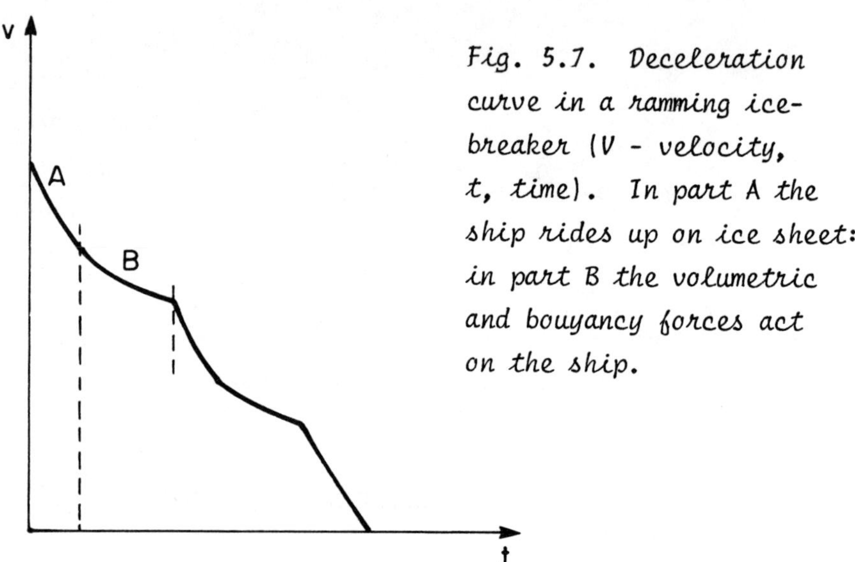


*Fig. 5.7. Deceleration curve in a ramming icebreaker (V - velocity, t, time). In part A the ship rides up on ice sheet: in part B the volumetric and bouyancy forces act on the ship.*

Shimansky (1938), Vinogradov (1940) and White (1965) have developed mathematical models for the process of an icebreaker impacting an ice field and the transformation of the kinetic energy available in the advancing vessel and thrust of its propellers into a force component normal to the ice sheet surface.

The analysis of the icebreaking process is discussed by Michel (1970). The principe of conservation of energy is applied to obtain the expression of the vertical component of the load acting on the ice. The energy expended is the ship's kinetic energy plus the work done by the propeller thrust acting through the distance traveled. The energy expended is directed into three channels: 1) energy dissipated by impact of the bow of the ship; 2) potential energy due to the ship being raised and changed in trim; 3) frictional loss caused by rubbing of the ship against the edge of the ice. In the case when the ship stops at the instant of ice collapse:

$$E_1 + E_2 = E_3 + E_4 + E_5 \tag{5-40}$$

where: $E_1$ - kinetic energy of the ship when the ice is first touched

$E_2$ - energy derived from propeller thrust

$E_3$ - energy dissipated by impact

$E_4$ - potential energy acquired by the ship

$E_5$ - energy loss by friction.

Let W represent the weight or displacement of the ship and $V_0$ the velocity of the ship when the ice is first touched as shown in Fig. 5.8. Kinetic energy absorbed during the operation (assuming negligible speed after impact) is:

$$E_1 = \frac{W}{2g} V_0^2 \tag{5-41}$$

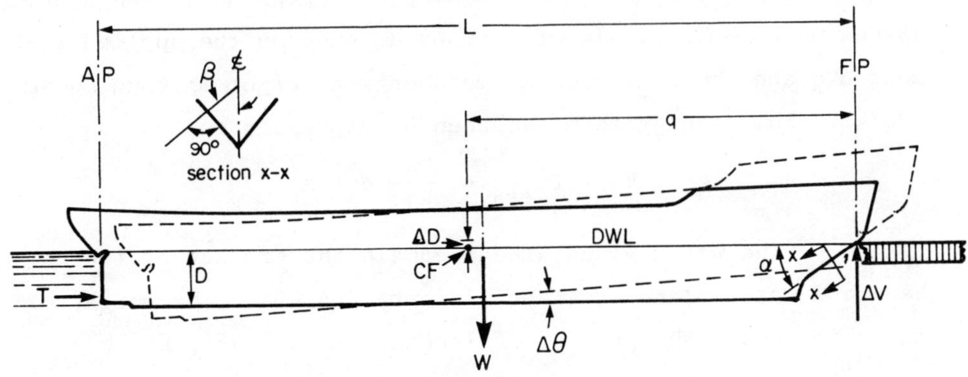


*Fig. 5.8. Variables and nomenclature used in ramming analysis.*

The next item considered is the energy delivered by the propellers to the ship while it is gliding up on the ice edge. During this interval there is a reduction in mean draft designated by $\Delta D$ and the ship assumes an angle of trim of $\Delta\theta$. The distance from the point of contact of the stem to the center of flotation is designated by q. The stem of the ship is sloped at angle $\alpha$ from horizontal. Then, from the instant of first contact until the ice collapses, the linear advance of the ship is given by x:

$$x = \Delta D \cot \alpha + q\Delta\theta \cot \alpha \tag{5-42}$$

Let T represent the average value of propeller thrust during this advance, then:

$$E_2 = T(\Delta D + q\Delta\theta) \cot \alpha \tag{5-43}$$

This formula should be expressed in different terms so as to include $\Delta V$ the maximum value of the vertical force developed at the stem. Assuming that $\Delta V$ is small compared with the displacement W, and that the change in draft and trim do not seriously change the properties of the water plane, we then have $\Delta D = V/\gamma' S$, S being the water plane area and $\gamma'$ the density of sea water. The angle of trim $\Delta\theta$ depends on the applied moment $\Delta Vq$ and the longitudinal metacentric height m: thus $\Delta\theta = \Delta Vq/Wm$. This energy term can then be expressed as:

$$E_2 = T \frac{\Delta V}{\gamma' S} + \frac{\Delta V q^2}{Wm} \cot \alpha \tag{5-44}$$

The water plane area S equals the product of length, beam and water plane area coefficient, S = LBa. The weight of the ship equals the product of length, beam, draft, block coefficient and density of sea water, or $W = LBD\delta\gamma'$. New non-dimensional coefficients $k_1$ and $k_2$ are arbitrarily set up by relationships $q = k_1$ (L/2) and $m = (k^2a^2L^2)/D\delta$. Substituing these new quantities in the last equation, we get:

$$E_2 = A \frac{D\Delta V}{W} T \cot \alpha \qquad (5\text{-}45)$$

where

$$A = \frac{\delta}{a}\left[1 + \left(\frac{k_1}{k_2}\right)^2 \frac{1}{4a}\right]$$

According to the theory of impact, when two bodies collide normally there is always a dissipation of energy, whose magnitude depends on the relative velocity and a physical constant e known as the coefficient of restitution. In fact the stem of the ship does not collide normally with the edge of the ice. The component of initial velocity $V_o$ which is directed normal to the edge of the ice is $V_o \sin \alpha$ and the energy dissipated by impact should be:

$$E_3 = \frac{W}{2g} (V_o \sin \alpha)^2 (1 - e^2) \qquad (6\text{-}46)$$

Michel (1970) does not think that there is such a type of elastic impact as given by Vinogradov and this term should instead take into account the crushing of an ice wedge in the ice shelf.

$$E_3 = \int \sigma_i A_n \, dn = \frac{\sigma_i x^3 \tan \alpha}{3} \tan \beta \qquad (5\text{-}47)$$

where $\sigma_i$ is the indentation strength of ice, $A_n$ the area of contact between hull and ice and dn the distance element normal to the shell through which the force moves and $1/3\ x^3 \tan \alpha \tan \beta$ is the volume of the crushed pyramid of ice whose length x is given by Eq. (5-42).

The value of x can also be expressed as in Eq. (5-44) and Eq. (5-45) by:

$$x = \frac{AD}{W} \Delta V \cot \alpha \qquad (5\text{-}48)$$

so that:

$$E_3 = \frac{\sigma_i A^3 D^3 \Delta V^3}{3W^3} \cot^2 \alpha \tan \beta \qquad (5\text{-}49)$$

The vertical force ΔV is a variable which keeps increasing as the ship slides up on the ice. The total rise of the point on the stem at which ΔV is first applied equals the reduction in draft ΔD plus the angle of trim, in radians, times the horizontal arm between the center of flotation and the stem Δθq. The potential energy set up by the force ΔV is therefore:

$$E_4 = \int_o^{\Delta D} \Delta V \, d \, \Delta D + \int_o^{\Delta\theta} \Delta V \, q \, d\Delta\theta \qquad (5\text{-}50)$$

Energy is dissipated by sliding friction between the shell plating and the ice. The coefficient of sliding friction μ must be applied to that component of the pressure which is normal to the plating. The resultant frictional force $F_f$ acts in a direction parallel to the stem of the ship and is a variable; half of it acts on one side of the stem and half on the other.

The energy dissipated by frictional force $F_f$ acting through a distance determined by the changes in draft and trim is given by:

$$E_5 = \frac{1}{\sin \alpha} \int_o^{\Delta D} F_f d \, \Delta D + \frac{1}{\sin \alpha} \int_o^{\Delta D} F_f q \, d\Delta\theta \qquad (5\text{-}51)$$

Consider an inclined plane intersecting the bow of the ship in a direction normal to the stem. This section of the bow appears as a wedge with essentially flat sides and the normal pressure on these sides makes an angle β with the centerline plane. As the bow rides up on the ice shelf it forms a wedge like groove. Pressure is developed normal to the faces of the groove, and friction along the faces of the groove directed parallel to the sloping stem.

Let R be the resultant force acting normal to the stem. On each side, then, the force acting normal to the plating is:

$$\frac{R}{2} \quad \frac{1}{\cos\beta}$$

So the frictional force is given by:

$$F_f = \mu R \frac{1}{\cos\beta} \qquad (5\text{-}52)$$

The magnitude of the force R is related to other forces acting on the ship as follows:

$$R = \Delta V \cos\alpha + T \sin\alpha$$

Eq. (5-52) can be rewritten as:

$$F_f = \mu \left(\Delta V \frac{\cos\alpha}{\cos\beta} + T \frac{\sin\alpha}{\cos\beta}\right) \qquad (5\text{-}53)$$

Using Eq. (5-53) in (5-51) and substituing the previously established values, $\Delta V = \gamma' S \Delta D$, $\Delta V = W m\Delta\theta$, for Eqs (5-50) and (5-51) together; we get:

$$E_4 + E_5 = \frac{AD}{W}\left[\left(1 + \mu \frac{\cot\alpha}{\cos\beta}\right)\frac{\Delta V^2}{2} + \mu \frac{T\Delta V}{\cos\beta}\right] \qquad (5\text{-}54)$$

Substituing all the values of component energies in Eq. (5-40) there finally results:

$$A\left[\frac{\Delta V^2}{W} - 2X\frac{\Delta V}{W}\frac{T}{W}\right] = Y\frac{V_0^2\,[1 - (1 - e^2)\sin^2\alpha]}{g\,D} \tag{5-55}$$

in which:

$$X = \frac{1 - \left(\frac{\mu}{\cos\beta}\right)\tan\alpha}{1 + \left(\frac{\mu}{\cos\beta}\right)\cot\alpha}\cot\alpha$$

$$Y = \frac{1}{1 + \left(\frac{\mu}{\cos\beta}\right)\cot\alpha}$$

From Eq. (5-54) we obtain the downward icebreaking force $\Delta V$ as given by Vinogradov:

$$\Delta V = XT + \left\{X^2T^2 + \frac{YW^2V_0^2}{AgD}\left[1 - (1 - e^2)\sin^2\alpha\right]\right\}^{\frac{1}{2}} \tag{5-56}$$

Taking the energy for crushing the ice instead of the theoretical energy of elastic impact (Eq. 5-49) we get:

$$\Delta V = \left\{-\frac{b}{2} + \left(\frac{b^2}{4} + \frac{a^3}{27}\right)^{1/2}\right\}^{1/3} + \left\{\frac{b}{2} - \left(\frac{b^2}{4} + \frac{a^3}{27}\right)^{1/2}\right\}^{1/3} - \frac{S}{3} \tag{5-57}$$

where

$$S = \frac{3W^2}{2\,\sigma_i\,A^2D^2Y\cot^2\alpha\tan\beta}$$

$$a = -\,2XTS - \frac{S^2}{2}$$

$$b = \frac{2}{27}S^3 + \frac{2}{3}XTS^2 - \frac{YW^2V_0^2S}{ADg}$$

The values of the coefficient X and Y depend on the dynamic friction of ice on the metal surface which is usually taken as 0.10 to 0.15 for fresh and brackish water ice and 0.20 for salt water and polar ice (Jansson, 1956). One of the important factors involved is the presence of a snow cover on the ice shelf that greatly increases the coefficient of dynamic friction and reduces the efficiency of the icebreaking operation. The parameters T, W, $V_0$, and A are characteristics of the ship and are discussed at length by Milano (1960).

The expression obtained for each of the terms of Eqs (5-56) or (5-57) is cumbersome. So is the value of these terms obtained by Milano (1961, 1973) in the case of continuous icebreaking. On the other hand, White (1965) used a similar mathematical model with many different ship sizes and shapes to examine the relationship between the downward force and various ship parameters. By curve fitting the results of his work, White developed a simplified expression for the downward force developed by an icebreaker.

$$\Delta V = 2.28\ V_o (IW)^{.845} \qquad (5\text{-}58)$$

$$I = 0.000234\ (10.72 + B/D)\ (0.1833 + a)$$
$$\times \quad (1.652 - \delta)\ (6.14 - \beta^2)\ (0.725 - \mu)\ (1.718 - \alpha) \qquad (5\text{-}59)$$

where $\Delta V$ - vertical force applied to the ice, kg

W - displacement, kg

$V_o$ - impact velocity m/s

B/D- beam-draft ratio

a - waterplane coefficient

$\delta$ - block coefficient

$\beta$ - spread angle (radians)

$\mu$ - friction factor

$\alpha$ - bow angle (radians)

The relationship between the vertical load and the maximum ice thickness which will not support such a load is dependent upon many variables. Some of these are: ice strength, loading area, characteristic length and geometry of the ice sheet in relation to the load.

The failure of an ice sheet due to a vertical load is discussed at length in Chapter 3. When an ice sheet is deflected to failure, radial cracks are formed, centered at the load. Ultimate failure occurs when cracks are formed across the wedges some distance from the load.

Fig. 5.9 shows the radial and circumferential cracks formed by a ramming icebreaker. Kasteljan *et al.*, (1968) have observed that full scale boundary conditions would be expressed by:

$$\theta o = 270^{\circ}, \ \theta = 67.5^{\circ}$$

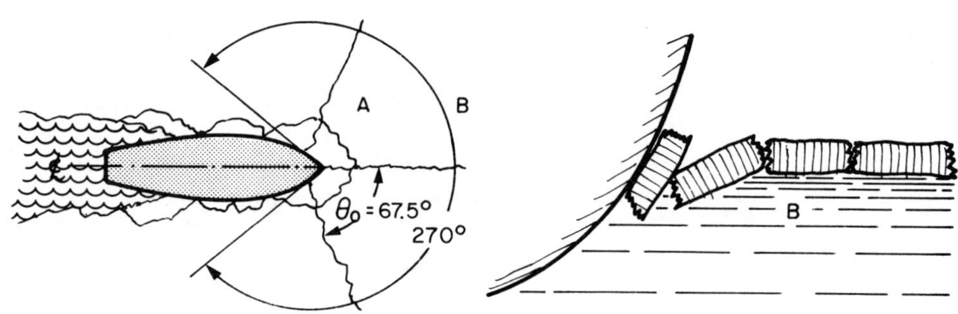


*Fig. 5.9. Ice fracture model by ramming. A) radial cracks B) circumferential crack.*

In Chapter 3, the bearing capacity of floating ice wedges is given from an analysis by Nevel (1961). Using this solution, Lewis *et al.*, (1970) have generalized the results to handle different boundary conditions by including the total sheet included angle and the wedge angle, with the following simplified result:

$$\Delta V = C_{\sigma} \ \sigma \ h^2 \tag{5-60}$$

with

$$C_{\sigma} = .174 \ \frac{\theta o}{\theta} \tan\left(\frac{\theta}{2}\right) \tag{5-61}$$

With the observed values of θo and θ this gives:

$$C_\sigma = 0.46$$

This compare with the values used by other authors (Chu, 1974):

| Authors | $C_\sigma$ |
|---|---|
| Panfilov | 0.45 |
| Kashteljan | 0.52 |
| Wageningen IMB | 0.53 |
| Norske Veritas | 0.67 |

All these values are derived for point loads at the tip of the wedges. Nothing has been done on real loading conditions representing an inclined triangular indentor.

The solution of the White's semi-empirical Eq. (5-58) with Eq. (5-60), will then give the limiting thickness that the icebreaker can break by ramming:

$$h = 1.51 \left(\frac{V_o}{C_\sigma \sigma}\right)^{0.5} (IW)^{.423} \qquad (5\text{-}62)$$

where h is in m and σ in $Kg/m^2$.

Vance (1975) presented some experimental data which indicate that the bow angle and spread angle have the effect describe by White through the term I in Eq. (5-59). Lewis *et al.*, (1970) believe, on the other hand, that insufficient reliable data have been analyzed with which to make firm assumptions about the effects of any parameters but ship mass and impact velocity on limiting ice thickness in the ramming process.

## 5.5 - MANEUVERABILITY IN ICE

Icebreaking ships should have good maneuvering and course keeping qualities, particularly icebreakers which needs them for rescue and escort operation.

For ordinary surface ships, the theory for predicting the important parameters affecting maneuvering have been developped to a satisfactory level. Very little has been done at this time for icebreaking ships except full-scale turning tests on the CCGS Labrador (German and Lawrence, 1975) and some model tests by Edwards *et al.*, (1976).

One of the main measure of maneuverability of a ship is the turning circle diameter. As the turning circle becomes smaller the ship is said to be more maneuverable; that is, the ship can change its course move quickly with rudder movement. In open water, the turning circle radius of a ship can be found using a linear theory of ship motion and is well developped (Abkowitz, 1964). In ice conditions, the conditions of analysis should include the component forces of the ice resistance and their moments. These equations cannot be solved at the present time (Kashtelyan *et al.*, 1972).

As the icebreaker sails in a curvilinear trajectory in the ice, asymmetry is observed in the force contact points between the ice at the starboard and port sides. The side that is external comes in contact with the ice cover almost all along its length, whereon the inside contacts the solid ice only with part of the bow. This can be seen in Fig. 5.10. Thus the ice forces produce a torque which is applied at some distance toward the bow from the center of gravity.

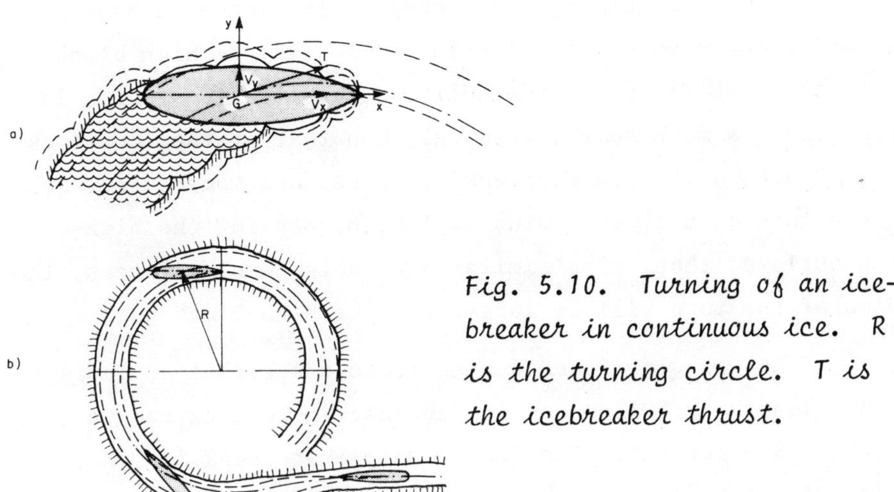


*Fig. 5.10. Turning of an ice-breaker in continuous ice. R is the turning circle. T is the icebreaker thrust.*

The CCGS Labrador tests were conducted south of Lowther Island in sea ice with an average thickness of 28 to 30 cm. The test were run with full power (approximately 8300 shp) with the rudder hard over, that is, a $37.5^{0}$ rudder angle. The results of the tests are shown on Fig. 5.11.

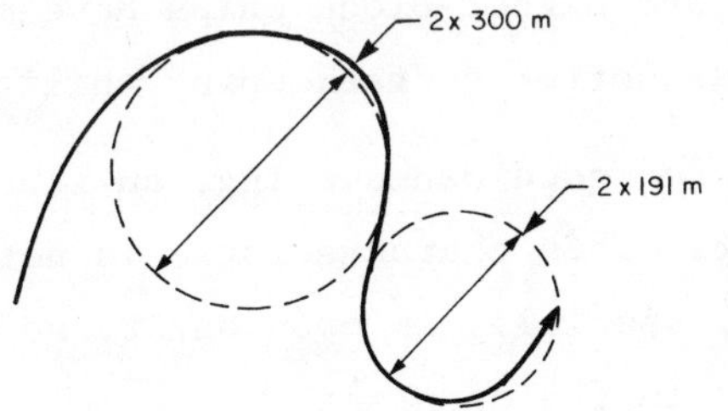


*Fig. 5.11. Labrador trial turning circles.*

Model tests in saline ice done by Edwards *et al.*, (1976) for the same ship gave very good agreement for the radius of these turning circles. Systematic modeling has shown that an increase in block coefficient increases the dimensionless turning radius R/L where R is the radius and L the ship's length.

The reason for this significant increase in dimensionless turning radius seems to be that for the hull with the high block coefficient, the slope of the hull, as the midship section is approached, is more nearly vertical than that of the low block hull. Consequently the icebreaking forces and moments developed in a turn of a given radius will be higher for the high-block hull, so that, with similar available turning forces, the radius of the turn will be larger for the high-block hull form.

It was also found during these tests that an increase in ship length of 20% caused an increase in turning radius per length of 300 percent. The longer ship must break ice on a greater length than must the shorter one, in a turn of equal radius.

Another characteristic of a ship is its stability, i.e. its ability to maintain a straight-line course without rudder action after a small disturbance (such as non-uniform ice conditions, cross winds and currents etc...). In general, all ships are directionally stable while continuously moving through level ice. The model tests (Edwards *et al.*, 1976) have shown that longer ships and higher-block ships have move directional stability and thus better coursekeeping ability.

On the other hand, for good maneuvering, an icebreaker needs a low block coefficient. This characteristic is not the one that makes for stability, specially in open water, so that icebreakers are heavy rollers in the open sea.

## 5.6 - CHARACTERISTICS OF ICEBREAKING SHIPS

### 5.6.1 - ICEBREAKERS

The normal task of an icebreaker is to assist other vessels through the ice. Other tasks include the destruction of river ice jams to prevent flooding (Fig. 5.12), winter ferry services, supply of isolated arctic harbors, patrol, buoy tending and hydrographic surveying.

*Fig. 5.12. Four icebreakers working together to break an ice jam across the St. Lawrence River.*

When assisting vessels in smooth waters the icebreaker makes an open track and the other vessels follow as closely as possible without risking collisions. At sea and in rivers this is often difficult because the track does not stay open because of currents and wind. One or more of the vessels may get stuck in the ice. To cut it loose the icebreaker must approach closely. Here good maneuverability is of great importance. The ideal is to have two icebreakers, with the wider of the two opening the way while the other offers loosening services if required, as done on the St. Lawrence River. In very difficult conditions it may even be necessary for the icebreaker to tow the ship it is assisting.

The characteristics of some recent icebreakers are shown in Table 5.2. The most important main dimension of the icebreaker is its beam, which determines the breadth of the vessel it is capable of assisting. The track that the icebreaker leaves is a little wider than the ship itself. For the majority of older icebreakers of the oceangoing variety the usual practice has been to adopt a water line length/water line beam ratio in the range of 4.0 to 4.5. For polar operations new ships have a length/beam ratio in the range of 5.0 to 6.0.

If there are no restriction on draft due to water depth, the draft should be large enough to accommodate large propellers placed as low as possible so as not to draw down air. As the stern propellers are usually designed for higher power and lower revolutions than the bow propellers, their diameter is larger. This difference is overcome by designing the icebreaker for a 0.6 m trim aft which also makes it easier to reverse out of ice walls. The large trim and tank capacities on icebreakers make it possible, within certain limits, to alter the draft and

| | |
|---|---|
| 1 Vessel | Murtaja |
| 2 Year of delivery | 1890 |
| 3 Shipyard | Bergsunds |
| 4 Nation | Finland |
| 5 Region of operation | Abo |
| 6 L oa (m) | 47.5 |
| 7 L wl (m) | 41.9 |
| 8 B max (m) | 10.98 |
| 9 B wl (m) | 10.77 |
| 10 Draft, normal (m) | 4.75 |
| 11 Draft, max (m) | 5.51 |
| 12 Moulded depth (m) | 7.60 |
| 13 Displacement normal metr. tons | 825 |
| 14 Displacement max. metr. tons | 900 |
| 15 Normal total power HP (metric) | 1200 |
| 16 Max. total power HP (metric) | 1793 |
| 17 Main machinery normal power HP (metric) | |
| 18 Main machinery max. power HP (metric) | |
| 19 Main machinery rot. speed R.P.M. | |
| 20 Stern propeller m/c normal power HP (metric) | 1x1200 |
| 21 Stern propeller m/c max. power HP (metric) | 1x1793 |
| 22 Stern propeller rot. speed R.P.M. | 83 |
| 23 Type of machinery | Steam |
| 24 Normal/Max. speed knots | -/12.5 |

Table 5.2 (Part 1) - Characteristics of icebreakers (Compiled from various sources).

| | | | | |
|---|---|---|---|---|
| 1 | Ermak | Stalin class | Kirov class | Wind Class |
| 2 | 1899 | 1937-39 | 1938-40 | |
| 3 | Annstrong | Leningrad Nikolajev | Leningrad | U.S.A. |
| 4 | U.S.S.R. | U.S.S.R. | U.S.S.R. | U.S.A. |
| 5 | U.S.S.R. | Arctic Ocean | Far East | Arctic and Antartic |
| 6 | 97.5 | 106.0 | 109.0 | 82.0 |
| 7 | 94.6 | 102.1 | 101.0 | 76.2 |
| 8 | 21.64 | 23.16 | 22.3 | 19.4 |
| 9 | 21.42 | 22.70 | 21.2 | 18.9 |
| 10 | 7.32 | 8.00 | 7.25 | 7.85 |
| 11 | 8.54 | 9.04 | 8.4 | 8.87 |
| 12 | 12.95 | 12.61 | | 5300 |
| 13 | 7875 | 9300 | 8330 | 6515 |
| 14 | 10000 | 11000 | 10250 | |
| 15 | 7500 | 10050 | | |
| 16 | 9000 | | 12000 | |
| 17 | | | | |
| 18 | | | 4x3000 | |
| 19 | | | 300 | |
| 20 | 3x2500 | 3x3350 | | 2x5200 |
| 21 | 3x3000 | | 2x2600<br>1x5200 | |
| 22 | 105 | 105 | | |
| 23 | Steam | Steam | Diesel el. | Diesel el. |
| 24 | 14/- | -/15.5 | -/- | 16 |

Table 5.2 (Part 2) - Characteristics of icebreakers (Compiled from various sources).

| | | | |
|---|---|---|---|
| 1 | d'Iberville | Labrador | General S. Martin |
| 2 | 1953 | 1953 | 1954 |
| 3 | Lauzon | Sorel | Weser |
| 4 | Canada | Canada | Argentina |
| 5 | Arctic and St.Lawrence | Arctic and East Coast | Antarctic |
| 6 | 94.6 | 82.0 | 84.0 |
| 7 | 92.0 | 76.2 | 76.95 |
| 8 | 20.27 | 19.44 | 19.00 |
| 9 | 19.68 | 18.90 | 18.60 |
| 10 | 7.92 | 9.30 | 6.50 |
| 11 | | | |
| 12 | 12.20 | 11.51 | 9.85/7.50 |
| 13 | 8840 | 5350 | |
| 14 | 9930 | 6940 | |
| 15 | 10800 | 10000 | 7500 |
| 16 | 15200 | 12000 | 8100 |
| 17 | | 6x1750 | 2x3750 |
| 18 | | 6x2000 | 2x4050 |
| 19 | | 810 | 275/300 |
| 20 | 2x5400 | | 2x3250 |
| 21 | 2x7600 | 2x5000 | 2x3550 |
| 22 | 145 | 145 | 138 |
| 23 | Skinner st. | Diesel el. | Diesel el. |
| 24 | 15/- | 16/- | -/16 |

Table 5.2 (Part 3) - Characteristics of icebreakers (Compiled from various sources).

| | | | |
|---|---|---|---|
| 1 | Moskva class | Glacier | Lenin |
| 2 | 1954 | 1955 | 1959 |
| 3 | Warsila | Ingalls | Leningrad |
| 4 | U.S.S.R. | U.S.A. | U.S.S.R. |
| 5 | Arctic Ocean | Antarctic | Arctic Ocean |
| 6 | 122.1 | 94.5 | 134 |
| 7 | 112.4 | 90.0 | 128 |
| 8 | 24.50 | 22.56 | 27.6 |
| 9 | 23.50 | 22.0 | 27.5 |
| 10 | 9.50 | 8.53 | 9.2 |
| 11 | 10.50 | | |
| 12 | 14.00/11.59 | | 16.1 |
| 13 | 13100 | 8763 | 16000 |
| 14 | 15200 | | |
| 15 | 26000 | | 44000 |
| 16 | | 24000 | |
| 17 | 8x3250 | | |
| 18 | | 10x24000 | |
| 19 | 330 | 720/810 | |
| 20 | 2x5500<br>1x11000 | 2x8450 | |
| 21 | | 2x10500 | |
| 22 | 105/120 | 120/175 | 185/205 |
| 23 | Diesel el. | Diesel el. | Atomic-steam turbine |
| 24 | 18/19.5 | 18/19 | 15.5/18 |

Table 5.2 (Part 4) - Characteristics of icebreakers (Compiled from various sources).

| | | | |
|---|---|---|---|
| 1 | John A. MacDonald | Louis St-Laurent | Polar Star |
| 2 | 1960 | 1969 | 1975 |
| 3 | Lauzon, Québec | Montréal | Lockheed - Seattle |
| 4 | Canada | Canada | U.S.A. |
| 5 | Arctic Ocean | Arctic Ocean | Arctic and Antarctic |
| 6 | 96 | 111.7 | 121.6 |
| 7 | 93.5 | 101.8 | 107.3 |
| 8 | 21.4 | 24.4 | 25.5 |
| 9 | 21.0 | 24.3 | 23.8 |
| 10 | 8.5 | 9.5 | 18.5 |
| 11 | 8.9 | | 9.7 |
| 12 | | | 11000 |
| 13 | | 13.1 | 13179 |
| 14 | 9160 | 14280 | |
| 15 | 15000 | 24000 | |
| 16 | | 31500 | |
| 17 | | | |
| 18 | | | |
| 19 | | | |
| 20 | | | 3x6200 D.E. |
| 21 | 3x | | |
| 22 | | | |
| 23 | Diesel el. | S.Turb. | Diesel el. |
| 24 | /15.5 | 15/17.7 | 17 |

Table 5.2 (Part 5) - Characteristic of icebreakers (Compiled from various sources)

the trim of the vessel.

The typical proportions and lines of an icebreaker (the C.C.G.S., "Louis S. St-Laurent") are shown in Fig. 5.13. The upper portion of the stem is nearly vertical. At water line and below, the stem slopes close to $30^{\circ}$. The midship sectional shape is close to a circular arc with a center above the water line. The virtue of the circular section appears to be that if facilitates working the ship loose by means of the heeling tanks when beset. The first effect of the heeling moment is to break the grip of the static friction and this is followed by a loosening of the ice adjacent to the hull. This shape tends to make the block coefficient very small, somewhat lower than 0.5 for typical icebreakers. It produces very little damping to any induced rolling motion and icebreakers are heavy rollers in the open sea if stabilizing fins are not fitted. The after-body form of an icebreaker usually narrows to a point, and a generous clearance is provided between propeller blades and shell so as not to let any ice lodge in between.

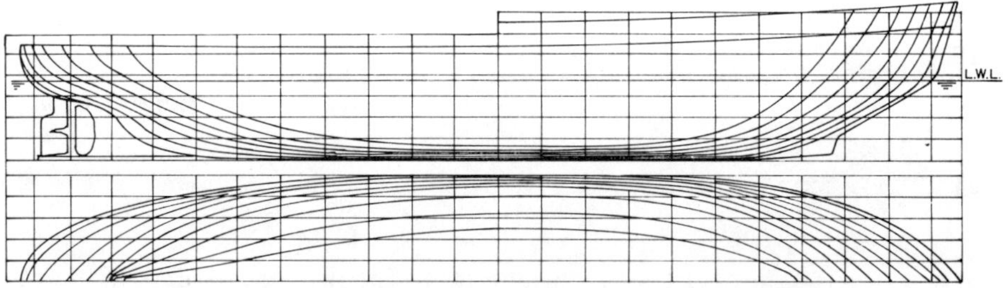


*Fig. 5.13. Proportions and lines of C.C.G.S. "Louis St-Laurent".*

All modern icebreakers have heeling tanks that can be put to use in one or two minutes. By pumping water from one side to the other a heel of 5 to $8^{\circ}$ to either side can be achieved, which is normally sufficient to loosen the vessel from the ice. The capacity of ballast tanks in icebreakers is high and may reach 15% of the displacement. In certain cases the icebreaking efficiency can be increased by pumping water between trim tanks in the fore and aft peaks, thus changing the trim.

## 5.6.2 - COMMERCIAL ICEBREAKING SHIPS

Although there is now only one existing commercial ship for transportation of material in bulk from Arctic regions, the examination of the feasibility of using them is well advanced in view of their enormous economic advantage on other trading routes.

We will summarize and cite directly, here, a paper presented by German and Dadachanyi (1975) on the possible hull forms for arctic bulk cargo transportation.

First the authors make it clear that the avoidance of unacceptable high costs can only be achieved by taking careful cognizance of the following steps in setting up an Arctic bulk carrier operation:

"1) Accurate data concerning the ice environment for the route, including:

a) Description of ice cover (pack, solid sheet, fast, open, first year, multi-year, etc.). This should be on a month-by-month basis or better and over the entire route.

b) Frequency, magnitude and extent of pressure hazards over the route.

c) Frequency, size and type of ridge formations and rafting which will be found over the route.

2) Accurate assessment of the powering and hull strength requirements to achieve a maximum return on investment. This requires the careful analysis of speed/power profile over the whole route and all operating seasons, the round trips possible varying, of course, with the extent and severity of the ice cover.

3) A correct derivation of the hull form to best suit the above. Choice of hull form is interrelated with the choice of machinery and power level, operating costs, ship handling, draft limitations, overall capital costs and many other considerations."

It is this latter aspect that is now considered by the authors, and that we cite textually:

Conventional icebreaking hullforms

"Experience and theory have shown that the principal ship parameter in minimizing the resistance in ice cover is the breadth at the waterline. Kashtelyan's formulation for resistance in ice indicates that the velocity dependent term (the significant one except at low speeds) is proportional to the 1.65 power of beam. Thus adapting a conventional hull form for Arctic ice means, primarily, reducing the beam as far as possible without losing stability. Present day tankers have a breadth to depth ratio of about 2.0. This ratio can be reduced to about 1.45 before the achievement of adequate stability creates difficulties.

In discussing bulk carrier hull forms for heavy ice service we shall not distinguish between different types of bulk carrier service. That is; ore carriers and tankers of comparable deadweight capacity will have similar hull form and proportions. This is due to the fact that efficient operation in heavy ice cover requires a hullform which should operate at a reasonably constant draft. The slopes of the hull at the waterline must be extended from above the loadline to below the ballast waterline, which imposes a real penalty on the hull size and capacity if the ballast and load waterlines are separated to any significant degree. Ore carriers must be fitted with large water ballast capacity which will effectively result in hull proportions quite similar to tankers.

Reduction of breadth at the waterline is of varying degrees of importance depending on the severity of the ice condition in which the ship will operate over a significant proportion of her service life. It carries with it the need to either lengthen or deepen the vessel, or both, to compensate for the reduced beam. In many potential areas of operation in the Arctic, deeper draft cannot be obtained due to the shallow water extending large distances offshore. For VLCC type ships the increase of draft might drastically reduce the number of unloading terminals available. Increasing length is expensive.

Within practical limits the effect of increasing severity of ice conditions, depicted by the Arctic Classification given in Table 5.3 and Fig. 5.14 increases the dimensions of the ships. The "conventional" forms are for surface vessels having a conventional icebreaking bow form. In other words they will have parallel body, full midship section, vertical sides at amidships and normal stern form, the principal effect of designing to Arctic conditions being the adoption of higher L/B ratios and lower B/H ratios.

| Column I Category | Column II Zone 1 | Column III Zone 2 | Column IV Zone 3 | Column V Zone 4 | Column VI Zone 5 |
|---|---|---|---|---|---|
| Arctic Class 10 1 | All Year | All Year | All Year | All Year | All Year |
| Arctic Class 8 2 | Jul. 1 to Oct. 15 | All Year | All Year | All Year | All Year |
| Arctic Class 7 3 | Aug. 1 to Sept. 30 | Aug. 1 to Nov. 30 | Jul. 1 to Dec. 31 | Jul. 1 to Dec. 15 | Jul. 1 to Dec. 15 |
| Arctic Class 6 4 | Aug. 15 to Sept. 15 | Aug. 1 to Oct. 31 | Jul. 15 to Nov. 30 | Jul. 15 to Nov. 30 | Aug. 1 to Oct. 15 |
| Arctic Class 4 5 | Aug. 15 to Sept. 15 | Aug. 15 to Oct. 15 | Jul. 15 to Oct. 31 | Jul. 15 to Nov. 15 | Aug. 15 to Sept. 30 |
| Arctic Class 3 6 | Aug. 20 to Sept. 15 | Aug. 20 to Sept. 30 | Jul. 25 to Oct. 15 | Jul. 20 to Nov. 5 | Aug. 20 to Sept. 25 |
| Arctic Class 2 7 | No Entry | No Entry | Aug. 15 to Sept. 30 | Aug. 1 to Oct. 31 | No Entry |
| Arctic Class 1A 8 | No Entry | No Entry | Aug. 20 to Sept. 15 | Aug. 20 to Sept. 30 | No Entry |
| Arctic Class 1 9 | No Entry | No Entry | No Entry | No Entry | No Entry |
| Type A 10 | No Entry | No Entry | Aug. 20 to Sept. 10 | Aug. 20 to Sept. 20 | No Entry |
| Type B 11 | No Entry | No Entry | Aug. 20 to Sept. 5 | Aug. 20 to Sept. 15 | No Entry |
| Type C 12 | No Entry | No Entry | No Entry | No Entry | No Entry |
| Type D 13 | No Entry | No Entry | No Entry | No Entry | No Entry |
| Type E 14 | No Entry | No Entry | No Entry | No Entry | No Entry |

Table 5.3 (Part 1) - Ice class vs zones & seasons (Canadian Arctic pollution prevention regulations).

| Column VII Zone 6 | Column VIII Zone 7 | Column IX Zone 8 | Column X Zone 9 | Column XI Zone 10 | Column XII Zone 11 |
|---|---|---|---|---|---|
| All Year | All Year | All Year | All Year | All Year | All Year |
| All Year | All Year | All Year | All Year | All Year | All Year |
| All Year | All Year | All Year | All Year | All Year | All Year |
| Jul. 15 to Feb. 28 | Jul. 1 to Mar. 31 | Jul. 1 to Mar. 31 | All Year | All Year | Jul. 1 to Mar. 31 |
| Jul. 20 to Dec. 31 | Jul. 15 to Jan. 15 | Jul. 15 to Jan. 15 | Jul. 10 to Mar. 31 | Jul. 10 to Feb. 28 | Jul. 5 to Jan. 15 |
| Aug. 1 to Nov. 20 | Jul. 20 to Dec. 15 | Jul. 20 to Dec. 31 | Jul. 20 to Jan. 20 | Jul. 15 to Jan. 25 | Jul. 5 to Dec. 15 |
| Aug. 15 to Nov. 20 | Aug. 1 to Nov. 20 | Aug. 1 to Nov. 20 | Aug. 1 to Dec. 20 | Jul. 25 to Dec. 20 | Jul. 10 to Nov. 20 |
| Aug. 25 to Oct. 31 | Aug. 10 to Nov. 5 | Aug. 10 to Nov. 20 | Aug. 10 to Dec. 10 | Aug. 1 to Dec. 10 | Jul. 15 to Nov. 10 |
| Aug. 25 to Sep. 30 | Aug. 10 to Oct. 15 | Aug. 10 to Oct. 31 | Aug. 10 to Oct. 31 | Aug. 1 to Oct. 31 | Jul. 15 to Oct. 20 |
| Aug. 15 to Oct. 15 | Aug. 1 to Oct. 25 | Aug. 1 to Nov. 10 | Aug. 1 to Nov. 20 | Jul. 25 to Nov. 20 | Jul. 10 to Oct.31 |
| Aug. 25 to Sep. 30 | Aug. 10 to Oct. 15 | Aug. 10 to Oct. 31 | Aug. 10 to Oct. 31 | Aug. 1 to Oct. 31 | Jul. 15 to Oct. 20 |
| Aug. 25 to Sep. 25 | Aug. 10 to Oct. 10 | Aug. 10 to Oct. 25 | Aug. 10 to Oct. 25 | Aug. 1 to Oct. 25 | Jul. 15 to Oct. 15 |
| No Entry | Aug. 10 to Oct. 5 | Aug. 15 to Oct. 20 | Aug. 15 to Oct. 20 | Aug. 5 to Oct.20 | Jul. 15 to Oct.10 |
| No Entry | Aug. 10 to Sep. 30 | Aug. 20 to Oct. 20 | Aug. 20 to Oct. 15 | Aug. 10 to Oct. 20 | Jul. 15 to Sep. 30 |

Table 5.3 (Part 2) - Ice class vs zones & seasons (Canadian Arctic pollution prevention regulations).

| Column XIII Zone 12 | Column XIV Zone 13 | Column XV Zone 14 | Column XVI Zone 15 | Column XVII Zone 16 |
|---|---|---|---|---|
| All Year | All Year | All Year | All Year | All Year |
| All Year | All Year | All Year | All Year | All Year |
| All Year | All Year | All Year | All Year | All Year |
| All Year | All Year | All Year | All Year | All Year |
| June 1 to Jan. 31 | June 1 to Feb. 15 | June 15 to Feb. 15 | June 15 to Mar. 15 | June 1 to Feb. 15 |
| June 10 to Dec. 31 | June 10 to Dec. 31 | June 20 to Jan. 10 | June 20 to Jan. 31 | June 5 to Jan. 10 |
| June 15 to Dec. 5 | June 25 to Nov. 15 | June 25 to Dec. 10 | June 25 to Dec. 20 | June 10 to Dec. 10 |
| Jul. 1 to Nov. 10 | Jul. 15 to Oct. 31 | Jul. 1 to Nov. 30 | Jul. 1 to Dec. 10 | June 20 to Nov. 30 |
| Jul. 1 to Oct. 31 | Jul. 15 to Oct. 15 | Jul. 1 to Nov. 30 | Jul. 1 to Nov. 30 | June 20 to Nov. 15 |
| June 15 to Nov. 10 | June 25 to Oct. 22 | June 25 to Nov. 30 | June 25 to Dec. 5 | June 20 to Nov. 20 |
| Jul. 1 to Oct. 25 | Jul. 15 to Oct. 15 | Jul. 1 to Nov. 30 | Jul. 1 to Nov. 30 | June 20 to Nov. 10 |
| Jul. 1 to Oct. 25 | Jul. 15 to Oct. 10 | Jul. 1 to Nov. 25 | Jul. 1 to Nov. 25 | June 25 to Nov. 10 |
| Jul. 1 to Oct. 20 | Jul. 30 to Sep. 30 | Jul. 10 to Nov. 10 | Jul. 5 to Nov. 10 | Jul. 1 to Oct. 31 |
| Jul. 1 to Oct. 20 | Aug. 15 to Sep. 20 | Jul. 20 to Oct. 31 | Jul. 20 to Nov. 5 | Jul. 1 to Oct. 31 |

Table 5.3 (Part 3) - Ice class vs zones & seasons (Canadian Arctic pollution prevention regulations).

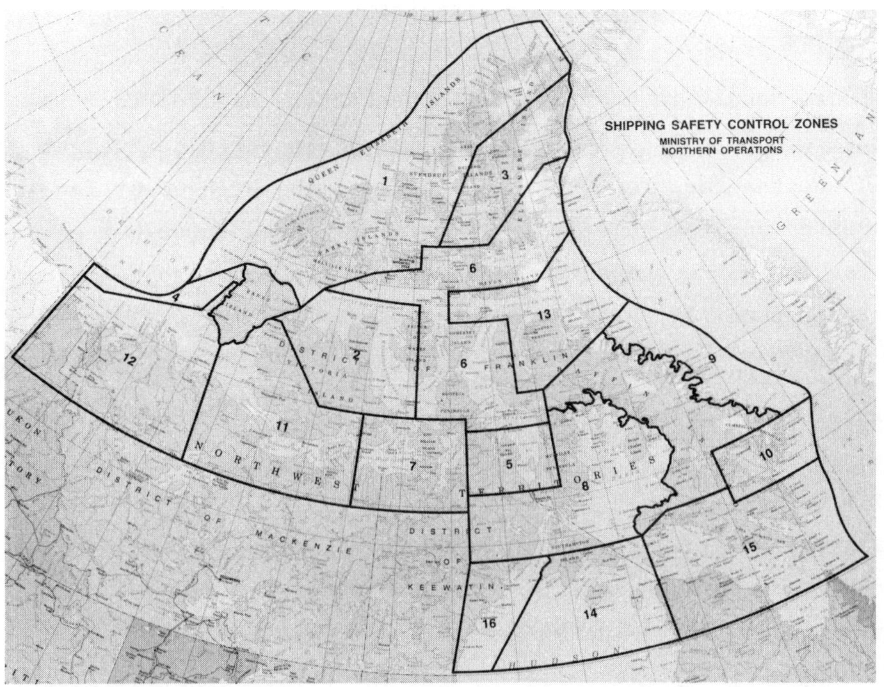


*Fig. 5.14. Zones for Canadian Arctic Pollution Prevention Regulations (CAPPR).*

Compared to conventional open water bulk carriers the forebody, designed for icebreaking, will necessarily be quite long and fine, as will the after body which will be designed in this manner to facilitate extraction when stuck in ice - in addition to providing a good flow of water to the screws in conditions approaching bollard. These considerations will cause a reduction in block coefficient to about 0.75 to 0.80 depending on the severity of the conditions.

Aside from the effect of ice class upon the so-called "conventional" icebreaking hull form, other effects upon the ship are apparent such as the propulsion machinery plant, design of hull structure, etc. to provide required power and strength respectively. For a given cargo deadweight requirement these aspects result in greater hull dimensions. An important influence upon the hull dimensions for given cargo deadweight will be the fuel requirement. Owing to the high power levels required and the relatively low transit speed through ice cover the bunker capacity requirements can prove to be much greater than for the open water vessel.

One minor advantage in ice operation is that continual friction will tend to keep the hull free of fouling, thereby reducing water resistance. Also, wavemaking losses will be reduced when moving among ice fields, although the gain would be negligible at other than low speeds or in thin ice. The minimum power requirement for an example vessel as set forth in the CAPPR Regulations would be about 74,000 SHP for a Class 6 vessel. However the power level stipulated in these Regulations is based upon safety considerations and for a commercial operation, the vessel should have sufficient power to maintain reasonnable headway in uniform ice of thickness equal to the average maximum expected (in this case 2 m). Assuming 5 knots as "reasonable headway" the installed power would need to be about 160,000 SHP on three screws.

From the above general discussion it will be understood that the most important single parameter of the hull form is the breadth. After that comes the shape of the forebody and midsection. These facts are by now widely known, but the manner in which they should be applied to the bulk carrier form is important because of the ease of making a serious economic error in judgment of the best form.

In a general discussion such as this it is difficult to cover all conditions likely to affect the design and economics. Some examples, therefore, may serve to illustrate the points to be made. One can start off by generalizing that the vessel's capabilities should be such that the vessel can make progress in continuous mode at an economical speed in pressure-free fast ice equal in thickness to the ice class (CAPPR). This means she would be operating in ice of lesser thickness most of the year and over the bulk of her journey out of the Arctic. The choice of "economical speed" in ice depends on many factors peculiar to a particular service and would include the percentage of her annual voyage mileage spent in the maximum ice conditions, the probable frequency of encountering ice pressure conditions and their expected magnitude and the probable mileage through such conditions when at a severity level in excess of the ship's capability to make continuous headway.

The "conventional" type hullforms (A1 & A2) represented in Fig. 5.15 incorporate reduced beam with increasing ice Class. In practice, the design of these hullforms will incorporate refinements designed to reduce as far as practicable the ice resistance and the tendency to become stuck under conditions of pressure. Flare in the midbody portion will be incorporated. The CAPPR, it should be noted, includes the introduction of flare as an integral part of the hull structural design, corrections being made to scantlings where designed flare angles are greater or less than the basic value set out in the Regulations. Other refinements would involve the bow form design and stern design."

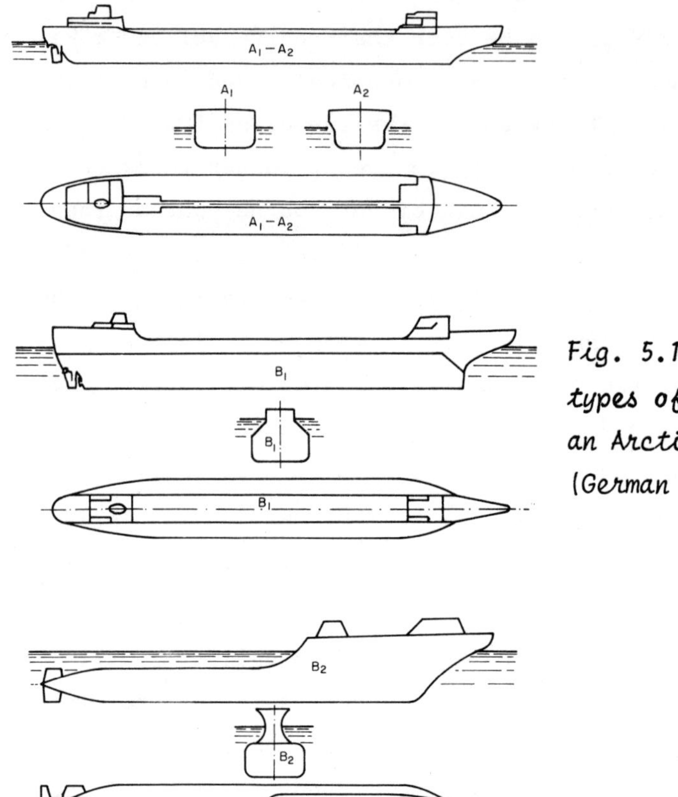


*Fig. 5.15. Different types of hullforms for an Arctic cargo ship (German et al., 1975).*

## Semi-submerged icebreaking hullforms

"The semi-submerged "B2" type form (see Fig. 5.15) is far less demanding for power in ice conditions than the "A" or "B1" forms. However, a correspondingly large draft is required which would limit its usefulness in most modern terminals. For regions of very heavy ice cover this form would be far less resistful than the wider waterline forms and would probably compete in such areas with the cargo submarine. In lesser ice formations it is clear the high draft and the structural arrangement would work against it compared to the more moderate form.

The semi-submerged type hullforms naturally start to manifest when ice conditions warrant exceptionally narrow breadths. These forms will achieve narrow breadths by incorporating waterlines, lowering the centre of gravity thereby allowing narrower beam at the waterline area. The B1 form is a reasonably moderate concept for the semi-submerged form and no difficulty is envisaged in maintaining trim. The pronounced reverse flare can be adapted to a downward-acting bow as shown in Fig. 5.15 and is, moreover, expecially well suited to adaptation to an upward-acting bow. Fig. 5.16 illustrates this type of bow. It should be mentioned, however, that development work must still be done on such a bow form for bulk carrier applications in order to reduce its resistance at open water service speeds. Furthermore, the evidence thus far from model and full scale tests is inconclusive but suggests it may be more resistful in ice compared to a well designed downward-acting form. The potential advantages of the upward-acting form are nonetheless quite real. These include:

1) additional deadweight lifting capacity for given ship length;

2) the removal of broken ice upward and sideways rather than beneath the vessel with attendant reduction of hazard to propellers and rudders;

3) possible reduction of side friction, especially in field pressure, due to the sharp angle of tumblehome and the fact that ice to ship contact is with the water lubricated underside of the ice.

4) This form may be well suited to the use of air blast systems for easing the ship passage through ice.

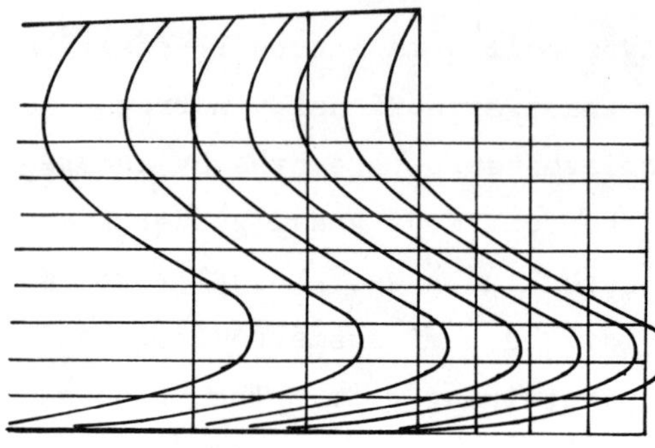

*Fig. 5.16. Upward acting bow.*

One important consideration for design of the B type forms is related to the requirement that the type must operate as nearly as practicable to a constant waterline. This is particularly true for the B2 types which are especially sensitive to changes in displacement, having relatively small waterline area. The low tons/cm immersion of the working waterline requires that these ships be capable of maintaining the load draft in the ballast condition. For tankers it would appear that the ballast water must be carried in the cargo tanks which would have to be pumped to a shore separating or disposal station when loading cargo. For ore carriers the problem may not be so severe as most of the ballast water could be carried in ballast tanks."

## 5.7 - SPECIAL METHODS TO BREAK ICE

Special methods are sometimes used to assist in the icebreaking process. One consists of spraying warm water from the circulating pumps onto the ice at the stem. Explosives, salt melting and even manual ice cutting may be used in exceptional circumstances. Large bow vibrators have been used in small river icebreakers in the U.S.S.R.

An air coating system was used on the Great Lake ore carrier Leon Fraser to reduce ice resistance (Levine *et al.*, 1974). Orifices set at 0.6 m center to center along two lines on the exterior portion of the bilge of the hull were fed with compressed air. The air bubble plumes from the orifices converged into a solid air sheet about 1.6 m above the manifold. This system was designed to provide lubrication between the broken pieces of ice and ship's hull but cannot help when the ship is pinched in moving ice fields. It has been found efficient for its purpose.

A new method to break ice was found in 1971 in northern Canada by Sun Oil Co. when they noticed that an air cushion vehicle, the ACT 100, could break 68 cm of ice at a speed of 8 km/hr. The principle is shown on Fig. 5.17. The air which comes from by the ACV is pushed under the ice which then collapse under a reduced weight because of lack of water support. In the high-speed method, a self-propelled ACV moves at such a speed that it generates a steep wave in the water, shattering the ice in its path.

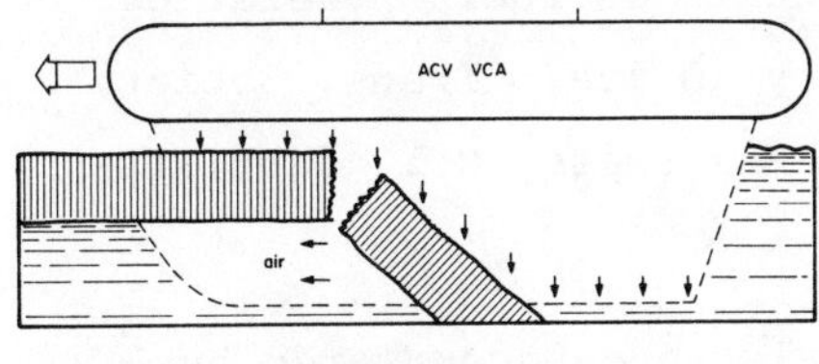

*Fig. 5.17. Icebreaking with an air cushion vehicle (ACV).*

## REFERENCES

Abkowitz, M.A. (1964) - "Lectures on Ship Hydrodynamics - steering and Maneuverability" Report No. Hy-5, Hydro and Aerodynamics Laboratory, Lyngby, Denmark.

Buzuev, A., Ryblin, A. (1961) - "Calculation of the resistance encountered by an icebreaker moving through ice cakes and brash" Morskoi Flot. Vol. 21, No. 8, p. 136-138.

Chu Fen-Dow (1974) - "Ship resistance in homogeneous ice fields" Thesis for the degree of Doctor of Technology, Helsinki University of Technology, 125p.

Edwards, R.Y. Jr, German, J.G., Lawrence, R.G.A. (1973) "Comparative Model Tests of the Icebreaker Performance of Two Canadian Coast Guard Icebreakers" Second International Conference on Port and Ocean Engineering Under Arctic Conditions (POAC), Reykavik, Iceland.

Edwards, R.Y., Lewis, J.W., Wheaton, J.W., Colburn, J. (1972) "Full Scale and Model Tests of a Great Lakes Icebreaker" Trans. SNAME, Vol. 80.

Edwards, R.Y., Major, R.A., Kirn, J.K., German, J.G., Lewis, J.W., Miller, D.R. (1976) - "Influence of major characteristics of icebreaker hulls on their powerering requirements and maneuverability in ice" Trans. Society of Naval Architects and Marine Engineers, N.Y. Sept. 11-13, 39p.

Enkvist, E. (1973) - "On the Ice Resistance Encountered by Ships Operating in the Continuous Mode of icebreaking" Swedish Academy of Engineering Sciences in Finland, Report No. 24.

German, J.G., Dadachanyi, N. (1975) - "Hullforms for arctic bulk cargo transportation" Presented to 'The Society Naval Architects and Marine Engineers' May 14-17, 1975 p. 53-64.

German, J.G., Lawrence, R.G.A. (1975) - "Full scale testing in ice of three icebreakers" Trans. Society of Naval Architects and Marine Eng. Montreal, April 1975, 67 p.

Jansson, J.E. (1956) - "Icebreakers and their design" European Ship building, Vol. 5.

Kathtelyan, V.I., Poznjak, I.I., Ryblin, A. Ya (1968) - "Ice Resistance to Motion of a Ship" (translation) Sudostroenie, Leningrad.

Kashtelyan, V.I., Ryblin, A.Y., Faddeyev, O.V., Yagodkin, V.Y. (1972) - "Icebreakers" Ledokoly, Leningrad, 287p. Translation CRREL Draft Transl. 418.

Levine, G., Voelker, R.P., Mentz, P.B. (1974) - "Advances in the development of commercial ice transiting ships" Trans. The Society of Naval Architects and Marine Engineers. Nov. 14-16, 1974, 25p.

Lewis, Jack W., Edwards, R.Y. Jr. (1970) - "Methods for predicting Icebreaking and Ice Resistance Characteristics of Icebreakers" Trans. SNAME Vol. 78.

Michel, B. (1970) - "Ice pressure on engineering structures" Cold Region Research and Engineering Lab. (CRREL) Monograph III-Blb, 71 p.

Michel, B., Lafleur, P. (1971) - "Ice management at marine terminal, Hershel Island" Report to Department of Public Works, Canada.

Milano, V.R. (1961) - "Notes on the Preliminary Design of Icebreakers" MS thesis, Webb Institute of Naval Architecture.

Milano, V.R. (1973) - "Ship Resistance to Continuous Motion in Ice" Trans. SNAME, Vol. 81.

Mookhoek, A.D., Bielstein, W.T. (1971) - "Problems associated with the design of an arctic marine transportation system" Offshore technology Conference; Houston, Texas.

Nevel, D.E. (1961) - "The Narrow Infinite Wedge on an Elastic Foundation" Research Report Number 79, Cold Regions Research and Engineering Laboratory, July 1961.

Runeberg, R. (1900) - "Steamers for Winter Navigation and Icebreaking" Proceedings of the Institution of Civil Engineers, Vol. CSL, Session 1899-1900, Part II.

Shimansky, J.A. (1938) - "Conditional Standards of Ice Qualities of a Ship" (translation) Arctic Institute of the Chief Administrator of the Northern Sea Route, Vol. 130, Leningrad.

Vinogradov, I.V. (1946) - "Ships for navigation in ice" Oborovgiz NKAP, 239 p.

Vance, G.P. (1975) - "A Scaling System for Vessels in Ice" SNAME Ice Tech. Symposium, Montreal, April 9-11, 1975, 26 p.

Watson, A. (1959) - "The design and building of icebreakers" Trans. of the Institute of Marine Engineers, Vol. 71, No. 2, p. 37-65.

White, R.M. (1965) - "Dynamically Developed Forces at the Bow of an Icebreaker" Ph.D. Dissertation, M.I.T.

# CHAPTER 6

# ICE MODELING

## 6.0 - INTRODUCTION

In the last few years knowledge of the mechanical behavior of ice has increased to the point where some basic laws can be formulated and scaled down for physical models to study engineering problems caused by ice. The first such models were built just after World War II in Canada, with paraffin to model the ice (Foundation Eng. Co, 1949) and in the U.S.S.R., with weakened ice and paraffin (Shvayshteyn, 1959).

The technique of ice modeling has evolved along two completely different lines. On one side studies were made on moving ice in rivers and the similitude was limited to representing the hydrodynamic conditions, the ice itself being considered a rigid, unbreakable body. This condition is well understood and can be accurately modeled.

On the other hand models were built to measure the forces caused by the impact of ice on structures and to obtain the resistance of icebreakers, and the similitude was aimed mainly at scaling down the strength properties of ice, somewhat neglecting other dynamic aspects. Because of incomplete similitude, modeling techniques for these problems usually differ from one application to the other.

## 6.1 - LAWS OF SIMILITUDE

### 6.1.1 - BASIC RELATIONS

The rules of similitude for ice models can be obtained by three different approaches; dimensional analysis, similitude of forces and the general method of comparing the equations of motion and ice resistance in the model and prototype.

In hydraulic textbooks, similitude is usually presented as a natural consequence of dimensional analysis. Because of the complexities involves in ice modeling, knowledge and understanding of the phenomena under study is generally necessary for deducing the rules of similitude, in order to be able to neglect the phenomena which are of secondary importance and for deciding the relative influence of scale effects. Thus, the basic laws of similitude are better derived by discussing geometric, kinematic and dynamic similitude.

The development of the laws of similitude for modeling ice has been presented by many authors such as Nogid (1959), White and Vance (1967, 68), Michel (1970, 1975), Edwards and Lewis (1970), Atkins (1975) and Vance (1975). Although the derivations differ in presentation, one common and major result is universal agreement that Froude scaling laws are applicable. The complete derivations are presented here (Michel, 1970):

A mechanical model is said to be in complete similitude if it has geometric, kinematic and dynamic similitude.

Geometric similarity will be satisfied when every linear dimension on the prototype is equal to constant times the corresponding linear dimension of the model. This can be expressed as:

$$L_p = \lambda L_m \tag{6-1}$$

Kinematic similarity will be satisfied when the velocity on the prototype is equal to the velocity on the model, times a constant, i.e.:

$$V_p = \lambda_k V_m \tag{6-2}$$

which gives for the time scale:

$$T_p = \frac{\lambda}{\lambda_k} T_m \tag{6-3}$$

Dynamic similarity is satisfied when the forces on the prototype are related to the forces on the model by a constant, i.e.:

$$F_p = \lambda_d F_m \tag{6-4}$$

In the above, $\lambda$, $\lambda_k$, and $\lambda_d$ are geometric, kinematic and dynamic scaling factors and L, V and F are length, velocity and force in the L, T, M dimension system. The subscript p refers to the prototype, while the subscript m refers to the model.

## Gravity and inertia forces

The gravity and inertia forces, $F_w$ and $F_i$, can be expressed as:

$$F_w = Mg = \rho g L^3 \tag{6-5}$$

$$F_i = Ma = \rho L^3 a \tag{6-6}$$

The ratio of weight forces between the model and prototype is expressed with (6-4) and (6-5) as:

$$\frac{F_{wp}}{F_{wm}} = \lambda_d = \frac{\rho_p}{\rho_m} \lambda^3 \tag{6-7}$$

this is the basic expression for the force scale in ice modeling. If water is used in the model, it then follows that all other substances should have the same density they have in the prototype:

$$\rho_p = \rho_m \tag{6-8}$$

So that the force scale is related to the geometric scale by (6-7):

$$\lambda_d = \lambda^3 \tag{6-9}$$

The ratio of prototype inertia forces to model inertia forces is then with Eqs (6-6) and (6-8):

$$\frac{F_{ip}}{F_{im}} = \lambda^3 \frac{a_p}{a_m} \tag{6-10}$$

The ratio of inertia forces will be the same as that of the gravity forces if Eq. (6-9) is equal to Eq. (6-10); that is:

$$\frac{a_p}{a_m} = \frac{\lambda_k}{T_p} T_m = 1 \tag{6-11}$$

with Eq. (6-3), it also gives:

$$\lambda_k = \frac{V_p}{V_m} = \sqrt{\lambda} \tag{6-12}$$

which corresponds to the Froude similitude where the Froude number $Fr = V/\sqrt{gL}$ is kept the same on model and prototype. It can easily be seen that the Froude similitude comes out by comparing only the force of inertia with the force of gravity. If there is any acceleration whatsoever in an ice model or change in velocity, it has to be scaled down with the Froude similitude.

## Forces of water friction

The water friction forces on different boundaries, including ice, can be represented by:

$$F_f = C \left(\frac{e}{L}, Re\right) \rho V^2 L^2 \qquad (6\text{-}13)$$

where $F_f$ is the tangential friction force, C is the friction coefficient depending on relative roughness $\frac{e}{L}$ and Reynolds number $Re = VL/\nu$; V is a representative velocity outside the boundary layer and $L^2$ is the surface on which $F_f$ is acting.

For rough turbulent flow, as is most generally found in natural rivers, the friction coefficient is independent of the Reynolds number. To get the same coefficient on the model and the same quadratic law it is necessary to have complete roughness similarity and a Reynolds number high enough to ensure rough turbulent flow at the reduced scale. For river models there is thus a limitation in the choice of the scale. It should be big enough to obtain this acceptable Reynolds number.

When a body is towed in water the friction force also has the same form as Eq. (6-13) and becomes independent of Re if the velocity is high enough. The friction force in the developing laminar boundary layer then becomes small before that of the fully developed turbulent layer and the same reasoning applies as for the previous case. In all cases the condition of similitude from Eqs (6-13) and (6-4) is:

$$\frac{F_{fp}}{F_{fm}} = \lambda_d = \lambda^2 \lambda_k^2 \qquad (6\text{-}14)$$

with $\lambda_d = \lambda^3$, in (6-9), this gives exactly the same as Eq. (6-12), so that Froude's similitude applies very well to simulate the friction forces in fully developed turbulent flow.

If the Reynolds number were not high enough and the flow laminar, Eq. (6-13) would become equal to:

$$F'_f = C'\rho V^2 L^2/Re \qquad (6\text{-}15)$$

and for the same fluid of viscosity $\nu$, we obtain for the same ratio of gravity forces:

$$\lambda_k = 1/\lambda \qquad (6\text{-}16)$$

which is incompatible with the Froude similitude given by Eq. (6-12).

### Forces of water pressure

The forces of water pressure which are always normal to a surface, can be expressed by:

$$F_n = pL^2 \qquad (6\text{-}17)$$

where p is the water pressure and $L^2$ is the representative surface. In similitude this gives:

$$\frac{F_{np}}{F_{nm}} = \frac{p_p}{p_m} \lambda^2 \qquad (6\text{-}18)$$

which, with Eq. (6-9), gives the pressure scale ratio:

$$\frac{p_p}{p_m} = \lambda \tag{6-19}$$

The ratio of inertia forces to pressure forces corresponds to the well known Euler number, $Eu = V/\sqrt{p/\rho}$.

## Ice friction forces

The friction of ice on other ice or on foreign substances can be expressed by:

$$F_t = \mu F_n \tag{6-20}$$

where $F_t$ and $F_n$ are the tangential and normal forces and $\mu$ is the static or dynamic friction factor. It can readily be seen, from similitude of the forces, that the friction factor has to be the same in the model as in the prototype.

## Internal ice forces at collapse of an ice sheet

If brittle fracture of a floating ice sheet of thickness h is considered for the prototype and the model and if also, in both cases, the failure can be predicted from elastic theory, three types of forces have to be taken into account corresponding to possible modes of failure:

Shear failure forces:

$$F_t = \tau h L C_1 \left(\frac{a}{L}, \frac{b}{L}, \frac{c}{L}, \nu\right) = C_1' G \gamma hL \tag{6-21}$$

Crushing failure forces:

$$F_c = \sigma_c h L C_2 \left(\frac{a}{L}, \frac{b}{L}, \frac{c}{L}, \nu\right) = C_2' E\varepsilon hL \tag{6-22}$$

Flexural failure forces:

$$F_b = \sigma_f \, h^2 \, C_3\left(\frac{a}{\ell}, \frac{b}{\ell}, \frac{c}{\ell}, \nu\right) \tag{6-23}$$

where $F_t$, $F_c$ and $F_b$ are the forces producing failure; $\tau$, $\sigma_c$ and $\sigma_f$ are the shear, compressive and flexural strength of ice: L, a, b and c are representative lengths (the last three for conditions of loading), $\nu$ is Poisson's ratio, G and E the rigidity and elastic modulus of the ice, $\gamma$ and $\dot{\varepsilon}$ the strains; and $\ell$ is a characteristic length of the ice sheet that is related to the position of the line of flexural fracture:

$$\ell = \sqrt[4]{\frac{E\,h^3}{12\,\rho\,g\,(1-\nu^2)}} \tag{6-24}$$

The ratio of crushing fracture forces or shear forces in the prototype and model should correspond to the ratio of gravity forces given by Eq. (6-9) if:

$$\lambda_d = \lambda^3 = \frac{E_p}{E_m}\lambda^2 = \frac{G_p}{G_m}\lambda^2 \tag{6-25}$$

The ratio of inertia forces to elastic forces leads to the Cauchy number, Ch:

$$Ch = \frac{V^2\rho}{E} \tag{6-26}$$

Thus, for the same fluid, with (6-25):

$$\frac{E_p}{E_m} = \frac{G_p}{G_m} = \lambda \tag{6-27}$$

and with Eq. (6-24):

$$\frac{\ell_p}{\ell_m} = \lambda \tag{6-28}$$

the correct simulation of the characteristic length is specially important because it corresponds to the correct reduced size of ice pieces broken by flexion in the model as in the prototype.

The mechanical similitude also requires with Eqs (6-21), (6-22) and (6-23) that:

$$\frac{\tau_p}{\tau_m} = \frac{\sigma_{cp}}{\sigma_{cm}} = \frac{\sigma_{fp}}{\sigma_{fm}} = \lambda \qquad (6\text{-}29)$$

Surface tension forces

Sometimes surface tension forces might become important in an ice model. These forces have the form:

$$F_s = \sigma_s L \qquad (6\text{-}30)$$

with basic Eq. (6-9), the similitude is then given by:

$$\frac{\sigma_{sp}}{\sigma_{sm}} = \lambda^2 \qquad (6\text{-}31)$$

The ratio of inertia forces to surface tension forces leads to the Weber number, $W = V^2L/\sigma_s$.

## 6.1.2 - COMPLETE SIMILITUDE

Complete similitude is possible for certain types of ice models, mainly those which requires only hydrodynamical similitude.

For complete hydromechanical similitude, Froude similitude will apply when all water friction forces and hydrodynamical forces are reduced by Eq. (6-12). The model scale should be large enough to have no Reynolds' effect. Furthermore the solid friction coefficient of ice in contact with boundaries should be the same in model and prototype. Finally the Weber effect should be negligible in the model as in the prototype, as at least be scaled down properly.

The basic relations for Froude similitude are then:

$$\left.\begin{aligned} \lambda_k &= \sqrt{\lambda} \quad & T_p/T_m &= \sqrt{\lambda} \\ a_p/a_m &= 1 \quad & F_p/F_m &= \lambda^3 \end{aligned}\right\} \qquad (6\text{-}32)$$

With a breakable ice in the model the complete hydro-mechanical similitude should give:

$$\left.\begin{aligned} &\frac{\sigma_{cp}}{\sigma_{cm}} = \frac{\sigma_{fp}}{\sigma_{fm}} = \frac{\tau_p}{\tau_m} = \\ &\frac{E_p}{E_m} = \lambda, \; \nu = 1, \; \mu = 1 \end{aligned}\right\} \qquad (6\text{-}33)$$

When only unconsolidated ice pieces have to be scaled down, it is possible to have practically full similitude. Two factors have to be closely evaluated: the important surface tension effect on the model ice pieces and the correct solid friction effect between these pieces.

In general, it has not been feasible to use in a practical manner continuous materials, simulating ice, with flexural strengths under $0.5 \times 10^5$ Pa. Below this value saline ice becomes too ductile (Schwarz, 1977) and synthetic ice very difficult to control. Because natural ice has a flexural strength between $5 \times 10^5$ to $15 \times 10^5$ Pa, very large models with scale $\lambda < 20$, would have to be built for complete similitude and may be often too costly to operate. Thus, generally, some distortion or special interpretation has then to be introduced for the use of models that include a solid ice cover.

## 6.1.3 - DISTORTED SIMILITUDE

### 6.1.3.1 - GENERAL EQUATIONS

Distortion is widely used in hydraulic modeling, largely for wide rivers when the ratio y/B is small, where y is the water depth and B the river width. Let us call the hydraulic distortion D:

$$D = \frac{y_m}{y_p} \cdot \lambda \tag{6-34}$$

We will add another type of distortion K, the ratio of the scale of ice thickness h to the horizontal length scale (Michel, 1975):

$$K = \frac{h_m}{h_p} \cdot \lambda \tag{6-35}$$

where both D and K are bigger than one.

We will now distinguish between the ratio of all forces discussed in 6.1.1, acting either on water or on ice. A prime sign will show the ice forces:

Gravity forces

$$\frac{F_{wp}}{F_{wm}} = \frac{\lambda^3}{D} \tag{6-36}$$

$$\frac{F'_{wp}}{F'_{wm}} = \frac{\lambda^3}{K} \tag{6-37}$$

Inertia forces

$$\frac{F_{ip}}{F_{im}} = \frac{\lambda^3}{D} \frac{a_p}{a_m} \qquad (6\text{-}38)$$

$$\frac{F'_{ip}}{F'_{im}} = \frac{\lambda^3}{K} \frac{a_p}{a_n} \qquad (6\text{-}39)$$

Friction forces and vertical hydrodynamic forces

$$\frac{F_{fp}}{F_{fm}} = \frac{F'_{fp}}{F'_{fm}} = \lambda^2 \ \lambda_k^2 \qquad (6\text{-}40)$$

Horizontal hydrodynamic forces

$$\frac{F_{hp}}{F_{hm}} = \lambda^2 \ \lambda_k^2/D \qquad (6\text{-}41)$$

$$\frac{F'_{hp}}{F'_{hm}} = \lambda^2 \ \lambda_k^2/K \qquad (6\text{-}42)$$

Flexural ice forces

$$F'_{bp}/F'_{bm} = \lambda^2 \left(\frac{\sigma_{fp}}{\sigma_{fm}}\right) \Big/ K^2 \qquad (6\text{-}43)$$

$$\ell_p/\ell_m = \lambda$$

Other ice forces

$$\frac{F'_{\tau p}}{F'_{\tau m}} = \lambda^2 \left(\frac{\tau_p}{\tau_m}\right) \Big/ K \qquad (6\text{-}44)$$

$$\frac{F'_{cp}}{F'_{cm}} = \lambda^2 \left(\frac{\sigma_{cp}}{\sigma_{cm}}\right) \Big/ K \qquad (6\text{-}45)$$

### 6.1.3.2 - DISTORTED HYDRAULIC MODELS

It is well known that for large rivers where the width is larger than five to ten times the water depth, distortion can be used (Yalin, 1971).

Relations (6-36), (6-38) and (6-40) will give again the Froude similitude on the vertical scale:

$$\lambda_k = \sqrt{\lambda/D} \qquad \frac{T_p}{T_m} = \sqrt{\lambda/D}$$

$$\frac{a_p}{a_m} = 1 \qquad \lambda_d = \lambda^3/D \tag{6-46}$$

It is interesting to note that relation (6-41) is not scaled as such. This only shows that the horizontal hydrodynamic thrust is not scaled down to the horizontal scale. A singular structure should be undistorted in the model if the hydrodynamic forces are measured, and the results should be interpreted using the vertical scale.

The distortion does impose however a reevaluation of the roughness to be used on the model to correctly simulate the head losses in nature (Yalin, 1971).

### 6.1.3.3 - UNCONSOLIDATED ICE ACCUMULATIONS

Because an unconsolidated ice accumulation takes into account the same forces as a simple hydraulic model it should then be possible to use the same distortion.

Let us consider the equilibrium of a jam of unconsolidated ice pieces given by Eq. (4.102):

$$\frac{d\sigma_n}{dx} + \frac{\sin 2\psi \; K\psi}{B} \sigma_n - \frac{\tau}{h} = 0 \qquad (4\text{-}102)$$

with
$$\sigma_n h \sim \frac{h^2}{2} \qquad (4\text{-}116)$$

where $\sigma_n$ is the normal stress inside the accumulation, $K\psi$, the coefficient of active thrust in the accumulation with an angle of internal friction $\psi$, B the width of the accumulation, $\tau$ the tangential stress given by the sum of Eqs (4-109) and (4-110):

$$\tau = \gamma \; (1 - e) \; hS + \gamma \; YS$$

where e is the porosity of the accumulation and S, the slope of the water line.

By writing these equations both for prototype and model it is most interesting to see that a distortion D is compatible with the solution of these equations, if the same distortion on the ice thickness is also used, such that K = D, then:

$$\left.\begin{aligned} \lambda_k &= \sqrt{\lambda/D} \qquad & \frac{T_p}{T_m} &= \sqrt{\lambda/D} \\ \frac{a_p}{a_m} &= 1 \qquad & \lambda_d &= \lambda^3/D \\ \frac{\sigma_{np}}{\sigma_{nm}} &= \lambda/D \qquad & \frac{\tau_p}{\tau_m} &= \frac{\lambda}{D^2} \end{aligned}\right\} \qquad (6\text{-}47)$$

For the same angle of internal friction on model and prototype, the accumulation on the model will establish itself on the model at a thickness smaller than that obtained with the horizontal scale only, by a factor equal to the distortion of the model.

## 6.1.3.4 - DISTORTION WITH A SOLID ICE COVER

Let us keep the basic hydraulic similitude of gravity forces, Eq. (6-36):

$$\lambda_d = \lambda^3/D$$

We then have the basic Froude similitude with Eqs (6-38) and (6-40):

$$\lambda_k = \sqrt{\lambda/D} \qquad a_p/a_m = 1$$

If K is different than D, there is a distortion K/D in the reproduction of forces of gravity and inertia on ice from Eqs (6-37) and (6-39), these forces being reduced by a factor K/D on the model. The similitude of horizontal hydrodynamic forces is also distorted by a factor of K/D, Eq. (6-42).

We do find that the failure forces give the following scales with Eqs (6-43), (6-44), (6-45) and (6-24):

$$\left(\frac{\sigma_{fp}}{\sigma_{fm}}\right) = \frac{\lambda K^2}{D}$$

$$\left(\frac{\sigma_{cp}}{\sigma_{cm}}\right) = \left(\frac{\tau_p}{\tau_m}\right) = \lambda\ K/D \qquad (6\text{-}48)$$

$$\frac{E_p}{E_m} = \lambda K^3 \frac{(1-\nu^2)_p}{(1-\nu^2)_m}$$

The scales are most interesting when K is the smallest. For K = 1:

$$\frac{\sigma_{fp}}{\sigma_{fm}} = \frac{\sigma_{cp}}{\sigma_{cm}} = \frac{\tau_p}{\tau_m} = \lambda/D \qquad (6\text{-}49a)$$

$$\frac{E_p}{E_m}\frac{(1-\nu_m^2)}{(1-\nu_p^2)} = \lambda \qquad (6\text{-}49b)$$

A study we made (Michel and Abdelnour, 1975) shows that a solid ice cover breaks for conditions such that the following non-dimensionnal numbers can be used:

$$\frac{V}{\sqrt{\sigma_f/\rho}} = A\left(\frac{h}{B}\right)^n \qquad (6\text{-}50)$$

Both numbers reduce to model numbers with the preceding scales:

$$\frac{V_m}{\sqrt{\sigma_{fm}/\rho}} \cdot \frac{\lambda/D}{\lambda/D} = A\left(\frac{h_m}{B_m}\right)^n \qquad (6\text{-}51)$$

Thus a distortion is possible and interesting for the behavior of the solid part of a river ice cover. The gravity forces on the ice have to be small compared to the failure forces, which is usually the case, and the acceleration forces are negligible on the ice when there is a steady state of movement. This could be accepted in most cases of model representation.

With a value of around $\sigma_{fm} = 0.5 \times 10^5$ Pa and a scale of about $\sigma_{fp}/\sigma_{fm} = 20$ a distortion of D = 5 would give an acceptable horizontal scale of 100 for the model. It is even more interesting to observe that the Young's modulus is easier to scale on the model, because the $E/\sigma_f$ then becomes distorted:

$$\frac{Em}{\sigma_{fm}} = \frac{Ep}{\sigma_{fp}} \cdot \frac{1}{D} \qquad (6\text{-}52)$$

## 6.2 - MODEL TECHNIQUES

### 6.2.0 - INTRODUCTION

Many ice models have given unreliable and unproven results and this has usually been caused by a very poor knowledge of field conditions including a lack of field data. There are still many ice phenomena that are not completely understood and if they are not observed very well at a particular site it is improbable that they will be well simulated in a model.

The first objective of the ice model is indeed to reproduce field conditions under more or less normal conditions, so that the model could then be used with a good degree of confidence to predict the effects of abnormal conditions or those produced by man-made changes. It is not sufficiant to adjust water levels and the bed roughness as in an hydraulic model. The ice phenomena has also to be correctly simulated.

The major variable in nature that can not be simulated in a model is that of thermal exchange. Except in the case of ice cover formation, where the thermal state of the ice varies quickly, the other thermal states during winter conditions or at breakup are constant for a long enough period so their variation could be eliminated in the model simulation. It is however of major importance to know well these ice conditions for a given thermal state so they could be correctly reproduced in the model.

The first and most important decision in river ice modeling is to decide if the simulation of the mechanical properties of ice is needed on top of the usual hydrodynamical similitude. This decision can not be taken lightly as full hydromechanical modeling calls for extended means and a longer testing time. However, in most cases, simple hydrodynamical simulation will not reproduce natural pehnomena and will lead to extensive errors in results and questionnable verification techniques.

For instance, there is little doubt that the simulation of the mechanism of ice cover progression, when the solid ice cover is not reproduced behind the progressing frontal edge, leads to extensive errors. The ice cover thickness and its state of stability is not correctly scaled down and even the mechanism of underhanging dam formation cannot be simulated.

The same is true if an ice accumulation is suddenly formed between large bands of shorefast ice. The thickness equilibrium of an unconsolidated ice pack varies quickly with the pack width, so the shorefast ice has a large influence on ice thickness and hydrodynamic conditions in the central part or the model. A river model reproducing middle winter conditions can hardly be made without full hydromechanical similitude.

Many ice jams, including dry jams, are caused by river singularities where a solid part of the ice cover is holding the unconsolidated ice pack. It is hard to see how this can be simulated without scaling down the breakable ice cover.

When ice forces are measured in a model the mechanical properties of ice should obviously be scaled down.

There are indeed some cases where only the hydrodynamical simulation is needed. This is the case of a pure ice drift problem, the study of the discharge of ice floes, the question of stability of long unconsolidated jams at breakup and the movement of ships in brash ice.

Once a modeling technique has been decided, the model should then be verified. This process usually consist in the following steps:

- adjustment of roughness and verification of water lines without ice
- verification of head losses with simulated ice for known field conditions
- verification of the similitude of ice processes for known field conditions. This last verification will usually be only qualitative but will ascertain that the model is really simulating the natural phenomena.

## 6.2.1 - MODEL MATERIALS

We will review the various materials that are now in use for modeling ice.

### 6.2.1.1 - SOLID PARTICLES AND BLOCKS

The technique of using solid artificial materials to simulate ice floes and ice accumulations has been widely exploited. Many materials have been used for ice modeling: polyethylene, polypropylene, wood, paraffin etc... The density, roughness and average form of the drifting ice can be reproduced but it is impossible to break the ice pieces in the model; so none of its internal properties are modeled.

It would be theoretically possible to have complete hydrodynamical similitude with these materials in an undistorted model if care is taken to destroy the surface tension effect in the model. The model ice pieces would have the same velocity of submergence on the frontal edge, the same velocity of entrainment underneath the cover and even the ice discharge would be simulated to scale.

However the thickness and internal stresses in the model ice accumulations would not be to scale, in the general case, because of the difference in internal friction properties of model and real ice, the accumulations of the latter being usually much more consolidated and strong. The closest acceptable simulation would happen at breakup when there is very little consolidation left between the ice pieces in nature.

A distortion may be useful in many cases, as it reduces the ice thickness accumulation in the model, makes it more stable, and closer to natural and real conditions.

This technique cannot be used if the solid part of the ice cover plays an important role in the process under study. In some cases the solid part is conserved during the process. It may be reproduced by a wooden panel or its edge be simulated by a floating cross-boom etc...

The solid part of the cover does not play an important role in a few phenomena. The study of drifting ice and ice drift control structures can be made in this manner. Many studies of ice at breakup can be made with that technique if long unconsolidated ice jams are formed. The stability of those jams, in itself, is then more important than the resistance of the solid ice cover downstream of the jam. The movement of ships in slush or brash ice can also be simulated with this technique.

One limitation to this technique is obviously when the solid part plays an important role in the phenomenon under study. This is usually the case at freeze-up and during winter when the ice pieces consolidate by freezing behind the frontal edge of the cover and is true of many cases of breakup when the resistance of the solid ice cover plays a dominant role in jam formation.

#### 6.2.1.2 - REAL ICE

Floating real ice is the best material to study the forces exerted by ice sheets, without hydrodynamic similitude. The scales of all the mechanical properties of the material depends on the types of ice which is simulated and its comparison with the corresponding ice in nature. In general, the strength of ice is slighty higher in the laboratory than in nature.

Because there is then no water motion all relations pertaining to the hydrodynamic similitude disappear including the Froude number. There is only mechanical similitude where the geometric undistorted scale is the ice thickness scale:

$$h_p = \lambda \, h_m \qquad (6\text{-}53)$$

and the force scale is given by:

$$\lambda_d = \frac{\sigma_p}{\sigma_m} \lambda^2 \qquad (6\text{-}54)$$

where $\frac{\sigma_p}{\sigma_m}$ is the distorted strength scale.

All boundary conditions should be respected geometrically. Because the strength of ice varies widely with the rate of loading of ice sheets, this factor should also be taken into account in data interpretation.

Testing could also be done in the horizontal direction if the hydrodynamic forces are small and negligible compared to the breaking forces.

Even distorted similitude could not be used with real ice in the general case of hydromechanical similitude. With relation (6-48) we would have:

$$\frac{\sigma_{fp}}{\sigma_{fm}} = \frac{\sigma_{cp}}{\sigma_{cm}} = \frac{\tau_p}{\tau_m} \simeq 1 \simeq \frac{\lambda K}{D} \qquad (6\text{-}55)$$

This means that the vertical scale, as well as the horizontal scale, has to be unity to obtain the Froude similitude.

### 6.2.1.3 - WEAK BREAKABLE MATERIALS

Models in which a solid ice sheet has to be broken, have been used mainly in the past for the design of icebreakers and bridge piers. They are now also used for river studies (Michel *et al.*, 1973).

Various materials are actually in use. Saline ice is used mainly for model studies of icebreaking ships. Synthetic ice is used for river work and ice interaction with structures.

Saline ice was first used in the Leningrad ship model basin (Shvayshteyn, 1959) by freezing water of high salinity (ice: 10-20$^o$/oo). Depending on the rate of freezing this produces ice of very high salinity and low strength. This method is used by all ship model basins because of its convenience and low cost. Under normal freezing conditions the high salinity ice which is produced is very ductile. Thus the ratio $E/\sigma_f$ of this saline ice drops to 200 to 500, when $\sigma_f \leq 0.5 \times 10^5$ Pa, while in nature this ratio is in the range of 2000 to 5000 (Schwartz, 1977). This normally limits the scale to $\lambda \leq 20$, which leads to very large models and difficult testing problems. Otherwise some corrections have to be brought to the term taking into account the contribution of the failure strength to the resistance of the ship. Arctec Inc. has developped a cryogenic freezing system, with liquified N , which produces very fine size crystals of high salinity ice, which increases the range of use of saline ice (Edwards and Lewis, 1970).

At the origin, synthetic ice was made of paraffin weakened with oil which was melted and then poured over water This material was extremely ductile and the ratio $E/\sigma_f$ was so small that it was impossible to correctly simulate ice phenomena at failure. One synthetic ice (Michel 1969) is a wax made of various components that enables the adjustment of the ratios $E/\sigma_f$, $\sigma_o/\sigma_f$ and $\mu$ (solid friction) in a suitable range for simulating different types of ice. The wax is mixed hot and poured over a water bassin to produce ice sheets or ice pieces of different forms. The $\sigma_f$ value can be adjusted down to $0.3 \times 10^5$ Pa with correct similitude. The synthetic ice cannot be re-used many times and is costly to make and use. Another synthetic ice is a plaster of Paris weakened with various components (Tryde 1975). The material can be produced in a range starting from $\sigma_f \geq 1.0 \times 10^5$ Pa. It has to be poured into separate forms and then laid on water. It can be used only one time.

The best ice modeling is made if no distortion is used as may be the case for the study of a particular structure in ice. Because it is hardly possible to operate systematically with a weak material having a flexural strength smaller than 0.3 x $10^5$ Pa, the undistorted scale is then only in the range of 15 to 40. The most important mechanical property to scale next to the flexural strength is usually the Young's modulus because it determines the size of the broken pieces on the model. All the properties of weak materials are very difficult to measure because of their range of values and special testing techniques have to be used. One of the major factor affecting the behavior of weak materials is the large variation of properties with loading rates and temperatures. It should be ascertained that the scale values will be tested at the correct loading rate and temperature in the model.

Distorted models could also be used and the distortion helps in simulating the Young's modulus because the required value is smaller (Eq. 6-52). The properties of the weak material should be adjusted as close as possible to target values for real ice in nature. In many cases, one property will not be correctly scaled down. This might be acceptable if it has a secondary effect which is known. Simulation of a breakable ice cover in a river may be carried in conjunction with the use of unbreakable ice pieces feeding the cover to simulate drifting ice floes. All type of modeling could then be made including freeze-up, winter operation and breakup. Weak materials have first been used for icebreaker testing and interaction of ice with structures where they still have all their full advantage.

## 6.2.2 - MODELS OF ICE JAMS

In many models of ice jams, the internal collapse of the continuous ice sheet need not be reproduced and a simple hydrodynamic model can then be used. The ice is then essentially considered as individual unbreakable pieces whose hydrodynamic behavior is simulated. The Froude similitude applies well and the main difficulty is choosing model ice that represents the dimensions, forms and surface friction effects of the floes.

A large ice model of that type was built to represent a stretch of more than 5 miles of the St. Lawrence River at scales of 1/240 horizontally and 1/150 vertically, (Hausser, 1964). The aim of the study was to determine the effect of ice jams in Montreal Harbour if new land were reclaimed for Expo 67. This site was known to be subject to frazil and slush ice jams up to about 10 m in depth with water level rises approximately 6 to 8 m above summer levels. The frazil floes, slush and floes were represented on the model by small polyethylene grains whose surfaces were treated to get a wetting effect and increase the roughness. The model was built according to standard hydraulic laboratory procedures for fixed bed models and the bed roughness was adjusted for clear water conditions. The model was then fed with ice by automatic feeders whose rates were simulated according to computed values in nature. The ice runs were then made on the model, simulating natural conditions, and the results of water level measurements on the model and in the prototype are shown in Fig. 6.1. In nature the jams form during cold weather and there is always some bonding by thermal exchanges between ice floes at the water surface. On the model there is no cohesion at all between the ice pieces, and the pack is much less resistant to thrust and shoves than in nature. The ice accumulations are thus bigger on the model, the head losses increase and the water levels are higher. The model thus errs on the safe side and gives the worst limiting conditions.

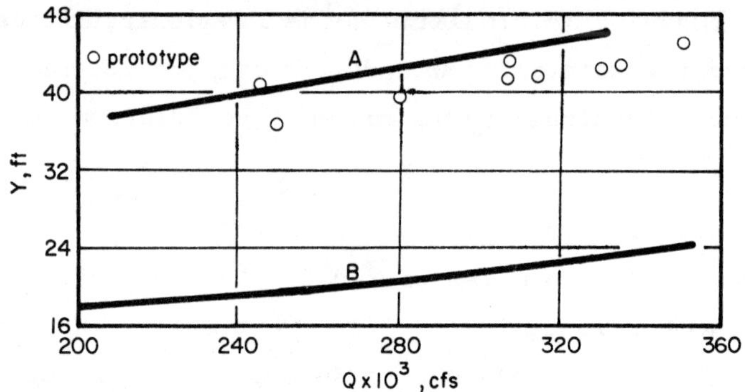


*Fig. 6.1. Water level and discharge for model and prototype. A) with ice, B) without ice.*

Another example is a model built to study the flow of frazil ice in a river in Iceland (Carstens, 1968). An intake to a power plant had to be designed that would get water free of ice while the heavy frazil ice run would be disposed of at the top of a regulating weir. Polyethylene foam of very small size was used with success to represent the heavy balls of aggregated particles of frazil.

Ice models are most useful to study floods caused by ice during the spring breakup of rivers. For those conditions there is pratically no thermal influence during the relatively short period of jam formation, and model results may be expected to give a good approximation of the truth. Fig. 6.2 shows such a model in operation where a temporary ice-delaying structure made of a boom and a control weir had to be studied. Conditions of ice accumulation and dry jam formation were studied in the model. Model ice consisted of 0.6 cm thick by 2.5 cm square polyethylene blocks, surface treated, which simulated the average characteristics of the ice floes at breakup for this site, at a scale of 1/120. On of the difficult aspect in this type of study was the representation of the solid ice sheet being broken up and pushed back under the thrust of the unconsolidated ice accumulation. On this model the problem was very local and the solid sheet in the pond formed by the weir was just simulated by a floating wooden boom. This boom was first placed at the upstream end of the pond. When the accumulation became unstable and was shoved under the boom it was moved artificially down, by trial and error, to a new section where the pack ice was again stable. The final retaining place was, of course, the position of the designed structures at the top of the weir. This process is indeed quite crude and there is no doubt that there is a lot of room of improvement in the techniques of ice model studies of this type.

*Fig. 6.2. Model of a jam fed by ice floes. Laval University, 1967.*

## 6.2.3 - MODELS OF ICE COVERS IN RIVERS

In some cases, river models have to be made where the effect of the solid ice cover has to be taken into account. This is the case, for example, of an ice jam which is primed by a solid ice cover retained by bridge piers. This might be the case where flooding, with large river discharges, occurs during the winter months when the ice cover is still very solid and strong. This is also the case when winter navigation is foreseen through the solid cover.

Such studies may be made by simulating the solid ice with synthetic ice reducing the strength properties to scale. A typical approach can be seen from a general study of the breakup of a solid ice cover made in a laboratory flume (Michel and Abdelnour, 1975).

A model ice wax was used for the tests which simulated to scales of 1/20 to 1/40, the properties of real ice. It had the same density as ice. Its flexural strength was measured with small cantilever beams, ten times for each test as shown in Fig. 6.3a. The average strength varied from 0.2 to $0.5 \times 10^5$ Pa from one test to the other. The Young's modulus was measured three times for each test, as shown on Fig. 6.3b, and it varied from $3 \times 10^7$ to $16 \times 10^7$ Pa. The ratio of the preceding properties varied from 700 to 5000 and this would cover a good range of possible combinations for the behavior of real ice. The wax was mixed in a tank and poured on warm water in a flume so it would spread evenly and form a uniform ice cover, 8 m long. The standard deviation on measured thickness was 5 to 10% in any one test. The Manning's hydraulic roughness of the ice and canal varied between 0.022 and 0.033.

Each test consisted essentially in increasing the discharge until complete failure of the ice cover, leading to accumulation of the pieces in front of the retaining structure. The failed ice in one test is shown in Fig. 6.3c.

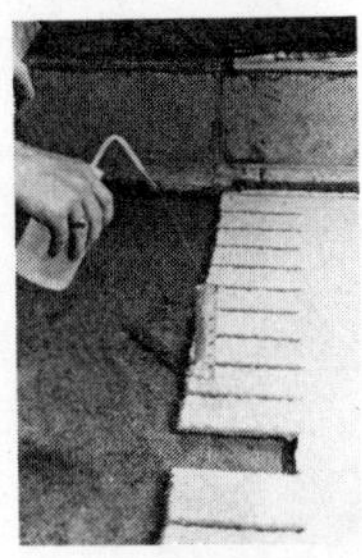

*Fig. 6.3. Model simulation of the breakup of a solid ice sheet in a flume with synthetic ice. a) measurement of flexural strength b) measurement of elastic modulus c) pieces of ice after breakup (Michel and Abdelnour, 1975).*

A scale model including the solid ice cover was built to study the effect of the construction of a bridge on the Bécancour River, in Quebec, on flooding by ice jams (Michel *et al.*, 1973). The bridge was designed with many piers and was located in a natural section for the formation of jams. This model had broad objectives because it also included a movable bed in the area of the bridge, to study scouring action at breakup. The horizontal scale was choosen as 1/160 and the vertical scale as 1/40. The distortion of 4 was found more appropriate to simulate the $E/\sigma_f$ ratio as shown by Eq. 6-52.

The model ice was produced very weak, with $\sigma_f = 0.15 \times 10^5$ Pa in order to simulate real ice of $6 \times 10^5$ Pa, flexural strength, according to Eq. 6-52.

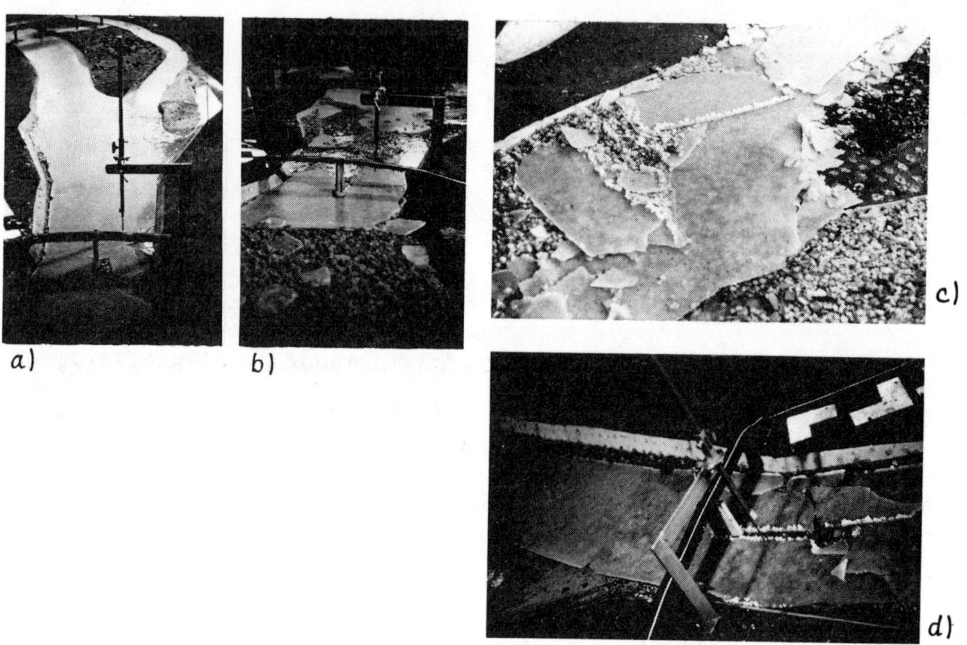

*Fig. 6.4. Hydraulic model study of breakup of ice cover on the Bécancour river (Michel et al., 1973). a) ice cover simulated with synthetic ice b) upstream edge fed with ice run made of simulated ice pieces c) breakup of the solid part of the cover d) ice push through a bridge pier.*

The operation of the model is shown in Fig. 6.4. A section of the solid ice cover is simulated with synthetic model ice in the reach of low velocity flow, as is found in nature at breakup time (Fig. 6.4a). The ice run coming from rapid flow sections of the river, upstream, is then let to move to the solid cover. The ice pieces are simulated by solid polyethylene blocks and large pieces of breakable model ice (Fig. 6.4b). Breakup then occurs as in nature, and flood levels are recorded at the upstream village on the model (Fig. 6.4c). Finally, bridge pier of various forms and location were set up in the model to study their effect on flood levels. On Fig. 6.4d is shown the failure of the ice cover by its crushing through one bridge pier.

## 6.2.4 - MODELS OF STRUCTURES IN ICE

Very simple structures can be tested in the laboratory with natural ice and the scale effect can be evaluated. This is the case of different types of simple indentors, either perpendicular or inclined when related to the ice sheet, and this type of testing was discussed in Chapter 4.

However for complex structures and special ice configurations, like pressure ridges, the effect of interaction of the ice with the structure cannot be modeled at a scale with natural ice. It is for those conditions that synthetic model ice is a most useful tool.

Fig. 6.5 shows a sequence of tests made to obtain the thrust exerted by a moving ice field against a conical structure (Michel, 1972). The synthetic ice sheet is 4.7 cm thick pushed against a cone 72 cm in diameter at water level. The material has a flexural strength of $0.2 \times 10^5$ Pa which corresponds to $10 \times 10^5$ Pa for real ice at a scale of 50. The model ice behaves very much like real ice in the ductile range. Fig.

6.6 shows the yield strength in compression of one mix of model ice compare to real ice at a scale of 125. The density of the model ice is 0.9, its static friction coefficient close to 0.2 on plexiglass and 0.11 on a teflon coated surface. The yield point of the model material occured at a strain of about 0.6% which is quite identical to most types of real ice. The results showed that the horizontal thrust was higher than that which could be computed by Korzhavin's formula (4-81) and this could easily be explained by the added force needed to ride-up the broken ice pieces on the cone.

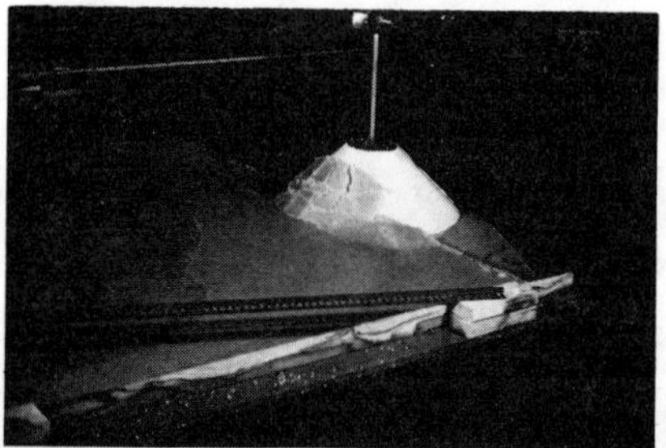

*Fig. 6.5. Sequences of model tests showing ride-up of ice sheet made of synthetic ice, on a cone.*

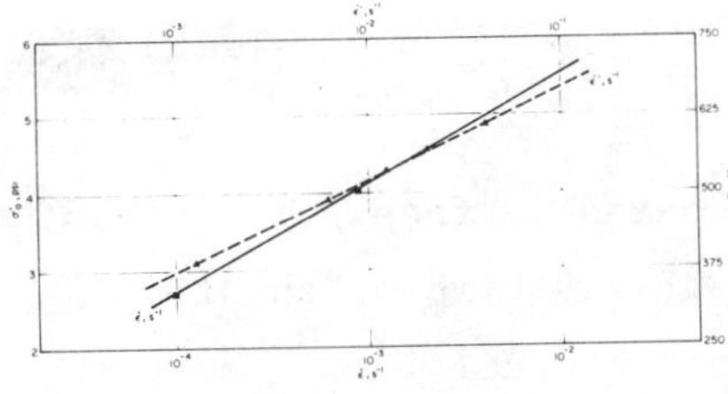

*Fig. 6.6. Comparison of yield strength of one mix of synthetic ice with one type of real ice in function of strain rates. Full lines (S2 ice at -10°C), dash line (model ice X-37).*

Fig. 6.7 shows one application of the use of model ice to study the design of a wharf (Michel and Ouellet, 1969). This wharf had three icebreaking piers as can be seen in Fig. 6.7a, serving as deflector for the ice to protect the berthing area. Fig. 6.7b shows the impact of the model ice on the outward protecting pier.

*Fig. 6.7. Model of a wharf having three icebreaking cells. Model ice interaction with the ship (a) and the structure (b) is shown.*

Synthetic model ice has also been used to study ice pile-up on the beaches of the artificial sand islands built in the MacKenzie Delta. Fig. 6.8 shows such a pile-up. The model was used to design a system to generate pile-up.

*Fig. 6.8. Model study of an ice pile-up on a sand beach (Courtesy of Arctec Canada Ltd).*

6.2.5 - MODELS OF SHIPS IN ICE

At the present time, saline ice is most generally used to model the action of ships in ice. Its advantage over synthetic ice is its lower cost for repetitive testing, more uniform ice properties in large tanks and automatic adjustment of friction properties. Its main drawback is the relatively low value of the $E/\sigma_f$ ratio, as compare to natural ice or synthetic ice.

In the mode of continuous ice breaking, there are many phenomena which intervene during the icebreaking processes and which must be modeled. Thus the resistance of the ship, Eq. (5-25), is given by:

$$R = C_b \sigma_f Bh + C_s \rho g Bh^2 + C_i \rho BhV^2 + C_f \rho \sqrt{g} BVh^{3/2} \quad (5\text{-}25)$$

The first term represents the force needed to break the ice, the second that needed to submerge the pieces and overcome frictional forces, the third is the inertia term and the last a solid type of friction of ice on the hull.

With the dimensionless resistance.

$$r = \frac{R}{\rho g \; Bh^2} \tag{5-10}$$

The equation reduces to:

$$r = C_b \frac{\sigma f}{\rho gh} + C_s + C_i \; F_r^{\;2} + C_f \; Fr \tag{5-26}$$

This equation is the basis for ice modeling in ship tanks. A pragmatic approach has shown that if the regression of r on both the strength term $\sigma_f / \rho gh$ and the Froude number $F_r$ is the same in the model and prototype than the modeling is correct, even if some properties of the model material are not perfectly to scale (Edwards and Lewis, 1970).

An extensive study of ice modeling techniques was done by Edwards *et al.*, 1976. He compared model studies at scales of 1/48 and 1/36 with full scale data of the Louis St-Laurent. The regression analysis for the full scale data obtained in the Canadian Arctic, gave:

$$r = 4.242 + 0.05033 \frac{\sigma_f}{\rho gh} + 8.948 \; F_r \tag{6-56}$$

This equation is shown in dash line in Fig. 6.9. The results of the model study can be discussed in the following manner. Let us cite directly Edwards *et al.*, (1976): "The purpose of the comparison was to examine whether there was a sufficiently better comparison with the full-scale data using the smaller-scale (larger) model to justify the reduced number of data points which can be obtained using it rather than the larger-scale (smaller) model. All of the full-scale data were collected using the method of ramming into an ice sheet the thickness of which was greater than the maximum continuous thickness capability of the propulsion plant. Therefore, it was deemed necessary to test the Louis S. St-Laurent in ramming-type tests in model scale similar to those from which the full-scale data were collected, as well as in routine constant-velocity towing tests. The full-scale range of the test data in thickness was from 0.3 to 2.0 m full-scale, in strength from $3.2 \times 10^5$ to $16.8 \times 10^5$ Pa, in speed from 0.2 to 10.7 knots, and in hull-ice friction from 0.08 to 0.48.

The first series conducted was constant-velocity towing tests. There are two scale models. These data from both of the scale models have been plotted on a dimensionless basis in Fig. 6.9. The data have been grouped according to eight separate friction groups, four for each model. Black data points represent the 1/36 th scale model while open data points are for the 1/48 th model. On the ordinate the resistance is plotted while along the abscissa is plotted the product of dimensionless strength and the thickness Froude number, which, during regression analysis, was found to be the only term that contributed to resistance with a statistical level of confidence greater than 95 percent. The predictor equations is plotted in full lines on Fig. 6.9 for three separate values of friction factor. Also plotted against the data and the predictor equation is the full-scale regression equation obtained from the full-scale test data in dash line.

The data collected during these tests were combine into dimensionless groups of elastic modulus over strength, $E/\sigma_f$; the thickness Froude number $F_r$; the flexural strength divided by weight density times thickness, $\sigma_f/\rho gh$; and the dimensionless resistance r. The data were first separated into groups of like hull-ice friction. Each of these groups was subjected to stepwise multiple regression analysis to determine what terms might be important in predicting the full-scale resistance. It was found in this multiple regression analysis that only two terms were significant at the 90 percent level of confidence. These are the thickness Froude number times dimensionless strength and the thickness Froude number squared. All of the eight data sets collected were force-fit to an equation containing both of these groups. A coefficient was obtained for each term in the hypothetical predictor equation, giving a total of eight separate values for each coefficient, corresponding to eight values of friction. Each of these coefficients, predictor equations can be constructed which are good for each of the individual models and one which represents both models combined, and three are shown on Fig. 6.9 for three values of friction coefficients.

Following this program, a series of non-self-propelled ramming tests was held. It was found from these tests that the results obtained from reducing the inertial ramming tests produced data equivalent to those collected during constant-velocity towing tests. The predictor equation includes values for hull-ice friction and ice strength measured during the ramming tests. At this one strength and one ice thickness, to have as much as 50 percent scatter in the magnitude of resistance indicates much poorer quality than has routinely been obtained through the constant-velocity method of testing. We therefore see that, although the results agree on a general basis and indicate that there are no strange phenomena happening during the ramming process, this method of collecting data does not hold as much promise in the model scale as the normally used constant-velocity towing method.

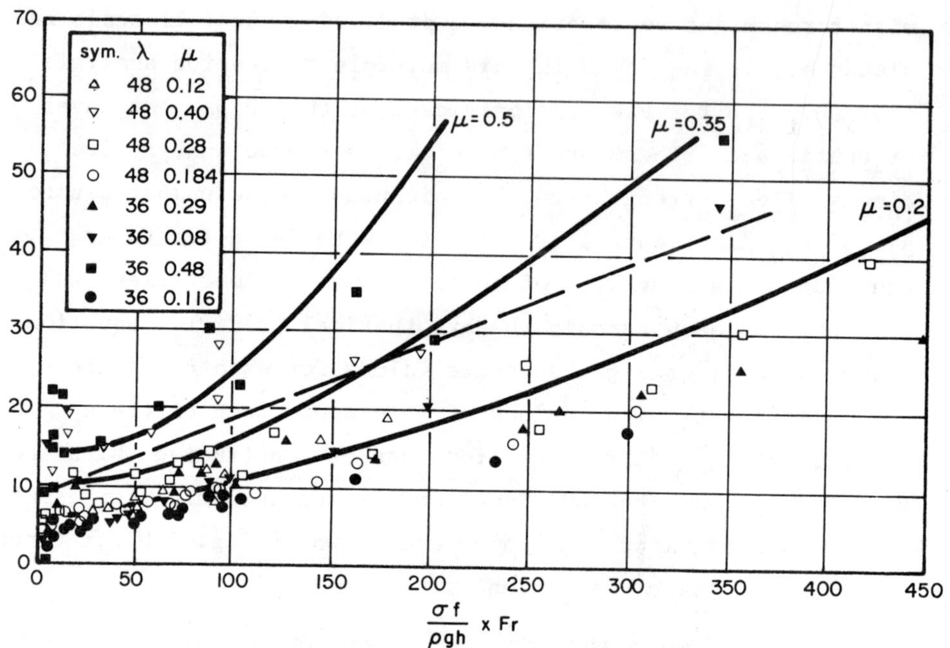


*Fig. 6.9. Dimensionless resistance model test data in saline ice. The full lines are predictor equation for model data. The dash line is the full-scale regression line.*

The results of these tests show that, at the Froude numbers normally associated with the limiting capability of the ship, the results given by the 36 th and 48 th scale models are nearly identical. As speed increases and we move toward thinner ice thicknesses, the predictor given by the smaller test model (1/48 th scale) gives a more conservative resistance. The effect of strength can be seen by this figure to be practically negligible with regard to the prediction of either model."

Another model analysis in saline ice by Schwartz (1977), indicates that when the flexural strength $\sigma_f$ is smaller than $0.5 \times 10^5$ Pa, Young's modulus is not scaled properly and this causes a much too small breaking portion of the total resistance.

*Fig. 6.10. Study of berthing of a ship along an Arctic terminal. a) ship penetrates ahead b) ship penetrates astern. (Courtesy of Arctec Canada Ltd)*

Not only standard tests of icebreaking resistance can be carried on in an ice model tank, but also other types of testing. Fig. 6.10 shows a study of berthing of a ship for the design of "Brithport inlet terminal" in the Canadian Arctic. In Fig. 6.10a the ship penetrates ahead, in Fig. 6.10 b it penetrates astern when the ice is very thick at the end of winter.

An interesting development in Canada is the use of air cushion vehicles to break ice; either to form a broken ice path for a ship or to destroy river ice jams. Such vehicles may by used in tandem like icebreakers. Fig. 6.11 shows a model study of ACV vehicles used to break saline ice.

*Fig. 6.11. Model study of ACV vehicles working in tandem to break saline ice (Courtesy of Arctec Canada Ltd).*

## REFERENCES

Atkins, A.G. (1975) - "Icebreaking modeling" J. of Ship Reearch, Vol. 19, No.1.

Carstens, T. (1968) - "Hydraulics of river ice" La Houille Blanche, Vol. 8, p. 271-84.

Edwards, R.Y., Lewis, J.W. (1970) - "Modeling the motion of ships through polar ice fields using unconstrained, self-propelled models" Proc. I.A.H.R. Symposium "Ice and its action on hydraulic structures" Reykjavik, Iceland, paper 4.14.

Edwards, R.Y., Major, R.A., Kirn, J.K., German, J.G., Lewis, J.W., Miller, D.R. (1976) - "Influence of major characteristics of icebreakers hulls on their powering requirements and maneuvrability in ice" Paper presented to 'The Society of Naval Architects and Marine Engineers' Ann. meeting, Nov. 11-13.

Foundation Eng. Co. of Canada (1949) - "Model study of the Bromptonville ice jam" Internal report.

Hausser, R. (1964) - "Hydraulic studies for enlarging Notre-Dame and St. Helen Islands" L'Ingénieur, June, p. 44-50.

Lewis, J.W. (1972) - "Ship model ice resistance experiments" Techn. Report 0029, Arctec Inc., Oct.

Michel, B. (1969) - "Nouvelle technique de simulation totale des glaces flottantes" L'Ingénieur, 248, pp. 16-20.

Michel, B. (1970) - "Ice Modeling in Hydraulic Engineering" Proc. I.A.H.R. Ice Symposium, Reykiavik, Paper 4.13.

Michel, B. (1970) - "Ice pressure on engineering structures" U.S.A. CRREL Monograph III-Blb, 71p.

Michel, B. (1972) - "The use of model ice to simulate ice action on offshore structures" Internal report to A.P.O.A.

Michel, B. (1975) - "Techniques of ice modeling including distortion" Report GCT-75-01-01, Civil Eng. Dept, Laval University.

Michel, B., Ouellet, Y. (1969) - "Hydraulic model study of Golden Eagle Wharf" Report T-11, Civil Eng. Dpt. Laval University.

Michel, B., Llamas, J., Verrette, J.L. (1973) - "Modèle de la rivière Bécancour" Report GCT-73-05-0, Civil Eng. Dpt., Laval University.

Michel, B., Abdelnour, R. (1975) - "Break-up of a river solid ice cover" Proc. Third Int. Symposium on Ice Problems, I.A.H.R. Hanover, U.S.A., p. 253-261.

Nogid, L.M. (1959) - "Model representation of a ship going through a continuous ice field or pack ice" Trans. of the Leningrad Ship building Institute, No. 45.

Schwartz, J. (1977) - "New developments in modeling ice problems" Proc. POAC 77, p. 45-61.

Shvayshteyn, Z.I. (1959) - "Laboratory for studying ice and testing models of icebreakers and ice-strengthened vessels" Problemi Arktiki, No. 2.

Tryde, P. (1975) - "Intermittent ice forces acting on inclined wedges" Proc. Third Int. Symposium on Ice Problems, I.A.H.R., p. 339-344.

Vance, G.P. (1968) - "Model testing in ice" Naval Engineers Journal, April, pp. 259-264.

Vance, G.P. (1975) - "A scaling system for vessels modeled in ice" Paper presented to 'The Society of Naval Architects and Marine Engineers' Ann. Meet., Ap. 9-11.

White, R.M., Vance, G.P. (1967) - "Icebreaker model tests" Naval Engineers Journal, August.

Yalin, M.S. (1970) - "Theory of Hydraulic Models" MacMillan Press Ltd, 266p.

NAME INDEX


Abdelnour, R., 452, 464, 465
Abkowitz, M.A., 410
Ackley, S., 47, 48
Altberg, W.G., 21
Anderson, D.L., 109
Assur, A., 32, 39, 42, 44, 108, 109, 192, 224
Atkins, A.G., 438
Atkinson, C.H., 326

Bader, H., 53
Banke, E.G., 279
Barnes, H.T., 249
Barnes, W., 2
Bauer, A., 61
Beaudais, D., 227
Bercha, F.G., 316
Bergdahl, L., 326,335
Bielstein, W.T., 398
Bigg, E.K., 18
Blenkarm, K.A., 324, 326
Bragg, W.H., 2
Brill, R., 149
Brown, J.H., 109
Buckley, 275
Butkovitch, T.R., 109, 170, 251
Buzuev, A., 387

Camp, P.R., 149
Caquot, A., 343
Carstens, T., 462
Carter, D., 87, 92, 93, 94, 95, 104, 111, 131, 166, 289, 307, 316
Chankin, P.A., 362
Chen, W.F., 317
Chu, F.D., 409
Colburn, J., 386, 387, 393
Cook, H.J., 273
Coon, M.D., 355
Corrucini, R.J., 8
Cox, G.F., 46, 47
Croasdale, K.R., 317, 319, 322, 326, 336
Cronin, D.L.R., 326

Dadachanyi, N., 382, 421, 430
Dantl, G., 84
Danys, J.V., 276, 316, 326, 332
Delagrave, M., 347, 348
Devik, O., 17
Dorsey, N.E., 18, 254, 267

Drouin, M., 104, 106, 107, 160, 161, 191, 194, 215, 233, 237, 239, 240, 249, 253, 262, 263, 264, 268, 283
Durelli, A.J., 296
Duval, P., 60, 130, 152, 163

Eckhard, G.F., 276
Edwards, R.Y., 317, 319, 321, 322, 326, 385, 386, 387, 389, 393, 394, 396, 399, 408, 409, 410, 411, 412, 438, 459, 471
Eide, L.I., 41, 43
Engineering Record, 249
Enkvist, E., 390
Ernstsons, E., 326, 335
Estifeev, A.M., 354

Fabre, B., 51, 59, 60
Faddayev, O.V., 410
Farrell, D.R., 328
Fletcher, N.H., 14, 85, 86
Flinn, A.E., 250
Foundation Eng. Co., 437
Frankenstein, G., 39
Frederking, R.F., 196, 224, 225, 226, 288, 289, 290, 365
Freiberger, A., 360
Friedel, J., 122, 123
Fukuda, A., 123

Garner, R., 39
Geiger, R., 17
Gere, J.M., 300
German, J.G., 382, 394, 396, 398, 410, 411, 412, 421, 430, 471
Ginnings, D.C., 8
Glen, J.W., 120, 121, 124, 146
Godbout, Y., 193
Gold, L.W., 87, 125, 126, 169, 192, 194, 196, 220, 221, 224, 225, 226, 229, 288, 290, 363, 365

Haggard, S., 326, 335
Hamanaka, K., 290, 291, 292
Hamza, H.E., 299, 301
Hausser, R., 461
Hawkes, I., 167
Hayes, C.E., 118
Haynes, F.D., 328
Hertz, H., 203
Hétényi, M., 300, 319, 391
Heverly, J.R., 19
Hibler III, W.B., 47, 48
Higashi, A., 118, 123, 146, 157, 158, 159
Hill, H.M., 249, 288
Hirayama, K., 289, 290, 295, 299, 321

Hobbs, P.V., 5, 90

Iakunin, A.E., 223
Isnard, J.L., 291, 292, 296

Jansson, J.E., 406
Jellinek, H.H.G., 359
Johannessen, O.M., 279
Johnston, W.G., 146
Jones, S.J., 121, 146

Kashtelian, V.I., 212, 390, 394, 408, 410
Katsaros, K.B., 17
Kennedy, J.F., 36
Kennedy, R.J., 347
Kérisel, J., 343
Kerr, A.D., 194, 195, 209
Ketcham, K.M., 90
Kheishin, D.E., 218, 223
Kingery, W.D., 146, 191, 192, 225, 226, 227, 238
Kirkham, I.E., 276, 340
Kirn, J.K., 394, 396, 410, 411, 412, 471
Kivisild, H.R., 36
Korzhavin, K.N., 170, 250, 291, 303, 310, 312, 314, 323, 326, 334, 339, 340
Kovacs, A., 47, 48
Krausz, A.S., 146
Kry, P.R., 356
Kumai, M., 6
Kuroiwa, D., 158

Lacks, H., 360
Lafleur, P. 178, 188, 383
Lagutin, B.L., 192
Langleben, M.P., 100, 279
Latyshenkov, A.M., 341, 347, 348
Lavrov, V.V., 321
Lawrence, R.G.A., 382, 398, 410, 430
Lefebvre, L.M., 194, 215, 233, 237, 239, 240
Levine, G., 387, 433
Lewis, J.W., 385, 386, 387, 389, 393, 394, 396, 399, 408, 409, 410, 411, 412, 438, 459, 471
Lewis, W.V., 53, 54
Liu, W.T., 17
Llamas, J., 458, 465, 466
Lliboutry, L., 48, 49, 51, 57, 58, 61, 126, 148, 161
Lofquist, B., 264, 274, 358, 362

Major, R.A., 394, 396, 411, 412, 471
Mamaieff, N.M., 341
Marshall, E.W., 275
Martin, S., 41
Masterson, D.M., 227
Megaw, H.D., 6
Mellor, M., 151, 167
Menely, W.A., 196
Mentz, P.B., 387, 433
Meyerhof, G.G., 224
Michel, B., 17, 19, 23, 26, 27, 28, 31, 34, 36, 61, 62, 82, 87, 92, 93, 94, 95, 100, 104, 106, 107, 111, 123, 126, 131, 133, 160, 161, 166, 191, 194, 215, 233, 237, 239, 240, 253, 262, 263, 264, 265, 266, 267, 268, 279, 283, 284, 288, 289, 294, 299, 310, 321, 323, 326, 341, 342, 354, 361, 383, 400, 403, 438, 447, 452, 458, 459, 464, 465, 466, 467, 469
Milano, V.R., 385, 407
Miller, D.R., 394, 396, 410, 411, 412, 471
Monfore, G.E., 249, 253, 257, 258, 259, 261, 262, 263
Mookhoek, A.D., 398
Murray, R., 194, 215, 233, 237, 239, 240

Naborro, F.R.N., 123
Nadai, A., 207
Nadreau, J.P., 178, 179, 183, 184, 185, 190
Nakaya, V., 99, 100
Neill, C.R., 323, 326, 327, 328, 330, 331, 335, 337
Nevel, D.E., 207, 208, 209, 210, 211, 212, 213, 214, 217, 218, 220, 223, 328, 408
Nogid, L.M., 438
Nye, J.F., 54, 83

Orowan, E., 120, 123
Osterkamp, T.E., 28
Ouellet, Y., 469
Ozadi, A., 290, 291, 292

Panc, V., 208
Panfilov, D.F., 196, 219, 220, 221, 223, 295
Paradis, M., 153, 154, 164, 166, 289
Pariset, E., 36
Parmerter, R.R., 355

Pauling, L., 6
Persson, B., 224
Peschanskii, I.S., 75
Petit, J.R., 51, 59, 60
Petrunichev, N.N., 271, 283, 341
Peyton, H.R., 324, 325, 326
Philipps, E.A., 296
Pinigrin, M.I., 340
Poirier, J.P., 122
Pounder, E.R., 43, 99, 308
Poznjak, I.I., 390, 393, 408
Pratley, P.L., 276
Proskuriakoff, B.V., 341

Ralston, T.D., 317, 319, 321
Ramseier, R., 62, 107, 166
Ready, D.W., 146
Reinius, E., 326, 335
Riehl, N., 7
Rigsby, G.P., 125, 162
Roads and Bridges, 276
Robert, S., 17
Rose, E., 265, 272, 274
Rosenberg, P., 194, 215, 233, 237, 239, 240
Ryblin, A., 388, 390, 393, 408, 410

Sadowsky, M.A., 91
Saeki, H., 290, 291, 292
Sanden, E.J., 326, 331
Schaefer, V.J., 33
Schnell, R.C., 17
Schwarz, J., 289, 290, 295, 299, 321, 326, 332, 333, 334, 446, 459, 475
Schwarzacher, W., 46, 47
Shadrin, F.C., 295
Shenchan, F.C., 249
Shimansky, J.A., 400
Shulman, A.P., 192
Shumskii, P.A., 25, 45, 69
Shvayshteyn, Z.I., 437, 458
Smith, S.D., 279
Sodhi, D.S., 299, 301
Steineman, S., 125, 146, 147, 161
Sternberg, E., 91

Taylor, F.W., 253, 254, 257, 258, 261, 262, 263
Telechev, V.I., 340
Testa, R., 151
Timoshenko, Woinowsky-Krieger, 197, 209, 300, 366
Tolokno, V.V., 340
Toussaint, N., 284, 288, 289, 294, 299
Transhe, N.A., 45
Transport Canada, 229, 235
Tryde, P., 376, 459

Tsao, C., 296

Uzuner, M.S., 36
U.S.S.R., 229

Vali, G., 17
Vallon, M., 51, 58, 59, 60
Vance, G.P., 388, 390, 396, 409, 438
Verrette, J.L., 458, 465, 466
Vinogradov, I.V., 400, 403, 406
Voelker, R.P., 387, 433
Voytkavskii, K.F., 170

Wakahama, B., 121, 125
Watson, A., 384
Watts, J.S., 227
Webb, W.W., 118
Weeks, W.F., 44, 46, 47, 108, 109
Weertman, J., 123
Whalley, E., 1
Wheaton, J.W., 321, 322, 386, 387, 393
White, R.M., 396, 400, 407, 438
Willmot, J.G., 249
Wilson, J.T., 275
Wu, H., 289, 290, 295, 299, 321
Wyman, M., 204

Yagodkin, U.Y., 410
Yalin, M.S., 449

Zubov, N.N., 341
Zumberge, J.H., 275

# INDEX OF SUBJECTS


Ablation zone, 49, 50, 51, 58
Absorptivity of ice, 256, 257
Activation energy, 158
Accumulation basin, 49
Adhesive strength of ice, 358
Air bubble content, 58, 75, 87, 99
Allotropic varieties, 1
Alpine glacier, 49, 51
Amorphous form, 1
Anisotropy, 58, 117
Apparent yield stress, 156
Arctic ice, 69, 99, 377
Arctic pack, 45, 47
Autonomous hydraulic regime, 57
Axial plane, 343

Basal plane, 25, 86, 87, 88, 89, 90, 91, 92, 95, 101, 108, 114, 116, 126, 133, 134, 136, 162
Basal slip, 60
Bearing capacity:
- highest, 222
- of floating ice, 408, 410, 412, 414
- of ice covers, 178, 193
- for stationary load, 196, 226

Bending splitting failure, 331
Bending strength, 363
Biaxial effect, 287
Biharmonic equation, 201, 217
Block coefficient, 411, 412, 420, 427
Boundary crack, 133
Boundary crack formation, 161
Boundary slip, 133, 161
Breakthrough load, 194
Brine:
- cells, 41, 43, 76
- channels, 42
- drainage & migration, 42, 43, 44, 47, 238
- inclusion, 42
- pockets, 39, 41
- volume, 42, 99, 108, 109, 230

Brittle behaviour, 81, 95
Brittle conditions, 81, 102, 111, 170
Brittle crack formation, 100
Brittle failure, 359
Brittle fracture, 93, 170, 171, 290, 292, 294, 295, 296, 298, 325, 443

Brittle strength, 81, 111, 131, 132, 164, 170, 246, 294
Brittle zone, 298
Built up ice apron, 238
Built up ice bridge, 239, 241
Buckling: 276, 299, 340
  conditions, 301
  load, 300
Bulk modulus, 83, 98

C-axis, 2, 4, 5, 60, 64, 68, 69, 71, 84, 86, 97, 107
Cavity formation, 125
Centers of crystallization, 8
Centrosymmetry, 4
Circumferential failure, 220
Cleavage process, 295
Climb:
  dislocation, 122, 123, 125
  process of, 121
Coefficient of:
  active thrust, 343, 450
  active resistance, 349
  active pressure, 355
  contact, 310, 311, 316, 334, 336, 390
  diffusion, 122
  dynamic friction, 406, 407
  heat transfer, 266
  ice friction, 393
  restitution, 403
  thermal expansion, 250, 251
Cohesive strength, 110, 111, 115, 159, 160
Columnar crystal, 87
Columnar structure, 44
Complete affinity (non), 15, 17, 18
Complete biaxial constraint, 264, 268
Compressibility, 85
Compression ridge, 276
Compressive strain energy, 110
Compressive strength of ice, 444
Compressive stress, 94, 110, 115, 131
Concentration:
  of crystals, 24
  ionic of Na+ Cl-, 38
  of salts, 38
Concentrated elastic energy, 86
Condition of equilibrium, 36
Condition of non-submersion, 35
Constant:
  limit plastic tangential stress, 55
  of proportionality, 386

velocity towing method, 473
velocity towing test, 472
Continuity equation, 34
Cooling:
curve, 9
rate of, 24
Crack:
boundary, 133, 161
circumferential, 192, 213, 219, 312, 389
cleavage, 295
contraction, 274
dry, 233
failure, 302
formation, 90, 93, 128, 110, 194, 195, 408
hair, 233, 234
hinge, formation, 319
lateral, 235
of oblate spheroid form, 90
radial, 192, 195, 219, 236 312, 313, 389, 408
shear, 302
tension, 302
transcrystalline, 313
transverse, 313
wet, 233, 234, 237
Creep: 120, 122, 123
accelerated, 128, 130, 144, 146
behavior, 365
constants, 227
curves, 124, 125, 126, 129, 131, 144, 147, 148, 149, 152, 153, 157, 160, 191
function, 137, 142, 156
laws of, 123, 124, 139, 148
model, 154, 155
permanent, 128, 130, 131, 132, 140, 141, 143, 144, 145, 147, 148, 154, 163, 166, 182, 183, 184, 191, 194
primary, secondary, tertiary, 126, 127, 128, 130, 151, 152, 153, 196,
process, 125, 127, 139, 161
properties, 290
rate, 164
time-recovery, 150
transitory, 126, 129, 130, 131, 140, 141, 144, 182, 183, 184,
steady, 130, 143, 145, 165
visco-elastic, 128
viscous, 128
Crevasses, 50

Critical velocity:
of resonance, 217, 218
of wave propagation, 236
Crushing: 47, 249, 295, 397
energy of, 281, 406
failure force, 443
fracture force, 446
load, 321
local, 340, 389
oscillations, 299
strength, 82, 95, 109, 112, 115, 282, 284, 329, 336, 339
Cryogenic freezing system, 459
Crystal nucleus, 10
Crystal, spherical, 10
Crystalline structure, 23, 58
Crystallization process, 25, 67
Crystallographic:
analysis, 230
orientation, 100, 163
preferred orientation, 63, 66, 67, 68, 69, 72
plane, 120
structure, 1, 3, 230, 259
Cube strength, 332, 335
Cubic structure, 1
Cylindrical shape bubbles, 75
Deflection, 361, 362, 363, 364, 391
Deflection vertical, 390
Deformation:
dish, 225
ductile, 179, 182
law of, 262
longitudinal, 149
narrow wedge, 212
pattern, 135
recovery curve, 149
Degree of opaqueness, 75
Delayed elasticity, 100, 130, 144, 156, 301
Delayed elastic deformation, 227
Dendrites, 29, 67
Densification, 53
Diffusion, 121
Dimensional analysis, 438
Dimensionless groups, 322
Direction of compression, 60
Discoids, 40
Dislocations, 116, 117, 118, 119, 121, 123, 124, 136, 146, 157, 158, 162
Dislocation:
edge, 117, 118, 120
loop, 119
mixed, 117, 119

mobile, 120, 123, 126, 131, 136, 142, 144, 145, 154, 166, 170, 269, 270
screw, 117, 118
velocity of, 123, 146
Distorted hydraulic model, 449
Distorted strength scale, 457
Distortion, 447, 449, 450, 452, 459, 460, 465
Dome, 50
Drag coefficient, 278, 279
Drift velocity, 280
Dry snow transition, 52
Ductile behaviour, 116, 134, 146, 183, 190, 258, 262, 284, 285, 287, 288, 301, 334
Ductile conditions, 8
Ductile failure, 170
Ductile fracture, 86, 183, 190
Dynamic friction force, 311, 313, 341
Dynamic loading conditions, 8
Dynamic modulii, 97, 99
Dynamic viscosity of water, 392

Effective compressive strength, 316
Effective ice pressure, 284, 285, 332, 333, 335, 336, 338, 339
Effective ice thickness, 231, 232
Effective strain rate, 286, 287
Effective stress, 356, 357
Elastic:
analysis, 215, 224
behaviour, 83, 227
behaviour formula, 191
full, behaviour, 216
conditions, 87, 89
constant, 84, 96, 308
deformation, 95, 125, 126, 129, 135, 138, 140, 142, 143, 145, 188, 196, 296, 312
full, deformation, 150, 153, 154
energy, 93, 94, 100
foundation, 197, 209, 212, 299, 300, 312, 319, 361
modulii, 82, 98, 99, 105, 112, 444
plate, 299, 314
stability, 96
stress distribution, 181, 186, 206
theory, 189, 193, 207, 231, 443

waves propagation, 87
yield moment, 363
Elastic-plastic response, 301
Elasto-ductile deformation process, 157
Elasto-plastic body, 157
Electron diffraction, 6
Elongated grains, 104, 107
Equation
biharmonic, 201, 217
of equilibrium, 200
Equiaxed grains, 68, 72, 73, 74
Equilibrium:
of the cover, 350
of hydrostatic forces, 55
thickness, 46
velocity, 281

Factor of safety, 195
Failure:
bending, 306, 311, 330
criterion, 195
by crushing, 307, 316
force, 331, 335, 451, 452
of ice covers, 192
by indentation, 301
mechanism, 284
mode, 339
Field strength, 328
Film conductance of convection, 255
Film conductance of radiation, 256
Firns, 49, 51, 52, 54, 58, 60, 99
Firn line, 51
Flexural:
failure force, 44
fracture, 444
ice force, 448
rigidity, 198
rigidity of the plate, 361
strength, 82, 109, 178, 187, 188, 190, 191, 193, 218, 314, 391, 444, 446, 460, 464, 466, 467, 475
Floe splitting, 304
Flow:
compressive, 56
extended, 56
model, 161, 162
Foliation and non-foliated ice, 58
Foliation of ice, 58
Forced oscillation, 325, 327, 330
Forced nucleation, 21, 25
Form coefficient, 290

Formation of sea ice, 40, 41, 46
Foundation modulus, 201, 300, 361, 362
Fracture plane, 10
Fracture strength, 81, 93, 95, 100, 104, 114
Frazil:
  congealed slush, 71, 72
  evolution, 32
  flocks, 33
  formation, 25, 26, 27, 40
  ice, 106, 168, 113, 384
  nucleation, 21, 23
  particles, 20, 21, 33, 36
  production front, 26
Frequency of oscillation, 327
Friction:
  angle, 308
  coefficient, 303, 308, 310, 441
  effect, 397
  force, 303, 344, 347, 441
  law, 57
  water, force, 345, 346, 448
Frictional force, 385, 393, 404, 405

Geometric similitude, 438
Glacier:
  cold polar, 53
  composite, 49
  equilibrium limit, 51, 53
  flow, 53, 55, 56
  front, 60, 61
  ice, 58
  morphology, 49, 53
  multiple emissaries, 49
  surges, 57
  tongues, 49
  velocity, 56, 57
Glide lines, 288
Glide planes, 4
Grain arrangements, 166
Grain boundary migration, 125
Grain boundary shape, 63, 74
Grain boundary slip, 125, 126
Grain size and shape, 63, 66, 67, 68, 69, 72, 73, 74
Gravity force, 435, 441, 447, 451, 452
Gravity-type dam, 272

Heat transfer equation, 30
Heeling moment, 420
Height limiting mechanism, 356
Heterogeneous nucleation, 13, 14, 17, 23,
Hexagonal crystal, 84
Hexagonal structure, 1

Hexagonal symmetry, 2
Highland plateau, 50
Homogeneous nucleation, 10, 15, 16
Horizontal layering, 74
Hull-ice friction, 472, 473
Hummocks, 45
Hydraulic:
  conditions, 353
  distortion, 447
  similitude, 451
  thrust, 350, 352
Hydrodynamic:
  conditions, 65, 437
  force, 342, 344, 445, 451
  thrust, 348, 349
Hydrodynamical model, 461
Hydrodynamical simulation, 454
Hydrodynamical similitude, 445, 453, 455, 457
Hydromechanical modelling, 454
Hydromechanical similitude, 445, 446, 454, 457, 458

Ice:
  accumulation, 56
  advection, 28
  agglomerate, 66, 74
  anchor, 29
  accumulation equilibrium, 342, 351
  bearing capacity, 218, 224, 226, 230, 234, 237
  black-ice, 31, 32
  border, 20, 28
  breaking force, 412
  breaking process, 388, 400
  bridge, 34, 41, 177
  candle, 235
  caps, 248
  cohesive strength, 94
  cold, 248
  columnar, 62, 63, 68, 69, 71, 73, 97, 100, 107, 108, 113, 168
  compliance constant, 82, 83, 84, 85, 86, 97
  corn-snow, 235
  drifting, 45, 298, 456
  drifting, impact, 276, 277
  drift problem, 454
  drift control structure, 456
  dynamic formation, 32
  embryo, 14, 15, 21

erosion, 332
expanding, sheet, 274
fast, 45,46
field, 380, 381
floes, 61. 276, 280, 281, 282, 292, 312, 313, 323, 340, 342, 280, 382, 455, 460, 461
floe movement, 277
floe splitting, 302
force, 385, 410
formation, 62, 66, 273
friction, 384, 393
germs, 20, 21
granular snow-ice, 97
glacier, 152
glacier, density, 99
high salinity, 458, 459
impinging, sheet, 298
internal pressure, 331, 336
islands, 47, 246, 248
jam, 382, 383, 384, 388, 413, 449, 454, 464, 476
jams models, 461
mixture, 383
models, 462, 466, 367, 469. 470, 472, 473, 474, 475
modelling, 437, 453, 459
modelling techniques, 471
multi-year, 46
needles, 20
nucleation, 22, 24, 28, 33, 62
nuclei, 40
overhanging, block, 61
pack, 383
pancakes, 33
peeling, 295
plate, 20, 33, 40
platform, 177
platelets, 41, 76
polar, 44, 46
polycrystalline, 81, 87, 95, 96, 97, 98, 99, 101, 104, 109, 110, 111, 112, 114, 117, 126, 129, 133, 139, 144, 148, 149, 151, 152, 153, 160, 164, 166, 167, 170, 224
pressure, 249, 250, 253, 265, 272, 331, 336, 381
primary, 62, 63, 66, 69, 73

pseudo-monocrystalline, 100, 109, 110. 169, 259, 268
rafted or ridges, 74
regelation, 45
ramparts, 275
resistance, 380, 433, 438
ridge, 319, 381
sea-ice, 38, 39, 41, 43, 44, 45, 71, 76
sea-ice growth, 46
sea-ice mechanical behaviour, 76
secondary, 62, 66
sheets, 49, 71
shore, growth, 29
slush, 377, 378, 384
snow-ice, 31, 32, 72, 166
synthetic, 459, 464, 466, 467, 469, 470
drained snow-ice, 73
soaked layer, 52
stiffness constant, 83, 84
stream, 50
strength, 114, 286, 332, 334, 335, 337, 338, 395, 400, 407
structure, 62, 82, 95, 98
superimposed, 63, 66, 72
superimposed layered, 73
temperature curve, 258
tongue, 49
thrust, 249, 250, 264, 268, 272, 273
viscous properties, 116
Icebergs, 47, 246
calving, 61
formation, 60
production, 61
Ice cover, 63, 65
bearing capacity, 215, 220, 223
dynamic behaviour, 197
formation, 22, 34, 38
static, formation, 28, 29, 31
stability, 35
thickness, 36
Indentation:
coefficient, 298, 339
formula, 340
of ice, 281, 282, 283, 289, 290, 292, 314
pressure, 289, 298
rate, 284, 285, 286, 287, 293
strength, 403
test, 288
Indentors shapes, 291, 292

Indlandsis, 49
Infinite plate, 219, 334, 361
Infinite theoretical deflection, 217
Inflection point, 265
Influence surface, 209
Inorganic nucleators, 16
Ionic concentration of Na+, Cl-, 39, 41
Isotropic body, 83
Isotropic crystal, 83
Instantaneous elasticity, 130
Instantaneous deflection, 183
Instantaneous elastic recovery, 149
Interconnected hydraulic regime, 57
Internal friction, 450

Jam formation, 457, 462
Jam stability, 454, 456

Kinematic method, 324, 330
Kinematic similitude, 438, 439

Lattice constants, 7
Laws of similitude, 438
Limiting capability of the ship, 474
Limiting stability, 352
Limiting thickness, 399, 409
Limiting of stability of an ice jam, 351, 352
Linear rise of temperature, 252, 255, 257, 259, 262, 265, 268
Linear strain distribution, 184
Linear theory of ship motion, 410
Load reversal, 152
Loading conditions, 82, 131
Local ice crushing, 340
Local ice pressure, 340, 341
Longitudinal crack, 362
Longitudinal metacentric heigh, 402
Longitudinal strain, 180
Low angle boundaries, 125

Macro-cracks, 293, 295
Macro-cracks formation, 295
Manoeuverability in ice, 410, 412, 414
Mass transport, 121
Maximum axial force, 304
Maximum deflection, 196, 205
Maximum shear stress, 115
Maximum yield strength, 333
Mechanism of cavitation, 37

Micro-cracks, 87, 96, 126, 284, 285, 295
Micro-crack formation, 295
Micro-crack production, 286, 287
Mirror plane, 4
Model of ice cover in rivers, 464, 465
Modes of failure, 81
Molecular self-diffusion, 158
Moment of inertia, 231, 320
Moment-shear, 272
Monocrystals, 82, 86, 89, 116, 133, 135, 146, 157
Monocrystalline ice, 110, 139
Moving ice sheet impact, 248
Mud or debris content, 58
Multiplication of crystals, 21

Narrow infinite wedge, 212
Non-isotropic behaviour, 87
Normal tensile strength, 133
Nucleating treshold, 17, 20, 21, 22
Nucleus, critical, 11, 12, 21
Nucleation:
of a crack, 87, 91, 101
rate, 12
secondary, 27, 40
surface, 19, 22, 23, 40
temperature, 8
"Nutcracker" tests, 336

Octahedral shear stress, 171
Open cracks, 46
Operation of ice bridge, 237
Optical axis, 59
Organic and inorganic impurities, 75
Oscillograms, 331, 337

Parabolic arches, 343
Percolation, 53
Permanent regime, 280
Piedmont glacier, 50
Pier movement, 277
Piezometer pressure, 57
Planes of maximum shear stress, 55
Plane of permanent shear, 60
Plastic:
bending moment, 363
deformation, 95, 125, 1 126, 135, 170, 142
flexural moment, 224
flow, 54, 133, 138, 162
ice properties, 274
limit analysis, 316
resisting moment, 196
shear, 120
theory, 288

Plastification coefficient, 184
Plastification zone, 284, 287, 295
Plate failure, 295
Plate theory, 312
Points defects, 117
Polar floe, 45
Polycrystalline aggregate, 137
Polygonisation, 125
Porosity, 112
Porosity of ice, 36, 99
Porosity of snow-ice, 105
Porous flocks, 26
Positive moments, 204
Predictor equation, 473, 474
Preferential slip bands, 135
Pressure force, 443
Pressure ridges, 45, 46, 47, 248, 274, 275, 325
Propellar thrust, 402
Prototype inertia, 440

Ramming, 399, 408, 409, 471, 473
Radial moments, 198, 204, 213
Rate of cooling, 24
Recrystallization, 125, 131
Regular slope, 50
Residual contact coefficient, 339
Resistance:
- of the cover, 346
- mean, 385, 386
- total, 386

Retarded elaticity, 152
Rheological behaviour, 265
Ridge sails, 47
Ridge porosity, 356
Rigidity modulus, 83, 98, 444
Rotation, 134

Safe capacity of ice, 237
Salinity profile, 42, 44, 48
Saturation zone, 52
Secant modulus, 96
Seeding, 22
Semi-infinite beam, 320
Semi-infinite plate, 209, 220, 334
Semi-infinite solid, 255
Shear:
- failure, 303, 306, 310
- failure force, 443
- force, 365
- interaction, 47
- modulus of ice, 136
- strength, 361, 444
- surface, 310, 311

Shelfs, 50

Ship resistance, 387, 390, 393, 394, 396, 398
Ship, icebreaker, resistance, 437, 459, 470
Side friction, 431
Sinusoidal variation, 252, 268
Sliding velocity, 57
Sliding friction coefficient, 404
Slip plane, 118, 121
Snow-like crust, 33
Solidification, 25, 75
Solid friction coefficient, 446
Solid friction force, 391, 397
Solid type of friction of ice, 470
Spectral density plot, 331
Spectral, power, density analysis, 332
Splitting failure, 304
Standard sea water, 39
Static friction, 420
Static friction coefficient, 468
Static or dynamic friction factor, 443
Static ice pressure, 271
Static pressure, 271
Step function, 252, 257
Strain:
- compression, 366
- critical rate, 190
- hardening, 128, 130
- rate, 336
- real rate, 339
- relief, 364
- softening process, 126

Strength of adhesion of ice, 359, 360, 361
Strength properties of ice, 437
Stress:
- concentration, 293
- distribution, 185, 350, 367
- field, 122
- function, 187
- strain curve, 262

Structural loading ice pressure, 325
Structural stress risers, 87
Sub-vertical preferrential orientation, 59
Surface tension effect, 446, 455
Surface tension force, 445
Supercooling, 9, 12, 13, 16, 17, 18, 19, 25, 27, 40

Syntectonic recrystallization, 60, 128, 130, 161

Tangential:
force, 342, 345, 443
friction force, 441
Temperate glacier, 59
Tensile:
force, 83
strain energy, 91
strenght, 92, 102, 104, 105, 106, 107, 108, 109, 170, 230, 304
testing, 91
Tetrahedral pattern, 2
Textural change, 75
Theory:
test, 339
yield strength, 289

Uniform floes, 325
Unit cell, 3

Velocity of climb, 122
Velocity of entrainment, 456
Velocity of submergence, 455, 456
Vertical circumferential shear plane, 193
Vertical:
force, 358, 359, 361, 363, 398
pile test, 338
shear displacement, 365
uplift by water, 61
Visco-elastic:
creep, 128
plates, 223
properties of ice, 359
recovery, 149, 183
Viscous creep, 128
Viscous flow, 137

Wave propagation, 217, 236
Waterplane area, 402
Worst limiting conditions, 462

X-Ray diffraction, 6

Yield line theory, 224
Yield load, 361
Yield moment, 314
Yield point, 158, 165, 170
Yield strength, 81, 111, 131, 132, 158, 164, 165, 170, 246, 328, 468